Macroscale and Microscale Organic Experiments

Selected Experiments from the 4th and 5th edition

Williamson

CENGAGE
Learning™

Australia • Brazil • Japan • Korea • Mexico • Singapore • Spain • United Kingdom • United States

Macroscale and Microscale Organic Experiments: Selected Experiments from the 4th and 5th edition

Macroscale and Microscale Organic Experiments, 5th Edition
Kenneth L. Williamson | Robert Minard | Katherine M. Masters

© 2007 Cengage Learning. All rights reserved.

Macroscale and Microscale Organic Experiments, 4th Edition
Kenneth L. Williamson

Executive Editors:
 Maureen Staudt
 Michael Stranz

Senior Project Development Manager:
 Linda deStefano

Marketing Specialist:
 Courtney Sheldon

Senior Production/Manufacturing Manager:
 Donna M. Brown

PreMedia Manager:
 Joel Brennecke

Sr. Rights Acquisition Account Manager:
 Todd Osborne

Cover Image:
 Getty Images*

*Unless otherwise noted, all cover images used by Custom
Solutions, a part of Cengage Learning, have been supplied
courtesy of Getty Images with the exception of the Earthview
cover image, which has been supplied by the National
Aeronautics and Space Administration (NASA).

For product information and technology assistance, contact us at
Cengage Learning Customer & Sales Support, 1-800-354-9706

For permission to use material from this text or product,
submit all requests online at **cengage.com/permissions**
Further permissions questions can be emailed to
permissionrequest@cengage.com

This book contains select works from existing Cengage Learning resources and
was produced by Cengage Learning Custom Solutions for collegiate use. As such,
those adopting and/or contributing to this work are responsible for editorial
content accuracy, continuity and completeness.

Compilation © 2011 Cengage Learning
ISBN-13: 978-1-133-35879-4

ISBN-10: 1-133-35879-9

Cengage Learning
5191 Natorp Boulevard
Mason, Ohio 45040
USA
Cengage Learning is a leading provider of customized learning solutions with
office locations around the globe, including Singapore, the United Kingdom,
Australia, Mexico, Brazil, and Japan. Locate your local office at:
international.cengage.com/region.

Cengage Learning products are represented in Canada by Nelson Education, Ltd.
For your lifelong learning solutions, visit **www.cengage.com/custom.**
Visit our corporate website at **www.cengage.com.**

Printed in the United States of America

Macroscale and Microscale Organic Experiments
Table of Contents

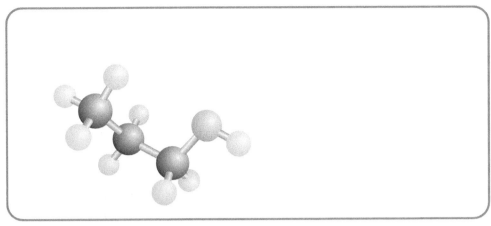

Introduction

PRELAB EXERCISE: Study the glassware diagrams presented in this chapter and be prepared to identify the reaction tube, the fractionating column, the distilling head, the filter adapter, and the Hirsch funnel.

Welcome to the organic chemistry laboratory! Here, the reactions that you learned in your organic lectures and studied in your textbook will come to life. The main goal of the laboratory course is for you to learn and carry out techniques for the synthesis, isolation, purification, and analysis of organic compounds, thus experiencing the experimental nature of organic chemistry. We want you to enjoy your laboratory experience and ask you to remember that safety always comes first.

Synthesis–The Big Picture

Synthesis is a pervasive process in the universe. Science has shown us that our world consists of energy and matter and that both of these are being transformed from one form to another throughout the cosmos (see Fig. 1.1).

- Cosmosynthesis started roughly 14 billion years ago when the Big Bang transformed subatomic particles into hydrogen and helium. These elements condense into stars and are fused by nucleosynthetic reactions to give us the periodic table of elements. Most stars ultimately expand or explode and disperse their product elements into space where they can recondense to form new stars, some with planets around them. For the early Earth, these elements combined into minerals and other substances such as water, carbon dioxide, and nitrogen.
- Biosynthesis on planet Earth has been going on for over 2 billion years. In the process of growth and reproduction, even the simplest single cell organisms can synthesize incredibly complex structural, catalytic, informational, and

■ FIG. 1.1
The three types of synthesis and
the steps by which organic
chemists study the resulting
products. Theory allows correlation
of structure to properties and
guides the cycle of synthetic
design, synthesis, separation/
purification, and analysis.

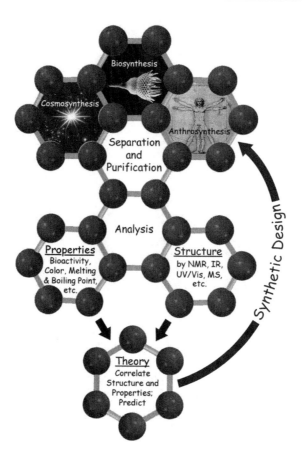

regulatory molecules from simple starting materials such as CO_2, H_2O, and N_2. Organic chemistry's beginnings are based on isolating and studying these natural products such as urea, ethanol, caffeine, or oil of wintergreen (methyl salicylate). The whole field of biochemistry is based on the isolation, structure determination, and reactivity of organic molecules found in living systems.

• Anthrosynthesis is a human activity that produces either known substances economically (e.g., iron, ammonia, and vitamin C) or new substances not found in the cosmos or biology (e.g., aspirin, Splenda, and Teflon). To date, chemists have synthesized or isolated from nature over 20 million different substances, and 99% of these are organic compounds. As you know, this great variety of organic compounds stems from carbon's unique ability to form chains that can be linear, branched, or cyclic and that can have other elements bonded to them to yield myriad structures.

Whether chemists study rocks from the earth, molecules in cells, or products of anthrosynthesis, they almost always separate and study the structure and properties of pure substances (e.g., silica, hemoglobin, or carbon nanotubes) rather than the mixtures (e.g., a shovel of dirt, blood, or soot) typically found in nature or synthetic reactions. Only pure substances will show reproducible data for physical, chemical, and biological properties. Chemists then try to understand the relationship between a compound's structure and its properties and devise a reasonable theory that relates these. If a theory is found to be true, it can be used to design the synthesis of a new substance with enhanced properties such as elasticity, antibiotic activity, or biodegradability. This cycle of Synthesis → Separation/Purification → Analysis → Theory → Synthetic Design is depicted in Figure 1.1. The cycle can be repeated to test and refine the theory and further improve the desired properties. This cycle is at the heart of progress in the pharmaceutical, polymer/plastic, and metallurgy industries.

Once you start your laboratory work, you will soon realize that synthesis involves much more than reacting chemicals. The techniques of separation, purification, and analysis often require considerable time and attention. Therefore, the first part of this lab text is designed to help you understand, apply, and master these techniques. They will then be used repeatedly in the dozens of synthesis and isolation experiments presented in later chapters.

Experimental Organic Chemistry– What Is in It for Me?

You are probably not a chemistry major. The vast majority of students in this laboratory course are majoring in the life sciences or engineering. Although you may never use the exact same techniques taught in this course, you will undoubtedly apply the skills taught here to whatever problem or question your ultimate career may present. Application of the scientific method involves the following steps:

1. Designing an experiment, therapy, or approach to solve a problem
2. Executing the plan or experiment
3. Observing the outcome to verify that you obtained the desired results
4. Recording the findings to communicate them both orally and in writing

You may not always get the desired result the first time when doing an experiment. Do not take this as a personal failure; it is evidence that carrying out an organic reaction can often be trickier than baking a cake (and we all probably had trouble baking our first cake!). Your laboratory instructor is there to provide guidance and advice but not to solve your problems. Learning to think for yourself is the most valuable skill you will take from this course. However, you should not be afraid to ask questions or bounce ideas off your instructor or fellow students. Just formulating the question or talking about your problem often helps you find the answer. To begin troubleshooting a problem, you will need to scrutinize carefully every detail of the approach, therapy, or experimental design; the execution of the experiment; and your observations to see where things might have gone wrong. This process will lead you to an idea or hypothesis about how to change things to make the experiment work. The sequence is then repeated, sometimes many times, until you solve the problem, answer the question, or prove/disprove the idea.

The teaching lab is more controlled than the real world. In this laboratory environment, you will be guided more than you would be on the job. Nevertheless, the experiments in this text are designed to be sufficiently challenging to give you a taste of experimental problem-solving methods practiced by professional scientists. We earnestly hope that you will find the techniques, the apparatus, and the experiments of just the right complexity, not too easy but not too hard, so that you can learn at a satisfying pace.

Macroscale and Microscale Experiments

This laboratory text presents a unique approach for carrying out organic experiments; they can be conducted on either a *macroscale* or a *microscale*. Macroscale was the traditional way of teaching the principles of experimental organic chemistry and is the basis for all the experiments in this book, a book that traces its history to 1934 when the late Louis Fieser, an outstanding organic chemist and professor at Harvard University, was its author. Macroscale experiments typically involve the use of a few grams of *starting material*, the chief reagent used in the reaction. Most teaching institutions are equipped to carry out traditional macroscale experiments. Instructors are familiar with these techniques and experiments, and much research in industry and academe is carried out on this scale. For these reasons, this book has macroscale versions of most experiments.

For reasons primarily related to safety and cost, there is a growing trend toward carrying out microscale laboratory work, on a scale one-tenth to one-thousandth of that previously used. Using smaller quantities of chemicals exposes the laboratory worker to smaller amounts of toxic, flammable, explosive, carcinogenic, and teratogenic material. Microscale experiments can be carried out more rapidly than macroscale experiments because of rapid heat transfer, filtration, and drying. Because the apparatus advocated by the authors is inexpensive, more than one reaction may be set up at once. The cost of chemicals is, of course, greatly reduced. A principal advantage of microscale experimentation is that the quantity of waste is one-tenth to one-thousandth of that formerly produced. To allow maximum flexibility in the conduct of organic experiments, this book presents both macroscale and microscale procedures for the vast majority of the experiments. As will be seen, some of the equipment and techniques differ. A careful reading of both the microscale and macroscale procedures will reveal which changes and precautions must be employed in going from one scale to the other.

Synthesis and Analysis

Synthesis and analysis are two major concerns of the organic chemist, and both are dealt with in this book. The typical sequence of activity in synthetic organic chemistry involves the following steps:

1. Designing the experiment based on knowledge of chemical reactivity, the equipment and techniques available, and full awareness of all safety issues
2. Setting up and running the reaction
3. Isolating the reaction product

4. Purifying the crude product, if necessary

5. Analyzing the product using chromatography or spectroscopy to verify purity and structure

6. Disposing of unwanted chemicals in a safe manner

Each of these steps uses certain laboratory techniques. The experiments at the beginning of this text are designed to teach you these techniques and to help you understand the molecular level principles that underlie them.

1. Designing the Experiment

Because the first step of experimental design often requires considerable experience, this part has already been done for you for most of the experiments in this introductory level book. Synthetic experimental design becomes increasingly important in an advanced course and in graduate research programs. Remember that safety is paramount, and therefore it is important to be aware of all possible personal and environmental hazards before running any reaction. These are more fully discussed in Chapter 2 and are highlighted in each experiment.

2. Running the Reaction

The rational synthesis of an organic compound, whether it involves the transformation of one functional group into another or a carbon-carbon bond-forming reaction, starts with a *reaction*. Organic reactions usually take place in the liquid phase and are *homogeneous*—the reactants are entirely in one phase. The reactants can be solids and/or liquids dissolved in an appropriate solvent to mediate the reaction. Some reactions are *heterogeneous*—that is, one of the reactants is a solid and requires stirring or shaking to bring the reactants in contact with one another. A few heterogeneous reactions involve the reaction of a gas, such as oxygen, carbon dioxide, or hydrogen, with material in solution. Examples of all these reactions are found in this book.

An *exothermic* reaction evolves heat. If it is highly exothermic with a low activation energy, one reactant is added slowly to the other, and heat is removed by external cooling. Most organic reactions are, however, mildly *endothermic*, which means the reaction mixture must be heated to overcome the activation barrier and to increase the rate of the reaction. A very useful rule of thumb is that *the rate of an organic reaction doubles with a 10°C rise in temperature*. Louis Fieser introduced the idea of changing the traditional solvents of many reactions to high-boiling solvents to reduce reaction times. Throughout this book we will use solvents such as triethylene glycol, with a boiling point (bp) of 290°C, to replace ethanol (bp 78°C) and triethylene glycol dimethyl ether (bp 222°C) to replace dimethoxyethane (bp 85°C). Using these high-boiling solvents can greatly increase the rates of many reactions.

The progress of a reaction can be followed by observing a change in color or pH, the evolution of a gas, or the separation of a solid product or a liquid layer. Quite often, the extent of reaction can be determined by withdrawing tiny samples at certain time intervals and analyzing them by *thin-layer chromatography* or *gas chromatography* to measure the amount of starting material remaining and/or the amount of product formed. The next step, product isolation, should not be carried out until one is confident that the desired amount of product has been formed.

Effect of temperature

Chapters 8–10: Chromatography

3. Product Isolation: Workup of the Reaction

Running an organic reaction is usually the easiest part of a synthesis. The real challenge lies in isolating and purifying the product from the reaction because organic reactions seldom give quantitative yields of a single pure substance. Any unwanted byproducts need to be removed.

Chapter 4: Recrystallization

In some cases the solvent and concentrations of reactants are chosen so that, after the reaction mixture has been cooled, the product will *crystallize* or *precipitate* if it is a solid. The product is then collected by *filtration*, and the crystals are washed with an appropriate solvent. If sufficiently pure at that point, the product is dried and collected; otherwise, it is purified by the process of *recrystallization* or, less commonly, by *sublimation*.

Chapter 7: Liquid/Liquid Extraction

More typically, the product of a reaction does not crystallize from the reaction mixture and is often isolated by the process of *liquid/liquid extraction*. This process involves two liquids, a water-insoluble organic liquid such as dichloromethane and a neutral, acidic, or basic aqueous solution. The two liquids do not mix, but when shaken together, the organic materials and inorganic byproducts go into the liquid layer, organic or aqueous, that they are the most soluble in. After shaking, two layers again form and can be separated. Most organic products remain in the organic liquid and can be isolated by evaporation of the organic solvent.

Chapter 5: Distillation

Chapter 6: Steam Distillation and Vacuum Distillation

If the product is a liquid, it is isolated by *distillation*, usually after extraction. Occasionally, the product can be isolated by the process of *steam distillation* from the reaction mixture.

4. Purification

When an organic product is first isolated, it will often contain significant impurities. This impure or crude product will need to be further purified or cleaned up before it can be analyzed or used in other reactions. Solids may be purified by recrystallization or sublimation and liquids by distillation or steam distillation. Small amounts of solids and liquids can also be purified by *chromatography*.

Chapters 11–14: Structure Analysis

5. Analysis to Verify Purity and Structure

The purity of the product can be determined by melting point analysis for solids, boiling point analysis or refractive index for liquids, and chromatographic analysis for either solids or liquids. Once the purity of the product has been verified, structure determination can be accomplished by using one of the various spectroscopic methods, such as ^{1}H and ^{13}C nuclear magnetic resonance (NMR), infrared (IR), and ultraviolet/visible (UV/Vis) spectroscopies. Mass spectrometry (MS) is another tool that can aid in the identification of a structure.

CAUTION: Never smell chemicals in an attempt to identify them.

6. Chemical Waste Disposal

All waste chemicals must be disposed of in their proper waste containers. Instructions on chemical disposal will appear at the end of each experiment. It is recommended that nothing be disposed of until you are sure of your product identity and purity; you do not want to accidentally throw out your product before the analysis is complete. Proper disposal of chemicals is essential for protecting the environment in accordance with local, state, and federal regulations.

Equipment for Experimental Organic Chemistry

A. Equipment for Running Reactions

Organic reactions are usually carried out by dissolving the reactants in a solvent and then heating the mixture to its boiling point, thus maintaining the reaction at that elevated temperature for as long as is necessary to complete the reaction. To keep the solvent from boiling away, the vapor is condensed to a liquid, which is allowed to run back into the boiling solvent.

Microscale reactions with volumes up to 4 mL can be carried out in a *reaction tube* (Fig. 1.2a). The mass of the reaction tube is so small and heat transfer is so rapid that 1 mL of nitrobenzene (bp 210°C) will boil in 10 s, and 1 mL of benzene (mp 5°C) will crystallize in the same period of time. Cooling is effected by simply agitating the tube in a small beaker of ice water, and heating is effected by immersing the reaction tube to an appropriate depth in an electrically heated sand bath. This sand bath usually consists of an electric 100-mL flask heater or heating mantle half filled with sand. The temperature is controlled by the setting on a variable voltage controller often called a Variac (Fig. 1.2b). The air above the heater is not hot. It is possible to hold a reaction tube containing refluxing solvents between the thumb and forefinger without the need for forceps or other protective device. Because sand is a fairly poor conductor of heat, there is a very large variation in temperature in the sand bath depending on its depth. The temperature of a 5-mL flask can be regulated by using a spatula to pile up or remove sand from near the flask's base. The heater is easily capable of producing temperatures in excess of 300°C; therefore, never leave the controller at its maximum setting. Ordinarily, it is set at 20%–40% of maximum.

Because the area of the tube exposed to heat is fairly small, it is difficult to transfer enough heat to the contents of the tube to cause the solvents to boil away. The reaction tube is 100 mm long, so the upper part of the tube can function as an efficient *air condenser* (Fig. 1.2a) because the area of glass is large and the volume of vapor is comparatively small. The air condenser can be made even longer by attaching the empty *distilling column* (Fig. 1.2c and 1.15o) to the reaction tube using the *connector with support rod* (Fig. 1.2c and Fig. 1.15m). The black connector is made of Viton, which is resistant to high-boiling aromatic solvents. The cream-colored connector is made of Santoprene, which is resistant to all but high-boiling aromatic solvents. As solvents such as water and ethanol boil, the hot vapor ascends to the upper part of the tube. These condense and run back down the tube. This process is called *refluxing* and is the most common method for conducting a reaction at a constant temperature, the boiling point of the solvent. For very low-boiling solvents such as diethyl ether (bp 35°C), a pipe cleaner dampened with water makes an efficient cooling device. A water-cooled condenser is also available (Fig. 1.3) but is seldom needed for microscale experiments.

A Petri dish containing sand and heated on a hot plate is not recommended for microscale experiments. It is too easy to burn oneself on the hot plate; too much heat wells up from the sand, so air condensers do not function well; the glass

Turn on the sand bath about 20 min before you intend to use it. The sand heats slowly.

CAUTION: Never put a mercury thermometer in a sand bath! It will break, releasing highly toxic mercury vapor.

Online Study Center

Photos: Williamson Microscale Kit, Refluxing a Liquid in a Reaction Tube on a Sand Bath; Video: The Reaction Tube in Use

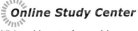

Online Study Center

Video: How to Assemble Apparatus

■ FIG. 1.2

(a) A reaction tube being heated on a hot sand bath in a flask heater. The area of the tube exposed to the heat is small. The liquid boils and condenses on the cool upper portion of the tube, which functions as an air condenser. (b) A variable voltage controller used to control the temperature of the sand bath. (c) The condensing area can be increased by adding a distilling column as an air condenser.

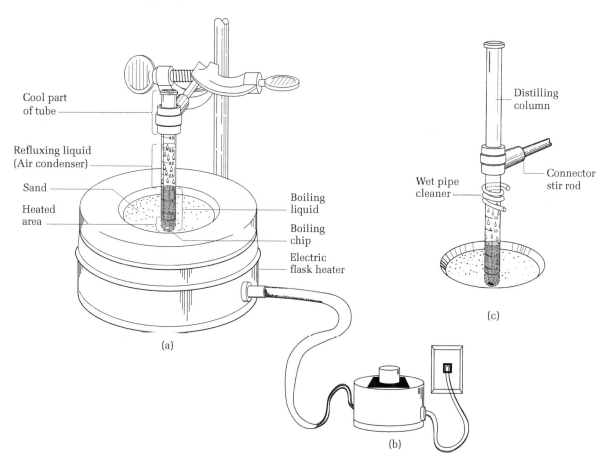

Cool part of tube

Refluxing liquid (Air condenser)

Sand

Heated area

Boiling liquid

Boiling chip

Electric flask heater

Distilling column

Wet pipe cleaner

Connector stir rod

(a)

(b)

(c)

CAUTION: Organic reactions should be conducted in a fume hood with the sash lowered.

dishes will break from thermal shock; and the ceramic coating on some hot plates will chip and come off.

Larger scale (macroscale) reactions involving volumes of tens to thousands of milliliters are usually carried out in large, round-bottom flasks that fit snugly (without sand!) into the appropriately sized flask heater or heating mantle (Fig. 1.4). The round shape can be heated more evenly than a flat-bottom flask or beaker. Heat transfer is slower than in microscale because of the smaller ratio of surface area to volume in a round-bottomed flask. Cooling is again conducted using an ice bath, but heating is sometimes done on a steam bath for low-boiling liquids. The narrow neck is necessary for connection via a *standard-taper ground glass joint* to a water-cooled *reflux condenser*, where the water flows in a jacket around the central tube.

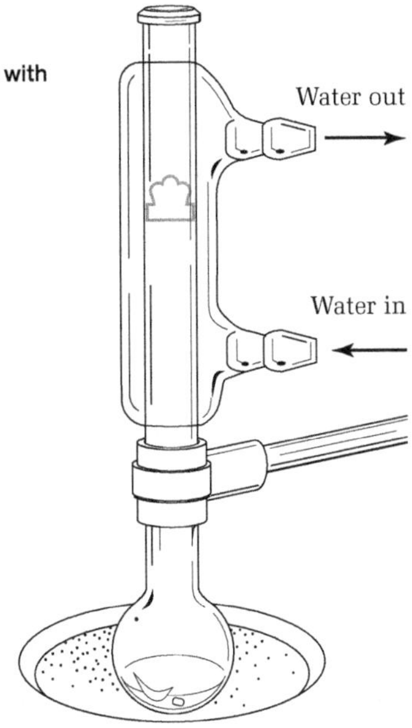

■ FIG. 1.3
Refluxing solvent in a 5-mL
round-bottomed flask fitted with
a water-cooled condenser.

Water out →

Water in →

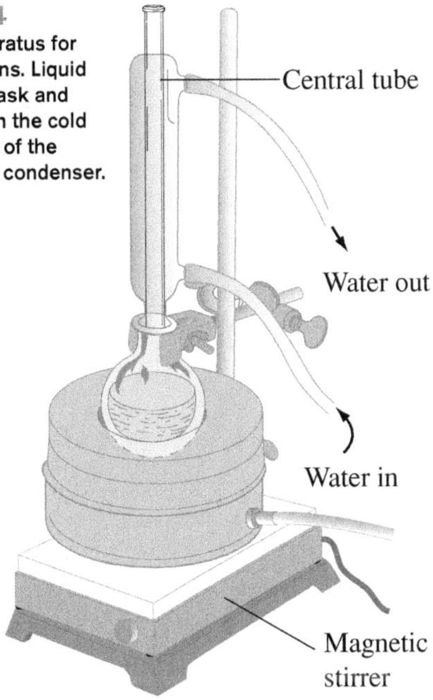

■ FIG. 1.4
A reflux apparatus for
larger reactions. Liquid
boils in the flask and
condenses on the cold
inner surface of the
water-cooled condenser.

Central tube

Water out

Water in

Magnetic
stirrer

The high heat capacity of water makes it possible to remove a large amount of heat
in the larger volume of refluxing vapor (Fig. 1.4).

Heating and Stirring

In modern organic laboratories, electric flask heaters (heating mantles), used
alone or as sand baths, are used exclusively for heating. Bunsen burners are almost never used because of the danger of igniting flammable organic vapors. For
solvents that boil below 90°C, the most common method for heating macroscale
flasks is the *steam bath*.

Reactions are often stirred using a *magnetic stirrer* (Fig. 1.4) to help mix
reagents and to promote smooth boiling. A Teflon-coated bar magnet (*stirring
bar*) is placed in the reaction flask, and a magnetic stirrer is placed under the flask
and flask heater. The stirrer contains a large, horizontally rotating bar magnet that
attracts the Teflon-coated stirring bar magnet and causes it to turn. The speed of
stirring can be adjusted on the front of the magnetic stirrer.

B. Equipment for the Isolation of Products

Filtration

Online Study Center

Videos: The Reaction Tube in
Use, Filtration of Crystals Using
the Pasteur Pipette

If the product of a reaction crystallizes from the reaction mixture on cooling, the
solid crystals are isolated by *filtration*. This can be done in several ways when
using microscale techniques. If the crystals are large enough and in a reaction
tube, insert a *Pasteur pipette* to the bottom of the tube while expelling the air and

withdrawing the solvent (Fig. 1.5). Highly effective filtration occurs between the square, flat tip of the pipette and the bottom of the tube. This method of filtration has several advantages over the alternatives. The mixture of crystals and solvent can be kept on ice during the entire process. This minimizes the solubility of the crystals in the solvent. There are no transfer losses of material because an external filtration device is not used. This technique allows several recrystallizations to be carried out in the same tube with final drying of the product under vacuum. If you know the *tare* (the weight of the empty tube), the weight of the product can be determined without removing it from the tube. In this manner a compound can be synthesized, purified by crystallization, and dried all in the same reaction tube. After removal of material for analysis, the compound in the tube can then be used for the next reaction. This technique is used in many of this book's microscale experiments. When the crystals are dry, they are easily removed from the reaction tube. When they are wet, it is difficult to scrape them out. If the crystals are in more than about 2 mL of solvent, they can be isolated by filtration with a *Hirsch funnel*. The one that is in the microscale kit of apparatus is particularly easy to use because the funnel fits into the *filter flask* with no adapter and is equipped with a *polyethylene frit* for the capture of the filtered crystals (Fig. 1.6). The Wilfilter is especially good for collecting small quantities of crystals (Fig. 1.7).

FIG. 1.5
Filtration using a Pasteur pipette and a reaction tube.

FIG. 1.7
A Wilfilter is placed upside down in a centrifuge tube and spun in a centrifuge.

FIG. 1.6
A Hirsch funnel with an integral adapter, a polyethylene frit, and a 25-mL filter flask.

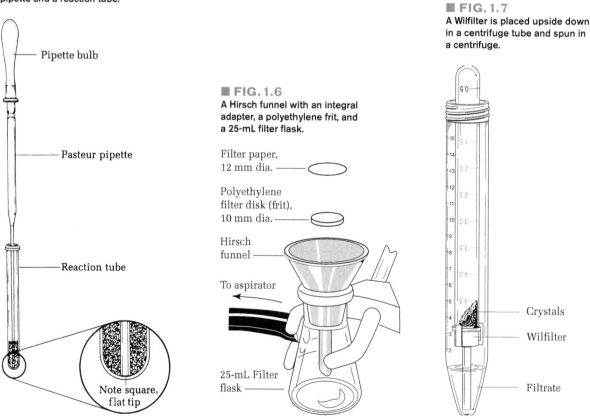

Pipette bulb

Pasteur pipette

Reaction tube

Note square, flat tip

Filter paper, 12 mm dia.

Polyethylene filter disk (frit), 10 mm dia.

Hirsch funnel

To aspirator

25-mL Filter flask

Crystals

Wilfilter

Filtrate

Online Study Center

Video: Macroscale Crystallization

Macroscale quantities of material can be recrystallized in conical *Erlenmeyer flasks* of the appropriate size. The crystals are collected in porcelain or plastic *Büchner funnels* fit with pieces of filter paper covering the holes in the bottom of the funnel (Fig. 1.8). A rubber *filter adapter* (*Filtervac*) is used to form a vacuum tight seal between the flask and the funnel.

Extraction

Chapter 7: Extraction

Online Study Center

Photos: Extraction with Ether, Extraction with Dichloromethane; Videos: Extraction with Ether, Extraction with Dichloromethane

The product of a reaction will often not crystallize. It may be a liquid or a viscous oil, it may be a mixture of compounds, or it may be too soluble in the reaction solvent being used. In this case, an immiscible solvent is added, the two layers are shaken to effect *extraction*, and after the layers separate, one layer is removed. On a microscale, this can be done with a Pasteur pipette. The extraction process is repeated if necessary. A tall, thin column of liquid, such as that produced in a reaction tube, makes it easy to selectively remove one layer by pipette. This is more difficult to do in the usual test tube because the height/diameter ratio is small.

On a larger scale, a *separatory funnel* is used for extraction (Fig. 1.9a). The mixture can be shaken in the funnel and then the lower layer removed through the

■ FIG. 1.9
(a) A separatory funnel with a Teflon stopcock. (b) A microscale separatory funnel. Remove the polyethylene frit from the micro Büchner funnel before using.

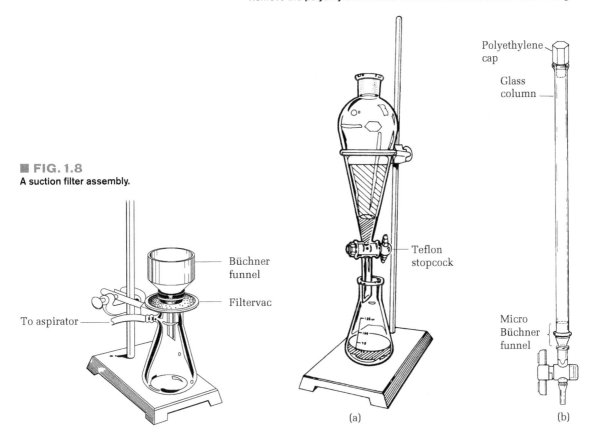

■ FIG. 1.8
A suction filter assembly.

Büchner funnel

Filtervac

To aspirator

Teflon stopcock

Polyethylene cap

Glass column

Micro Büchner funnel

(a)   (b)

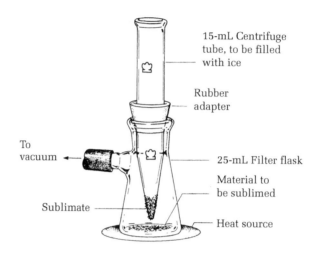

15-mL Centrifuge
tube, to be filled
with ice

Rubber
adapter

To
vacuum ←

25-mL Filter flask

Material to
be sublimed

Sublimate

Heat source

stopcock after the stopper is removed. These funnels are available in sizes from 10 mL to 5000 mL. The chromatography column in the apparatus kit is also a *micro separatory funnel* (Fig. 1.9b). Remember to remove the frit at the column base of the micro Büchner funnel and to close the valve before adding liquid.

Online Study Center

Photo: Sublimation Apparatus

Online Study Center

Photo: Column Chromatography;
Videos: Extraction with Ether,
Extraction with Dichloromethane

Chapter 9: Column
Chromatography

⚠
CAUTION: Use mercury-free
thermometers whenever
possible.

Online Study Center

Photo: Simple Distillation
Apparatus; Video: How to
Assemble Apparatus; Photo:
Fractional Distillation Apparatus

C. Equipment for Purification

Many solids can be purified by the process of *sublimation*. The solid is heated, and the vapor of the solid condenses on a cold surface to form crystals in an apparatus constructed from a *centrifuge tube* fitted with a rubber adapter and pushed into a *filter flask* (Fig. 1.10). Caffeine can be purified in this manner. This is primarily a microscale technique, although sublimers holding several grams of solid are available.

Mixtures of solids and, occasionally, of liquids can be separated and purified by *column chromatography*. The *chromatography column* for both microscale and macroscale work is very similar (Fig. 1.11).

Some of the compounds to be synthesized in these experiments are liquids. On a very small scale, the best way to separate and purify a mixture of liquids is by *gas chromatography*, but this technique is limited to less than 100 mg of material on the usual gas chromatograph. For larger quantities of material, *distillation* is used. For this purpose, small distilling flasks are used. These flasks have a large surface area to allow sufficient heat input to cause the liquid to vaporize rapidly so that it can be distilled and then condensed for collection in a *receiver*. The apparatus (Fig. 1.12) consists of a *distilling flask*, a *distilling adapter* (which also functions as an air condenser on a microscale), a *thermometer adapter*, and a *thermometer*; for macroscale, a water-cooled *condenser* and *distilling adapter* are added to the apparatus (Fig. 1.13). *Fractional distillation* is carried out using a small, packed *fractionating column* (Fig. 1.14). The apparatus is very similar for both microscale and macroscale. On a microscale, 2 mL to 4 mL of a liquid can be

A chromatography column consisting of a funnel, a tube, a base fitted with a polyethylene frit, and a Leur valve.

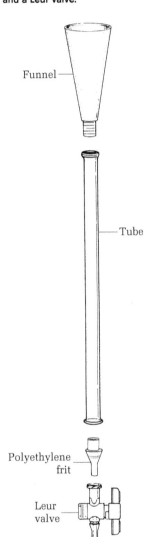

Funnel

Tube

Polyethylene frit

Leur valve

A small-scale simple distillation apparatus. Note that the entire thermometer bulb is below the side arm of the distilling adapter.

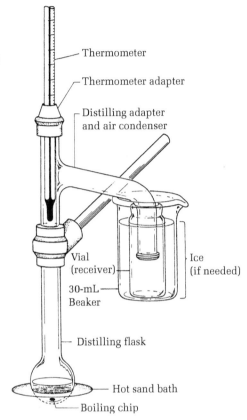

Thermometer

Thermometer adapter

Distilling adapter and air condenser

Vial (receiver)

Ice (if needed)

30-mL Beaker

Distilling flask

Hot sand bath

Boiling chip

fractionally distilled, and 1 mL or more can be simply distilled. The usual scale in these experiments for macroscale distillation is about 25 mL.

Some liquids with a relatively high vapor pressure can be isolated and purified by *steam distillation*, a process in which the organic compound codistills with water at the boiling point of water. The microscale and macroscale apparatus for this process are shown in Chapter 6.

The collection of typical equipment used for microscale experimentation is shown in Figure 1.15 and for macroscale experimentation in Figure 1.16. Other equipment commonly used in the organic laboratory is shown in Figure 1.17.

Check-in of Lab Equipment

Your first duty will be to check in to your assigned lab desk. The identity of much of the apparatus should already be apparent from the preceding outline of the experimental processes used in the organic laboratory.

Check to see that your thermometer reads about 22°C–25°C (71.6°F–77°F), which is normal room temperature. Examine the fluid column to see that it is

■ FIG. 1.13
An apparatus for simple distillation.

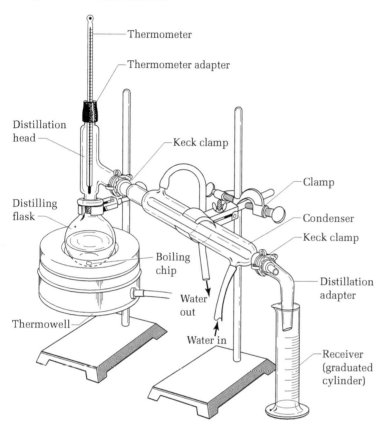

Thermometer

Thermometer adapter

Distillation
head

Keck clamp

Clamp

Condenser

Keck clamp

Distillation
adapter

Distilling
flask

Boiling
chip

Water
out

Thermowell

Water in

Receiver
(graduated
cylinder)

■ FIG. 1.14
A microscale fractional distillation
apparatus. The thermometer adapter is
to be fitted with a thermometer.

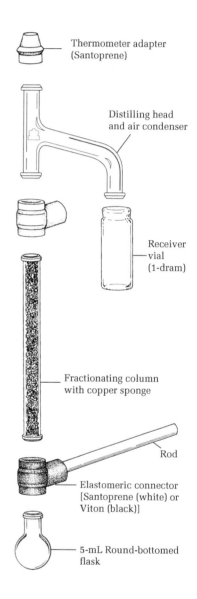

Thermometer adapter
(Santoprene)

Distilling head
and air condenser

Receiver
vial
(1-dram)

Fractionating column
with copper sponge

Rod

Elastomeric connector
[Santoprene (white) or
Viton (black)]

5-mL Round-bottomed
flask

■ FIG. 1.15
Microscale apparatus kit.

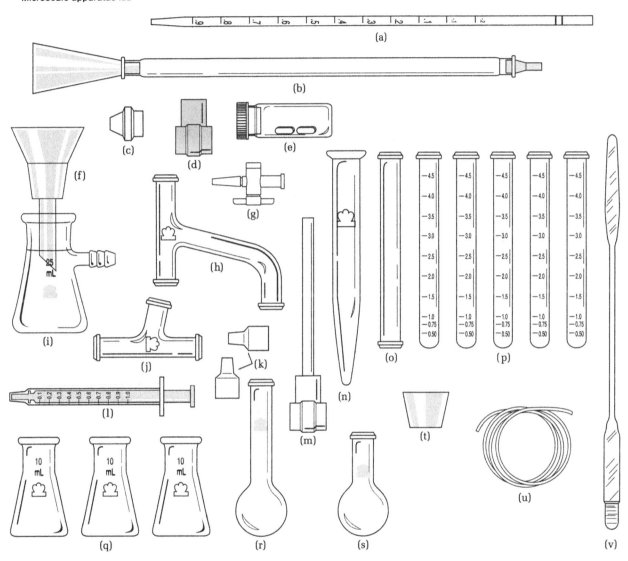

(a) Pipette (1 mL), graduated in 1/1000ths.
(b) Chromatography column (glass) with polypropylene funnel and 20-μm polyethylene frit in base, which doubles as a micro Büchner funnel. The column, base, and stopcock are also used as a separatory funnel.
(c) Thermometer adapter.
(d) Connector only (Viton).
(e) Magnetic stirring bars (4 × 12 mm) in distillation receiver vial.
(f) Hirsch funnel (polypropylene) with 20-μm fritted polyethylene disk.
(g) Stopcock for chromatography column and separatory funnel.
(h) Claisen adapter/distillation head with air condenser.
(i) Filter flask, 25 mL.
(j) Distillation head, 105° connecting adapter.
(k) Rubber septa/sleeve stoppers, 8 mm.
(l) Syringe (polypropylene).
(m) Connector with support rod.
(n) Centrifuge tube (15 mL)/sublimation receiver, with cap.
(o) Distillation column/air condenser.
(p) Reaction tube, calibrated, 10 × 100 mm.
(q) Erlenmeyer flasks, 10 mL.
(r) Long-necked flask, 5 mL.
(s) Short-necked flask, 5 mL.
(t) Rubber adapter for sublimation apparatus.
(u) Tubing (polyethylene), 1/16-in. diameter.
(v) Spatula (stainless steel) with scoop end.

■ **FIG. 1.16**
Macroscale apparatus kit with 14/20 standard-taper ground-glass joints.

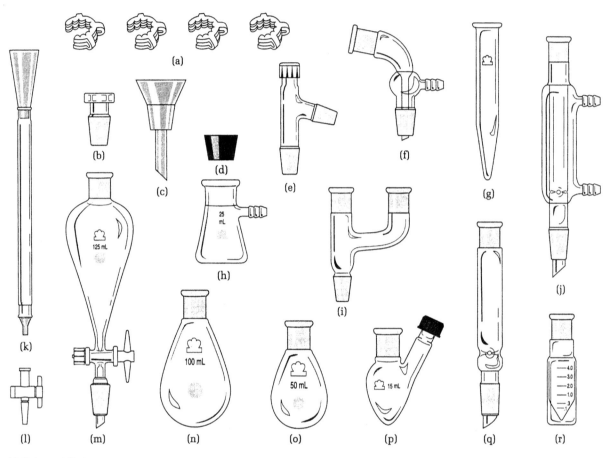

(a) Polyacetal Keck clamps, size 14.
(b) Hex-head glass stopper, 14/20 standard taper.
(c) Hirsch funnel (polypropylene) with 20-μm fritted polyethylene disk.
(d) Filter adapter for sublimation apparatus.
(e) Distilling head with O-ring thermometer adapter.
(f) Vacuum adapter.

(g) Centrifuge tube (15 mL)/sublimation receiver.
(h) Filter flask, 25 mL.
(i) Claisen adapter.
(j) Water-jacketed condenser.
(k) Chromatography column (glass) with polypropylene funnel and 20-μm polyethylene frit in base, which doubles as a micro Büchner funnel.

(l) Stopcock for chromatography column.
(m) Separatory funnel, 125 mL.
(n) Pear-shaped flask, 100 mL.
(o) Pear-shaped flask, 50 mL.
(p) Conical flask (15 mL) with side arm for inlet tube.
(q) Distilling column/air condenser.
(r) Conical reaction vial (5 mL)/distillation receiver.

unbroken and continuous from the bulb up. Replace any flasks that have star-shaped cracks. Remember that apparatus with graduations, stopcocks, or ground glass joints and anything porcelain are expensive. Erlenmeyer flasks, beakers, and test tubes are, by comparison, fairly cheap.

■ **FIG. 1.17**
Miscellaneous apparatus.
(a) 1.0 ± 0.01 mL graduated
pipette. (b) Septum. (c) 1.0-mL
syringe with blunt needle.
(d) Calibrated Pasteur pipette.
(e) Pipette pump. (f) Glass scorer.
(g) Filtervac. (h) Set of neoprene
filter adapters.

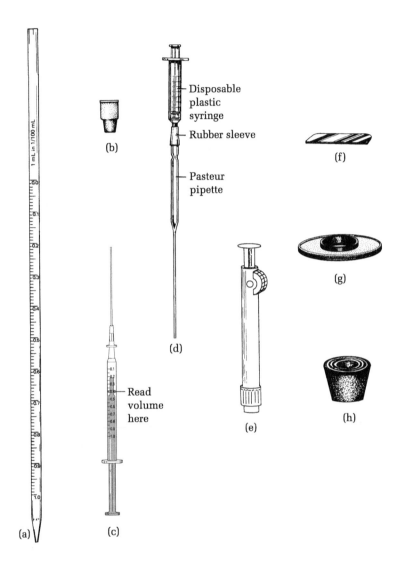

Transfer of Liquids and Solids

Pasteur pipettes (Fig. 1.18) are very useful for transferring small quantities of liquid, adding reagents dropwise, and carrying out recrystallizations. Discard used Pasteur pipettes in the special container for waste glass. Surprisingly, the acetone used to wash out a dirty Pasteur pipette usually costs more than the pipette itself.

A plastic funnel that fits on the top of the reaction tube is very convenient for the transfer of solids to reaction tubes or small Erlenmeyer flasks for microscale experiments (Fig. 1.19). It is also the top of the chromatography column (Fig. 1.11). A special spatula with a scoop end (Fig. 1.17p) is used to remove solid material

■ **FIG. 1.17** (continued)
(i) Hirsch funnel with perforated plate in place. (j) Rubber thermometer adapter. (k) Powder funnel. (l) Polyethylene wash bottle. (m) Single-pan electronic balance with automatic zeroing and 0.001 g digital readout; 100 g capacity. (n) Electric flask heater. (o) Solid-state control for electric flask heater. (p) Stainless steel spatula. (q) Stirring bar. (r) Keck clamp. (s) Wilfilter.

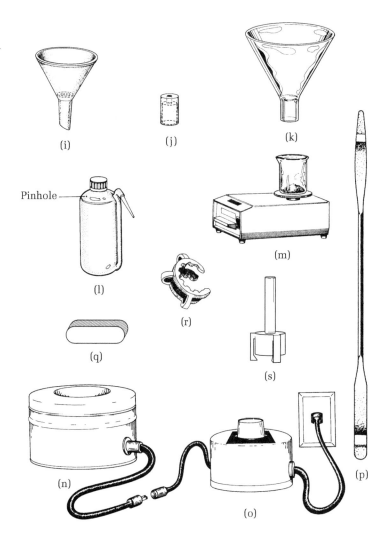

(i)   (j)   (k)

Pinhole

(l)   (m)

(r)

(q)   (s)

(n)   (o)   (p)

from the reaction tube. On a large scale, a powder funnel is useful for adding solids to a flask (Fig. 1.17k). A funnel can also be fashioned from a sheet of weighing paper.

Weighing and Measuring

The single-pan electronic balance (Fig. 1.17m), which is capable of weighing to ±0.001 g and having a capacity of at least 100 g, is the single most important instrument making microscale organic experiments possible. Most of the weighing measurements made in microscale experiments will use this type of balance. Weighing is fast and accurate with these balances. There should be one electronic balance for every 12 students. For macroscale experiments, a balance of such

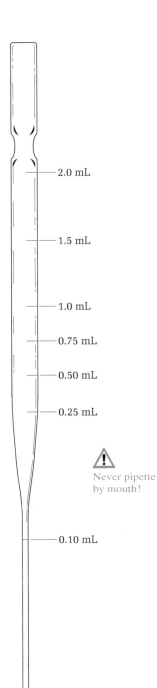

high accuracy is not necessary. Here, a balance with ±0.01 g accuracy would be satisfactory.

A container such as a reaction tube standing in a beaker or flask is placed on the pan. Set the digital readout to register zero and then add the desired quantity of the reagent to the reaction tube as the weight is measured periodically to the nearest milligram. Even liquids are weighed when accuracy is needed. It is much easier to weigh a liquid to 0.001 g than it is to measure it volumetrically to 0.001 mL.

It is often convenient to weigh reagents on glossy weighing paper and then transfer the chemical to the reaction container. The success of an experiment often depends on using just the right amount of starting materials and reagents. Inexperienced workers might think that if 1 mL of a reagent will do the job, then 2 mL will do the job twice as well. Such assumptions are usually erroneous.

Liquids can be measured by either volume or weight according to the following relationship:

$$\text{Volume (mL)} = \frac{\text{weight (g)}}{\text{density (g/mL)}}$$

Modern Erlenmeyer flasks and beakers have approximate volume calibrations fused into the glass, but these are *very* approximate. Better graduations are found on the microscale *reaction tube*. Somewhat more accurate volumetric measurements are made in 10-mL graduated cylinders. For volumes less than 4 mL, use a graduated pipette. **Never** apply suction to a pipette by mouth. The pipette can be fitted with a small plastic syringe using appropriately sized rubber tubing. A Pasteur pipette can be converted into a calibrated pipette with the addition of a plastic syringe (Fig. 1.17d). Figure 1.18 also shows the calibration marks for a 9-in. Pasteur pipette. You will find among your equipment a 1-mL pipette, calibrated in hundredths of a milliliter (Fig. 1.17a). Determine whether it is designed to *deliver* 1 mL or *contain* 1 mL between the top and bottom calibration marks. For our purposes, the latter is the better pipette.

Because the viscosity, surface tension, vapor pressure, and wetting characteristics of organic liquids are different from those of water, the so-called automatic pipette (designed for aqueous solutions) gives poor accuracy in measuring organic liquids. Syringes (Fig. 1.17c and Fig. 1.17d) and pipette pumps (Fig. 1.20), on the other hand, are quite useful, and these will be used frequently. Do not use a syringe that is equipped with a metal needle to measure corrosive reagents because these reagents will dissolve the metal in the needle. Because many reactions are "killed" by traces of moisture, many students' experiments are ruined by damp or wet apparatus. Several reactions that require especially dry or oxygen-free atmospheres will be run in systems sealed with a rubber septum (Fig. 1.17b). Reagents can be added to the system via syringe through this septum to minimize exposure to oxygen or atmospheric moisture.

Careful measurements of weights and volumes take more time than less accurate measurements. Think carefully about which measurements need to be made with accuracy and which do not.

■ **FIG. 1.18**
The approximate calibration of Kimble 9" Pasteur pipette.

■ **FIG. 1.19**
A funnel for adding solids and
liquids to a reaction tube.

■ **FIG. 1.20**
Using a pipette pump to measure
liquids to ±0.01 mL.

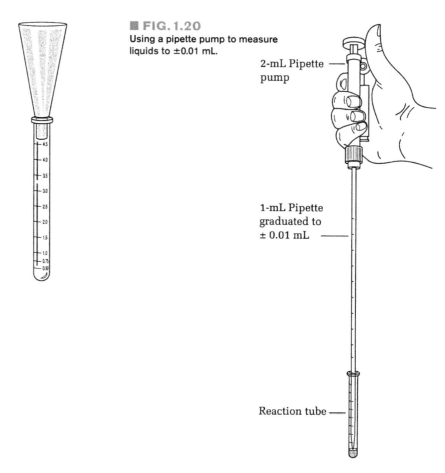

2-mL Pipette
pump

1-mL Pipette
graduated to
± 0.01 mL

Reaction tube

Tares

Tare = weight of empty
container

The tare of a container is its weight when empty. Throughout this laboratory
course, it will be necessary to know the tares of containers so that the weights of
the compounds within can be calculated. If identifying marks can be placed on the
containers (e.g., with a diamond stylus), you may want to record tares for frequently used containers in your laboratory notebook.

To be strictly correct, we should use the word *mass* instead of *weight* because
gravitational acceleration is not constant at all places on earth. But electronic balances record weights, unlike two-pan or triple-beam balances, which record masses.

Washing and Drying
Laboratory Equipment

Washing

Clean apparatus immediately.

Considerable time may be saved by cleaning each piece of equipment soon after use, for you will know at that point which contaminant is present and be
able to select the proper method for its removal. A residue is easier to remove

Both ethanol and acetone are very flammable.

Wash acetone goes in the organic solvents waste container; halogenated solvents go in the halogenated solvents waste container.

■ **FIG. 1.21**
A recrystallization solvent bottle and dispenser.

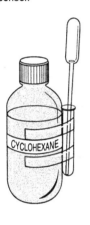

CYCLOHEXANE

before it has dried and hardened. A small amount of organic residue can usually be dissolved with a few milliliters of an appropriate organic solvent. Acetone (bp 56.1°C) has great solvent power and is often effective, but it is extremely flammable and somewhat expensive. Because it is miscible with water and vaporizes readily, it is easy to remove from the vessel. Detergent and water can also be used to clean dirty glassware if an appropriate solvent is not found. Cleaning after an operation may often be carried out while another experiment is in process.

A *polyethylene bottle* (Fig. 1.17l) is a convenient wash bottle for acetone. Be careful not to store solvent bottles in the vicinity of the reaction where they can provide additional fuel for an accidental fire. The name, symbol, and formula of a solvent is written on a bottle with a marker or a wax pencil. For macroscale crystallizations, extractions, and quick cleaning of apparatus, it is convenient to have a bottle for each frequently used solvent—95% ethanol, ligroin or hexanes, dichloromethane, ether, and ethyl acetate. A pinhole opposite the spout, which is covered with the finger when in use, will prevent the spout from dribbling the solvent. For microscale work, these solvents are best dispensed from 25-mL or 50-mL bottles with an attached test tube containing a graduated (1-mL) polypropylene pipette (Fig. 1.21). Be aware of any potential hazards stemming from the reactivity of these wash solvents with chemical residues in flasks. Also, be sure to dispose of wash solvents in the proper container. Acetone and most other organic solvents do not contain halogens and can therefore go in the regular organic solvents waste container. However, if dichloromethane or another halogen-containing solvent is used, it must be disposed of in the halogenated solvents waste container.

Sometimes a flask will not be clean after a washing with detergent and acetone. At that point try an abrasive household cleaner. If still no success, try adding dilute acid or base to the dirty glassware, let it soak for a few minutes, and rinse with plenty of water and acetone.

Drying

To dry a piece of apparatus rapidly, rinse with a few milliliters of acetone and invert over a beaker to drain. **Do not use compressed air**, which contains droplets of oil, water, and particles of rust. Instead, draw a slow stream of air through the apparatus using the suction of your water aspirator or house vacuum line.

Miscellaneous Cleanup

If a glass tube or thermometer becomes stuck to a rubber connector, it can be removed by painting on glycerol and forcing the pointed tip of a small spatula between the rubber and the glass. Another method is to select a cork borer that fits snugly over the glass tube, moisten it with glycerol, and slowly work it through the connector. If the stuck object is valuable, such as a thermometer, the best policy is to cut the rubber with a sharp knife. Care should be taken to avoid force that could potentially cause a thermometer to break, causing injury and the release of mercury.

The Laboratory Notebook

The laboratory notebook: What you did. How you did it. What you observed. Your conclusions.

A complete, accurate record is an essential part of all laboratory work. The failure to keep such a record means that laboratory labor is lost. An adequate record includes the procedure (what was done), observations (what happened), and conclusions (what the results mean).

Typically, laboratory personnel use a lined, 8.5 × 11 paperbound notebook and record all data in ink. Never record **anything** on scraps of paper to be recorded later in the notebook. Do not erase, remove, or obliterate notes. The use of whiteout is not good practice either. Simply draw a single line through incorrect entries.

When working in the laboratory, record everything you do and everything you observe **as it happens**. The recorded observations constitute the most important part of the laboratory record; they form the basis for the conclusions you will draw at the end of each experiment. One way to record observations is in a narrative form. Alternatively, the procedure can be written in outline form on the left-hand side of the page, with the observations recorded on the right-hand side.

In some colleges and universities, you will be expected to have all the relevant information about the running of an experiment entered in your notebook *before coming to the laboratory* so that your textbook will not be needed when you are conducting experiments. In industrial laboratories, a notebook may be designed so that carbon copies of all entries are kept. Notebooks are signed and dated by a supervisor, and the carbon copies are removed from the notebook each day and stored in a secure repository. Notebook records become critical legal documents, especially if a discovery turns out to be worth millions of dollars!

Record the physical properties of the product from your experiment, the yield in grams, and the percent yield. Analyze your results. When things did not turn out as expected, explain why. When your record of an experiment is complete, another chemist should be able to understand your account and determine what you did, how you did it, and what conclusions you reached. From the information in your notebook, this person should be able to repeat your work.

The detailed organization and format of your laboratory notebook will depend on your lab instructor's course design and varies from one instructor to another. On the publisher website, you will find an outline for preparing a typical laboratory record and two examples of completed laboratory records.

Online Study Center

General Resources
Laboratory Records

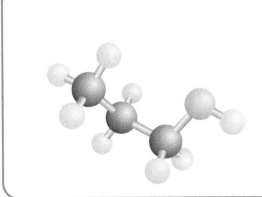

CHAPTER 2

Laboratory Safety, Courtesy, and Waste Disposal

PRELAB EXERCISE: Read this chapter carefully. Locate the emergency eyewash station, the safety shower, the fire extinguisher, and the emergency exits in your laboratory. Check your safety glasses or goggles for size and transparency. Learn which reactions must be carried out in the hood. Learn to use your laboratory fire extinguisher; learn how to summon help and how to put out a clothing fire. Learn first aid procedures for acid and alkali spills on the skin. Learn how to tell if your laboratory hood is working properly. Learn which operations under reduced pressure require special precautions. Check to see that compressed gas cylinders in your lab are firmly fastened to benches or walls. Learn the procedures for properly disposing of solid and liquid waste in your laboratory.

Small-scale (microscale) organic experiments are much safer to conduct than their macroscale counterparts that are run on a scale up to 100 times larger. However, for either microscale or macroscale experiments, the organic chemistry laboratory is an excellent place to learn and practice safety. The commonsense procedures practiced here also apply to other laboratories as well as to the shop, kitchen, and studio.

General laboratory safety information—particularly applicable to this organic chemistry laboratory course—is presented in this chapter. But it is not comprehensive. Throughout this text you will find specific cautions and safety information presented as margin notes printed in red. For a brief and thorough discussion of the topics in this chapter, you should read *Safety in Academic Chemistry Laboratories.*[1] There are also some specific admonitions regarding contact lenses (*see* "Eye Safety").

1. American Chemical Society Joint Board-Council Committee on Chemical Safety. *Safety in Academic Chemistry Laboratories, Vol. 1: Accident Prevention for College and University Students,* 7th ed.; American Chemical Society: Washington, DC, 2003.

Important General Rules

- Know the safety rules of your particular laboratory.
- Know the locations of emergency eyewashes, fire extinguishers, safety showers, and emergency exits.
- Never eat, drink, smoke, or apply cosmetics while in the laboratory.
- Wear gloves and aprons when handling corrosive materials.
- Never work alone.
- Perform no unauthorized experiments and do not distract your fellow workers; horseplay has no place in the laboratory.
- Dress properly for lab work. Do not wear open-toed shoes; your feet must be completely covered; wear shoes that have rubber soles and no heels or sneakers. Confine long hair and loose clothes. Do not wear shorts.
- Immediately report any accident to your instructor.
- If the fire extinguisher is used, report this to your instructor.
- Never use mouth suction to fill a pipette.
- Always wash your hands before leaving the laboratory.
- Do not use a solvent to remove a chemical from your skin. This will only hasten the absorption of the chemical through the skin.
- Do not use cell phones or tape, CD, MP3, or similar music players while working in the laboratory.
- Refer to the chemical supplier's hazard warning information or Material Safety Data Sheet (MSDS) when handling a new chemical for the first time.

Dress sensibly.

Eye protection

Eye Safety

Eye protection is extremely important. Safety glasses of some type must be worn at all times. It has been determined "that contact lenses can be worn in most work environments provided the same approved eye protection is worn as required of other workers in the area. Approved eye protection refers to safety glasses or goggles."[2] Chemical splash goggles are the preferred eye protection. One of the most important features of safety glasses/goggles is the brow bar. It is critical to have proper eye protection above the eyes; a brow bar satisfies this requirement for adequate splash protection.

Ordinary prescription eyeglasses do not offer adequate protection. Laboratory safety glasses should be constructed of plastic or tempered glass. If you do not have such glasses, wear goggles that afford protection from splashes and objects coming from the side as well as from the front. If plastic safety glasses are permitted in your laboratory, they should have side shields (Fig. 2.1). Eye safety cannot be overemphasized in the chemistry laboratory.

2. Ramsey, H.; and Breazeale, W. H. J. Jr. *Chem. Eng. News* **1998**, *76* (22), 6.

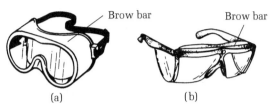

■ **FIG. 2.1**
(a) Chemical splash goggles.
(b) Safety glasses.

Brow bar   Brow bar

(a)   (b)

Laboratory Courtesy

Please show up on time and be prepared for the day's work. Clean up and leave promptly at the end of the lab period. Clean your desktop and sink before you leave the lab. Be certain that no items such as litmus paper, used filter papers, used cotton, or stir bars collect in the sink. Dispose of all trash properly. Please keep the balances clean. Always replace the caps on reagent bottles after use.

Working with Flammable Substances

Relative flammability of organic solvents.

Flammable substances are the most common hazard in the organic laboratory. Two factors can make today's organic laboratory much safer than its predecessor: (1) making the scale of the experiments as small as possible and (2) not using flames. Diethyl ether (bp 35°C), the most flammable substance you will usually work with in this course, has an ignition temperature of 160°C, which means that a hot plate at that temperature will cause it to burn. For comparison, *n*-hexane (bp 69°C), a constituent of gasoline, has an ignition temperature of 225°C. The flash points of these organic liquids—that is, the temperatures at which they will catch fire if exposed to a flame or spark—are below −20°C. These are very flammable liquids; however, if you are careful to eliminate all possible sources of ignition, they are not difficult to work with. Except for water, almost all of the liquids you will use in the laboratory are flammable.

■ **FIG. 2.2**
Solvent safety can.

Bulk solvents should be stored in and dispensed from safety cans (Fig. 2.2). These and other liquids will burn in the presence of the proper amount of their flammable vapors, oxygen, and a source of ignition (most commonly a flame or spark). It is usually difficult to remove oxygen from a fire, although it is possible to put out a fire in a beaker or a flask by simply covering the vessel with a flat object, thus cutting off the supply of air. Your lab notebook might do in an emergency. The best prevention is to pay close attention to sources of ignition—open flames, sparks, and hot surfaces. Remember, the vapors of flammable liquids are **always** heavier than air and thus will travel along bench tops, down drain troughs, and remain in sinks. For this reason all flames within the vicinity of a flammable liquid must be extinguished. Adequate ventilation is one of the best ways to prevent flammable vapors from accumulating. Work in an exhaust hood when manipulating large quantities of flammable liquids.

Flammable vapors travel along bench tops.

CAUTION: Keep ignition sources away from flammable liquids.

If a person's clothing catches fire and there is a safety shower close at hand, then shove the person under it and turn the shower on. Otherwise, push the person down and roll him or her over to extinguish the flames (stop, drop, and roll!). It is extremely important to prevent the victim from running or standing because the greatest harm comes from breathing the hot vapors that rise past the mouth. The safety shower

■ FIG. 2.3
Carbon dioxide fire
extinguisher.

⚠
Flammable vapors plus air
in a confined space are
explosive.

might then be used to extinguish glowing cloth that is no longer aflame. A so-called fire blanket should not be used because it tends to funnel flames past the victim's mouth, and clothing continues to char beneath it. It is, however, useful for retaining warmth to ward off shock after the flames are extinguished.

An organic chemistry laboratory should be equipped with a carbon dioxide or dry chemical (monoammonium phosphate) *fire extinguisher* (Fig. 2.3). To use this type of extinguisher, lift it from its support, pull the ring to break the seal, raise the horn, aim it at the base of the fire, and squeeze the handle. Do not hold on to the horn because it will become extremely cold. Do not replace the extinguisher; report the incident so the extinguisher can be refilled.

When disposing of certain chemicals, be alert for the possibility of *spontaneous combustion*. This may occur in oily rags; organic materials exposed to strong oxidizing agents such as nitric acid, permanganate ion, and peroxides; alkali metals such as sodium; or very finely divided metals such as zinc dust and platinum catalysts. Fires sometimes start when these chemicals are left in contact with filter paper.

Working with Hazardous Chemicals

If you do not know the properties of a chemical you will be working with, it is wise to regard the chemical as hazardous. The *flammability* of organic substances poses the most serious hazard in the organic laboratory. There is a possibility that storage containers in the laboratory may contribute to a fire. Large quantities of organic solvents should not be stored in glass bottles; they should be stored in solvent safety cans. Do not store chemicals on the floor.

A flammable liquid can often be vaporized to form, with air, a mixture that is explosive in a confined space. The beginning chemist is sometimes surprised to learn that diethyl ether is more likely to cause a laboratory fire or explosion than a worker's accidental anesthesia. The chances of being confined in a laboratory with a concentration of ether high enough to cause a loss of consciousness are extremely small, but a spark in such a room would probably destroy the building.

The probability of forming an explosive mixture of volatile organic liquids with air is far greater than that of producing an explosive solid or liquid. The chief functional groups that render compounds explosive are the *peroxide, acetylide, azide, diazonium, nitroso, nitro,* and *ozonide* groups (Fig. 2.4). Not all members of

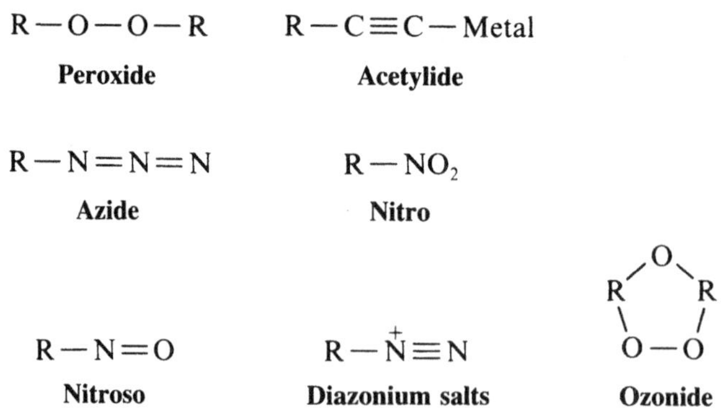

■ FIG. 2.4
Functional groups that can be
explosive in some compounds.

$$R-O-O-R \qquad R-C\equiv C-Metal$$
Peroxide **Acetylide**

$$R-N=N=N \qquad R-NO_2$$
Azide **Nitro**

$$R-N=O \qquad R-\overset{+}{N}\equiv N$$
Nitroso **Diazonium salts** **Ozonide**

these groups are equally sensitive to shock or heat. You would find it difficult to detonate trinitrotoluene (TNT) in the laboratory, but nitroglycerine is treacherously explosive. Peroxides present special problems that are discussed in the next section.

You will need to contend with the corrosiveness of many of the reagents you will handle. The principal danger here is to the eyes. Proper eye protection is mandatory, and even small-scale experiments can be hazardous to the eyes. It takes only a single drop of a corrosive reagent to do permanent damage. Handling concentrated acids and alkalis, dehydrating agents, and oxidizing agents calls for commonsense care to avoid spills and splashes and to avoid breathing the often corrosive vapors.

Certain organic chemicals present acute toxicity problems from short-duration exposure and chronic toxicity problems from long-term or repeated exposure. Exposure can result from ingestion, contact with the skin, or, most commonly, inhalation. Currently, great attention is being focused on chemicals that are teratogens (chemicals that often have no effect on a pregnant woman but cause abnormalities in a fetus), mutagens (chemicals causing changes in the structure of the DNA, which can lead to mutations in offspring), and carcinogens (cancer-causing chemicals). Small-scale experiments significantly reduce these hazards but do not completely eliminate them.

Working with Explosive Hazards

1. Peroxides

Certain functional groups can make an organic molecule become sensitive to heat and shock, such that it will explode. Chemists work with these functional groups only when there are no good alternatives. One of these functional groups, the peroxide group (R—O—O—R), is particularly insidious because it can form spontaneously when oxygen and light are present (Fig. 2.5). Ethers, especially cyclic ethers and those made from primary or secondary alcohols (such as tetrahydrofuran, diethyl ether, and diisopropyl ether), form peroxides. Other compounds that form peroxides are aldehydes, alkenes that have allylic hydrogen atoms (such as cyclohexene), compounds having benzylic hydrogens on a

Safety glasses or goggles must be worn at all times.

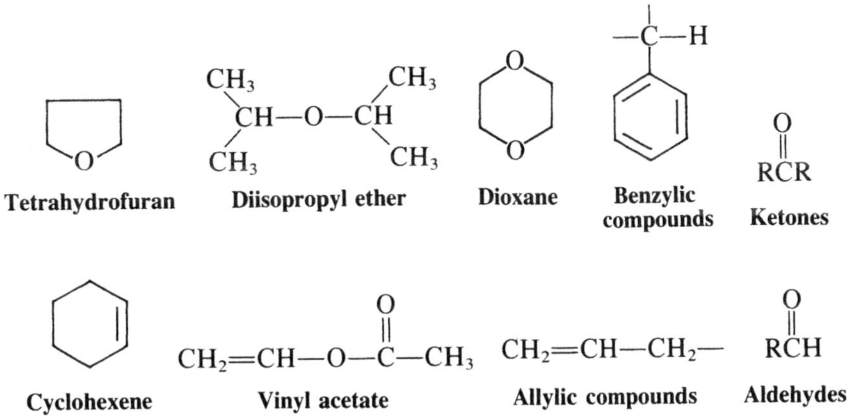

Ethers form explosive peroxides.

■ FIG. 2.5
Some compounds that form peroxides.

Tetrahydrofuran   Diisopropyl ether   Dioxane   Benzylic compounds   Ketones

Cyclohexene   Vinyl acetate   Allylic compounds   Aldehydes

tertiary carbon atom (such as isopropyl benzene), and vinyl compounds (such as vinyl acetate). Peroxides are low-power explosives but are extremely sensitive to shock, sparks, light, heat, friction, and impact. The greatest danger from peroxide impurities comes when the peroxide-forming compound is distilled. The peroxide has a higher boiling point than the parent compound and remains in the distilling flask as a residue that can become overheated and explode. For this reason, one should never distill a liquid until the distilling flask is completely dry, and the distillation should be run in a hood with the sash down to help contain a possible explosion.

Never distill to dryness.

The Detection of Peroxides	The Removal of Peroxides
To a solution of 0.01 g of sodium iodide in 0.1 mL of glacial acetic acid, add 0.1 mL of the liquid suspected of containing a peroxide. If the mixture turns brown, a high concentration of peroxide is present; if it turns yellow, a low concentration of peroxide is present.	Pouring the solvent through a column of activated alumina will remove peroxides and simultaneously dry the solvent. Do not allow the column to dry out while in use. When the alumina column is no longer effective, wash the column with 5% aqueous ferrous sulfate and discard the column as nonhazardous waste.

Problems with peroxide formation are especially critical for ethers. Ethers (R-O-R′) form peroxides readily. Because ethers are frequently used as solvents, they are often used in quantity and then removed to leave reaction products. Cans of diethyl ether should be dated when opened. If opened cans are not used within one month, they should be treated for peroxides and disposed of.

t-Butyl methyl ether, $(CH_3)_3C-O-CH_3$, with a primary carbon on one side of the oxygen and a tertiary carbon on the other does not form peroxides easily. It is highly desirable to use this in place of diethyl ether for extraction. Refer to the discussion in Chapter 7.

You may have occasion to use *30% hydrogen peroxide*. This material causes severe burns if it contacts the skin and decomposes violently if contaminated with metals or their salts. Be particularly careful not to contaminate the reagent bottle.

2. Closed Systems

A closed system is defined as not being open to the atmosphere. Any sealed system is a closed system. If a closed system is not properly prepared, an explosion may result from the system being under pressure, caused from gas or heat evolution from the reaction or from applied heat to the system. One way to prevent an explosion of a closed system is to use glassware that can withstand the pressure and to evacuate the system under vacuum before it is closed to the atmosphere.

Most reactions are run in open systems; that is, they are run in apparatus that are open to atmosphere, either directly or through a nitrogen line hooked up to a bubbler, which is open to atmosphere.

Working with Corrosive Substances

Handle strong acids, alkalis, dehydrating agents, and oxidizing agents carefully so as to avoid contact with the skin and eyes and to avoid breathing the corrosive vapors that attack the respiratory tract. All strong, concentrated acids attack the skin and eyes. *Concentrated sulfuric acid* is both a dehydrating agent and a strong acid and will cause very severe burns. *Nitric acid* and *chromic acid* (used in cleaning solutions) also cause bad burns. *Hydrofluoric acid* is especially harmful and causes deep, painful, and slow-healing wounds. It should be used only after thorough instruction. You should wear approved safety glasses or goggles, protective gloves, and an apron when handling these materials.

Add H_2SO4, P_2O_5, CaO, and NaOH to water, not the reverse.

Sodium, potassium, and ammonium hydroxides are common bases that you will encounter. Sodium and potassium hydroxides are extremely damaging to the eyes, and ammonium hydroxide is a severe bronchial irritant. Like sulfuric acid, sodium hydroxide, phosphorous pentoxide, and calcium oxide are powerful dehydrating agents. Their great affinity for water will cause burns to the skin. Because they release a great deal of heat when they react with water, to avoid spattering they should always be added to water rather than water being added to them. That is, the heavier substance should always be added to the lighter one so that rapid mixing results as a consequence of the law of gravity.

You will receive special instructions when it comes time to handle metallic sodium, lithium aluminum hydride, and sodium hydride, three substances that can react explosively with water.

Among the strong oxidizing agents, *perchloric acid* ($HClO_4$) is probably the most hazardous. It can form heavy metal and organic *perchlorates* that are *explosive*, and it can react explosively if it comes in contact with organic compounds.

If one of these substances gets on the skin or in the eyes, wash the affected area with very large quantities of water, using the safety shower and/or eyewash station (Fig. 2.6) until medical assistance arrives. Do not attempt to neutralize the reagent chemically. Remove contaminated clothing so that thorough washing can take place. Take care to wash the reagent from under the fingernails.

Wipe up spilled hydroxide pellets *rapidly.*

Take care not to let the reagents, such as sulfuric acid, run down the outside of a bottle or flask and come in contact with your fingers. Wipe up spills immediately with a very damp sponge, especially in the area around the balances. Pellets of sodium and potassium hydroxide are very hygroscopic and will dissolve in the water they pick up from the air; they should therefore be wiped up very quickly. When handling large quantities of corrosive chemicals, wear protective gloves, a face mask, and a neoprene apron. The corrosive vapors can be avoided by carrying out work in a good exhaust hood.

Do not use a plastic syringe with a metal needle to dispense corrosive inorganic reagents, such as concentrated acids or bases.

Working with Toxic Substances

Many chemicals have very specific toxic effects. They interfere with the body's metabolism in a known way. For example, the cyanide ion combines irreversibly with hemoglobin to form cyanometmyoglobin, which can no longer carry oxygen.

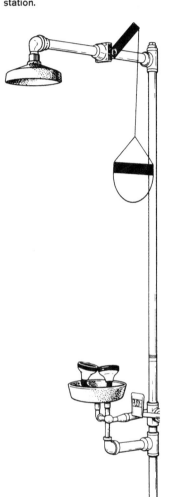

Aniline acts in the same way. Carbon tetrachloride and other halogenated compounds can cause liver and kidney failure. Carcinogenic and mutagenic substances deserve special attention because of their long-term insidious effects. The ability of certain carcinogens to cause cancer is very great; for example, special precautions are needed when handling aflatoxin B_1. In other cases, such as with dioxane, the hazard is so low that no special precautions are needed beyond reasonable, normal care in the laboratory.

Women of childbearing age should be careful when handling any substance of unknown properties. Certain substances are highly suspected as teratogens and will cause abnormalities in an embryo or fetus. Among these are benzene, toluene, xylene, aniline, nitrobenzene, phenol, formaldehyde, dimethylformamide (DMF), dimethyl sulfoxide (DMSO), polychlorinated biphenyls (PCBs), estradiol, hydrogen sulfide, carbon disulfide, carbon monoxide, nitrites, nitrous oxide, organolead and mercury compounds, and the notorious sedative thalidomide. Some of these substances will be used in subsequent experiments. Use care when working with these (and all) substances. Of course, the leading known cause of embryotoxic effects is ethyl alcohol in the form of maternal alcoholism. The amount of ethanol vapor inhaled in the laboratory or absorbed through the skin is so minute that it is unlikely to have morbid effects.

It is impossible to avoid handling every known or suspected toxic substance, so it is wise to know what measures should be taken. Because the eating of food or the consumption of beverages in the laboratory is strictly forbidden and because one should never taste material in the laboratory, the possibility of poisoning by mouth is remote. Be more careful than your predecessors—the hallucinogenic properties of LSD and **all** artificial sweeteners were discovered by accident. The two most important measures to be taken, then, are (1) avoiding skin contact by wearing the *proper* type of protective gloves (*see* "Gloves") and (2) avoiding inhalation by working in a good exhaust hood.

Many of the chemicals used in this course will be unfamiliar to you. Their properties can be looked up in reference books, a very useful one being the *Aldrich Handbook of Fine Chemicals*.[3] Note that 1,4-dichlorobenzene is listed as a "toxic irritant" and naphthalene is listed as an "irritant." Both are used as mothballs. Camphor, used in vaporizers, is classified as a "flammable solid irritant." Salicylic acid, which we will use to synthesize aspirin (Chapter 41), is listed as a "moisture-sensitive toxic." Aspirin (acetylsalicylic acid) is classified as an "irritant." Caffeine, which we will isolate from tea or cola syrup (Chapter 7), is classified as "toxic." Substances not so familiar to you, for example, 1-naphthol and benzoic acid, are classified, respectively, as "toxic irritant" and "irritant." To put things in perspective, nicotine is classified as "highly toxic." Pay attention to these health warnings. In laboratory quantities, common chemicals can be hazardous. Wash your hands carefully after coming in contact with laboratory chemicals. Consult the *Hazardous Laboratory Chemicals Disposal Guide*[4] for information on truly hazardous chemicals.

---

3. Free copies of this catalog can be obtained from http://www.sigmaaldrich.com/Brands/Aldrich.html.

4. Armour, M-A. *Hazardous Laboratory Chemicals Disposal Guide*, 3rd ed.; CRC Press LLC: Boca Raton, FL, 2003.

Because you have not had previous experience working with organic chemicals, most of the experiments you will carry out in this course will not involve the use of known carcinogens, although you will work routinely with flammable, corrosive, and toxic substances. A few experiments involve the use of substances that are suspected of being carcinogenic, such as hydrazine. If you pay proper attention to the rules of safety, you should find working with these substances no more hazardous than working with ammonia or nitric acid. The single, short-duration exposure you might receive from a suspected carcinogen, should an accident occur, would probably have no long-term consequences. The reason for taking the precautions noted in each experiment is to learn, from the beginning, good safety habits.

Gloves

Be aware that protective gloves in the organic laboratory may not offer much protection. Polyethylene and latex rubber gloves are very permeable to many organic liquids. An undetected pinhole may bring with it long-term contact with reagents. Disposable polyvinyl chloride (PVC) gloves offer reasonable protection from contact with aqueous solutions of acids, bases, and dyes, but no one type of glove is useful as a protection against all reagents. It is for this reason that no less than 25 different types of chemically resistant gloves are available from laboratory supply houses. Some gloves are quite expensive and will last for years.

If disposable gloves are available, fresh nitrile gloves can be worn whenever handling a corrosive substance and disposed of once the transfer is complete. When not wearing gloves, it is advised that you wash your hands every 15 min to remove any traces of chemicals that might be on them.

Using the Laboratory Hood

Modern practice dictates that in laboratories where workers spend most of their time working with chemicals, there should be one exhaust hood for every two people. This precaution is often not possible in the beginning organic chemistry laboratory, however. In this course you will find that for some experiments the hood must be used and for others it is advisable; in these instances, it may be necessary to schedule experimental work around access to the hoods. Many experiments formerly carried out in the hood can now be carried out at the lab desk because the concentration of vapors is significantly minimized when working at a microscale.

The hood offers a number of advantages when working with toxic and flammable substances. Not only does it draw off the toxic and flammable fumes, but it also affords an excellent physical barrier on all four sides of a reacting system when the sash is pulled down. If a chemical spill occurs, it may be contained within the hood.

Keep the hood sash closed.

It is your responsibility each time you use a hood to see that it is working properly. You should find some type of indicating device that will give you this information on the hood itself. A simple propeller on a cork works well. The hood is a backup device. Never use it alone to dispose of chemicals by evaporation; use an aspirator tube or carry out a distillation. Toxic and flammable fumes should be trapped or condensed in some way and disposed of in the prescribed manner. The sash should be pulled down unless you are actually carrying out manipulations on the experimental apparatus. The water, gas, and electrical controls should be on the outside of the hood so it is not necessary to open the hood to make adjustments. The ability of the hood to remove vapors is greatly enhanced if the apparatus is kept as close to the back of the hood as possible. Everything should be at least 15 cm back from the hood sash. Chemicals should not be permanently stored in the hood but should be removed to ventilated storage areas. If the hood is cluttered with chemicals, you will not achieve a good, smooth airflow or have adequate room for experiments.

Working at Reduced Pressure

Implosion

Whenever a vessel or system is evacuated, an implosion could result from atmospheric pressure on the empty vessel. It makes little difference whether the vacuum is perfect or just 10 mm Hg; the pressure difference is almost the same (760 versus 750 mm Hg). An implosion may occur if there is a star crack in the flask or if the flask is scratched or etched. Only with heavy-walled flasks specifically designed for vacuum filtration is the use of a safety shield (Fig. 2.7) ordinarily unnecessary. The chances of implosion of the apparatus used for microscale experiments are remote.

Dewar flasks (thermos bottles) are often found in the laboratory without shielding. These should be wrapped with friction tape or covered with a plastic net to prevent the glass from flying about in case of an implosion (Fig. 2.8). Similarly, vacuum desiccators should be wrapped with tape before being evacuated.

■ FIG. 2.7
Safety shield.

■ FIG. 2.8
Dewar flask with safety net in place.

Working with Compressed Gas Cylinders

Many reactions are carried out under an inert atmosphere so that the reactants and/or products will not react with oxygen or moisture in the air. Nitrogen and argon are the inert gases most frequently used. Oxygen is widely used both as a reactant and to provide a hot flame for glassblowing and welding. It is used in the oxidative coupling of alkynes (Chapter 24). Helium is the carrier gas used in gas chromatography. Other gases commonly used in the laboratory are ammonia, often used as a solvent; chlorine, used for chlorination reactions; acetylene, used in combination with oxygen for welding; and hydrogen, used for high- and low-pressure hydrogenation reactions.

The following rule applies to all compressed gases: Compressed gas cylinders should be firmly secured at all times. For temporary use, a clamp that attaches to the laboratory bench top and has a belt for the cylinder will suffice (Fig. 2.9). Eyebolts and chains should be used to secure cylinders in permanent installations. Flammable gases should be stored 20 ft from oxidizing gases.

A variety of outlet threads are used on gas cylinders to prevent incompatible gases from being mixed because of an interchange of connections. Both right- and left-handed external and internal threads are used. Left-handed nuts are notched to differentiate them from right-handed nuts. Right-handed threads are used on non-fuel and oxidizing gases, and left-handed threads are used on fuel gases, such as hydrogen. Never grease the threads on tank or regulator valves.

Cylinders come equipped with caps that should be left in place during storage and transportation. These caps can be removed by hand. Under these caps is a cylinder valve. It can be opened by turning the valve counterclockwise; however, because most compressed gases in full cylinders are under very high pressure (commonly up to 3000 lb/in.^{2}), a pressure regulator must be attached to the cylinder. This pressure regulator is almost always of the diaphragm type and has two gauges, one indicating the pressure in the cylinder, the other the outlet pressure (Fig. 2.10). On the outlet, low-pressure side of the regulator is a small needle valve and then the outlet connector. After connecting the regulator to the cylinder,

Always clamp gas cylinders.

█ FIG. 2.9
Gas cylinder clamp.

— Hand-wheel valve

█ FIG. 2.10
Gas pressure regulator. Turn two-flanged diaphragm valve clockwise to increase outlet pressure.

Clockwise movement of diaphragm valve handle *increases* pressure.

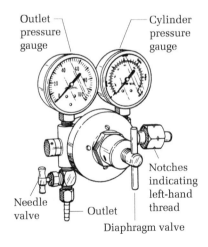

Outlet pressure gauge — Cylinder pressure gauge

Needle valve — Outlet

Notches indicating left-hand thread

Diaphragm valve

■ **FIG. 2.11**
Gas cylinder cart.

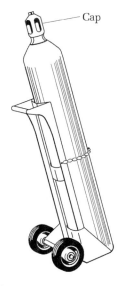

Never attempt to identify an
unknown organic compound by
smelling it.

Clean up spills rapidly.

Mercury requires special
measures—see your instructor.

unscrew the diaphragm valve (turn it counterclockwise) before opening the cylinder valve on the top of the cylinder. This valve should be opened only as far as necessary. For most gas flow rates in the laboratory, this will be a very small amount. The gas flow or pressure is increased by turning the two-flanged diaphragm valve **clockwise**. When the apparatus is not being used, turn off the cylinder valve (clockwise) on the top of the cylinder (Fig. 2.9). Before removing the regulator from the cylinder, reduce the flow or pressure to zero. Cylinders should never be emptied to zero pressure and left with the valve open because the residual contents will become contaminated with air. Empty cylinders should be labeled "empty." Their valves should be closed and capped, and the cylinders should be returned to the storage area and separated from full cylinders. Gas cylinders should never be dragged or rolled from place to place but should be fastened onto and moved in a cart designed for that purpose (Fig. 2.11). The cap should be in place whenever the cylinder is moved. If you detect even a small leak from any valve or connection, immediately seek the help of an instructor to remedy the problem. If there is a major leak of a corrosive or flammable gas, notify those around you to leave the area and seek help immediately.

Odoriferous Chemicals

Some organic chemicals just smell bad. Among these are the thiols (organic derivatives of hydrogen sulfide), isonitriles, many amines (e.g., cadaverine), and butyric acid. Washing apparatus and, if necessary, hands in a solution of a quaternary ammonium salt may solve the problem. These compounds apparently complex with many odoriferous substances, allowing them to be rinsed away. Commercial products (e.g., Zephiran, Roccal, San-O-Fec, and others) are available at pet and farm supply stores.

Waste Disposal–Cleaning Up

Spilled solids should simply be swept up and placed in the appropriate solid waste container. This should be done promptly because many solids are hygroscopic and become difficult if not impossible to sweep up in a short time. This is particularly true of sodium hydroxide and potassium hydroxide; these strong bases should be dissolved in water and neutralized with sodium bisulfate before disposal.

The method used to clean up spills depends on the type and amount of chemical spilled. If more than 1 or 2 g or mL of any chemical, particularly a corrosive or volatile one, is spilled, you should consult your instructor for the best way to clean up the spill. If a large amount of volatile or noxious liquid is spilled as might happen if a reagent bottle is dropped and broken, advise those in the area to leave the laboratory and contact your instructor immediately. If a spill involves a large amount of flammable liquid, be aware of any potential ignition sources and try to eliminate them. Large amounts of spilled acid can be neutralized with granular limestone or cement; large amounts of bases with solid sodium bisulfate, $NaHSO_4$. Large amounts of volatile liquids can be absorbed into materials such as vermiculite, diatomaceous earth, dry sand, kitty litter, or paper towels and these materials swept up and placed in a separate disposal container.

For spills of amounts less than 2 g of chemical, proceed as follows. Acid spills should be neutralized by dropping solid sodium carbonate onto them, testing the pH, wiping up the neutralized material with a sponge, and rinsing the neutral salt solution down the drain. Bases should be neutralized by sprinkling solid sodium bisulfate onto them, checking the pH, and wiping up with a sponge or towel. Do not use paper towels to wipe up spills of strong oxidizers such as dichromates or nitrates; the towels can ignite. Bits of sodium metal will also cause paper towels to ignite. Sodium metal is best destroyed with *n*-butyl alcohol. Always wear goves when cleaning up a spill.

Cleaning Up

In the not-too-distant past it was common practice to wash all unwanted liquids from the organic laboratory down the drain and to place all solid waste in the trash basket. For environmental reasons, this is never a wise practice and is no longer allowed by law.

Organic reactions usually employ a solvent and often involve the use of a strong acid, a strong base, an oxidant, a reductant, or a catalyst. None of these should be washed down the drain or placed in the wastebasket. Place the material, classified as waste, in containers labeled for nonhazardous solid waste, organic solvents, halogenated organic solvents, or hazardous wastes of various types.

Waste containers:
Nonhazardous solid waste
Organic solvents
Halogenated organic solvents
Hazardous waste (various types)

Nonhazardous waste encompasses such solids as paper, corks, TLC plates, solid chromatographic absorbents such as alumina or silica that are dry and free of residual organic solvents, and solid drying agents such as calcium chloride or sodium sulfate that are also dry and free of residual organic solvents. These will ultimately end up in a sanitary landfill (the local dump). Any chemicals that are leached by rainwater from this landfill must not be harmful to the environment. In the *organic solvents* container are placed the solvents that are used for recrystallization and for running reactions, cleaning apparatus, and so forth. These solvents can contain dissolved, solid, nonhazardous organic solids. This solution will go to an incinerator where it will be burned. If the solvent is halogenated (e.g., dichloromethane) or contains halogenated material, it must go in the *halogenated organic solvents* container. Ultimately, this will go to a special incinerator equipped with a scrubber to remove HCl from the combustion gases. The organic laboratory should also have several other waste disposal containers for special hazardous, reactive, and noncombustible wastes that would be incompatible with waste organic solvents and other materials. For example, it would be dangerous to place oxidants in lysts with many organics. In particular, separate waste containers should be provided for toxic heavy metal wastes containing mercury, chromium, or lead salts, and so forth. The cleaning up sections throughout this text will call your attention to these special hazards.

Hazardous wastes such as sodium hydrosulfite (a reducing agent), platinum catalysts, and Cr^{6+} (an oxidizing agent) cannot be burned and must be shipped to a secure landfill. To dispose of small quantities of a hazardous waste (e.g., solid mercury hydroxide), the material must be carefully packed in bottles and placed in a 55-gal ($\approx$208-L) drum called a lab pack, to which an inert material has been added. The lab pack is carefully documented and hauled off to a site where such waste is disposed of by a bonded, licensed, and heavily regulated waste disposal company. Formerly, many hazardous wastes were disposed of by burial in a secure landfill.

The kinds of hazardous waste that can be thus disposed of have become extremely limited in recent years, and much of the waste undergoes various kinds of treatment at the disposal site (e.g., neutralization, incineration, or reduction) to put it in a form that can be safely buried in a secure landfill or flushed to a sewer. There are relatively few places for approved disposal of hazardous waste. For example, there are none in New England, so most hazardous waste from this area is trucked to South Carolina. The charge to small generators of waste is usually based on the volume of waste. So, 1000 mL of a 2% cyanide solution would cost far more to dispose of than 20 g of solid cyanide, even though the total amount of this poisonous substance is the same. It now costs far more to dispose of most hazardous chemicals than it does to purchase them new.

American law states that a material is not a waste until the laboratory worker declares it a waste. So—for pedagogical and practical reasons—we want you to regard the chemical treatment of the byproducts of each reaction in this text as a part of the experiment.

In the section titled "Cleaning Up" at the end of each experiment, the goal is to reduce the volume of hazardous waste, to convert hazardous waste to less hazardous waste, or to convert it to nonhazardous waste. The simplest example is concentrated sulfuric acid. As a byproduct from a reaction, it is obviously hazardous. But after careful dilution with water and neutralization with sodium carbonate, the sulfuric acid becomes a dilute solution of sodium sulfate, which in almost every locale can be flushed down the drain with a large quantity of water. Anything flushed down the drain must be accompanied by a large excess of water. Similarly, concentrated bases can be neutralized, oxidants such as Cr^{6+} can be reduced, and reductants such as hydrosulfite can be oxidized (by hypochlorite or household bleach). Dilute solutions of heavy metal ions can be precipitated as their insoluble sulfides or hydroxides. The precipitate may still be a hazardous waste, but it will have a much smaller volume.

One type of hazardous waste is unique: a harmless solid that is damp with an organic solvent. Alumina from a chromatography column and calcium chloride used to dry an ether solution are examples. Being solids, they obviously cannot go in the organic solvents container, and being flammable they cannot go in the nonhazardous waste container. A solution to this problem is to spread the solid out in the hood to let the solvent evaporate. You can then place the solid in the nonhazardous waste container. The savings in waste disposal costs by this operation are enormous. However, be aware of the regulations in your area as they may not allow evaporation of small amounts of organic solvents in a hood. If this is the case, special containers should be available for disposal of these wet solids.

Our goal in "Cleaning Up" is to make you more aware of *all* aspects of an experiment. Waste disposal is now an extremely important aspect. Check to be sure the procedure you use is permitted by law in your location. Three sources of information have been used as the basis of the procedures at the end of each experiment: the *Aldrich Catalog Handbook of Fine Chemicals*,[5] which gives brief

Waste disposal is very expensive.

The law: A waste is not a waste until the laboratory worker declares it a waste.

Cleaning up: reducing the volume of hazardous waste or converting hazardous waste to less hazardous or nonhazardous waste.

Disposing of solids wet with organic solvents: alumina and anhydrous calcium chloride pellets.

---

5. See footnote 3 on page 30.

disposal procedures for every chemical in their catalog; *Prudent Practices in the Laboratory: Handling and Disposal of Chemicals*[6]; and the *Hazardous Laboratory Chemicals Disposal Guide*.[7] The last title listed here should be on the bookshelf of every laboratory. This 464-page book gives detailed information about hundreds of hazardous substances, including their physical properties, hazardous reactions, physiological properties, health hazards, spillage disposal, and waste disposal. Many of the treatment procedures in "Cleaning Up" are adaptations of these procedures. *Destruction of Hazardous Chemicals in the Laboratory*[8] complements this book.

The area of waste disposal is changing rapidly. Many levels of laws apply—local, state, and federal. What may be permissible to wash down the drain or evaporate in the hood in one jurisdiction may be illegal in another, so before carrying out any waste disposal, check with your college or university waste disposal officer.

Biohazards

The use of microbial growth bioassays is becoming common in chemistry laboratories. The use of infectious materials presents new hazards that must be recognized and addressed. The first step in reducing hazards when using these materials is to select infectious materials that are known not to cause illness in humans and are of minimal hazard to the environment. A number of procedures should be followed to make use of these materials safe: Individuals need to wash hands after they handle these materials and before they leave the laboratory; work surfaces need to be decontaminated at the end of each use; and all infectious materials need to be decontaminated before disposal.

QUESTIONS

1. Write a balanced equation for the reaction between the iodide ion, a peroxide, and the hydrogen ion. What causes the orange or brown color?
2. Why does the horn of the carbon dioxide fire extinguisher become cold when the extinguisher is used?
3. Why is water *not* used to extinguish most fires in an organic laboratory?

6. National Research Council. *Prudent Practices in the Laboratory: Handling and Disposal of Chemicals* National Academy Press: Washington, DC, 1995.

7. See footnote 4 on page 30.

8. Lunn, G.; Sansone, E. B. *Destruction of Hazardous Chemicals in the Laboratory*; Wiley: New York, 1994.

CHAPTER 3

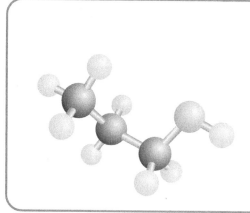

Melting Points and Boiling Points

Online Study Center

This icon will direct you to techniques, equipment setups, and online resources at http://college. hmco.com/PIC/ williamsonMMOE5e.

PRELAB EXERCISE: Draw the structure of each of the following organic compounds, identify the intermolecular attractive forces for each, and list them in order of increasing boiling point as predicted by the relative strength of those intermolecular forces: (a) acetaldehyde, (b) sodium formate, (c) ethanol, and (d) propane.

PART 1: Five Concepts for Predicting Physical Properties

In organic chemistry, structure is everything. A molecule's structure determines both its physical properties and its reactivity. Since the dawn of modern chemistry 200 years ago, over 20 million substances, most of them organic compounds, have been isolated, and their properties and reactions have been studied. It became apparent from these studies that certain structural features in organic molecules would affect the observed properties in a predictable way and that these millions of organic compounds could be organized into classes based on molecular size, composition, and the pattern of bonds between atoms. Chemists also saw trends in certain properties based on systematic changes in these structural features. This organized knowledge allows us to look at a compound's structure and to predict the physical properties of that compound.

Physical properties, such as melting point, boiling point, and solubility, are largely determined by *intermolecular attractive forces*. You learned about these properties in previous chemistry courses. Because a solid understanding of these concepts is critical to understanding organic chemistry, we will review the different types of forces in the context of structural organic chemistry. Using five simple concepts, you should be able to look at the structures of a group of different organic molecules and predict which might be liquids, gases, or solids and which might be soluble in water. You can often predict the boiling point, melting point, or solubility of one structure relative to other structures. In fact, as your knowledge grows, you may be able to predict a compound's approximate melting or boiling temperature based on its structure. Your understanding of intermolecular attractive forces will be very useful in this chapter's experiments on melting and boiling points and those in Chapters 5 and 6 that involve distillation and boiling points.

1. London Attractive Forces (Often Called Van der Waals Forces)

Organic molecules that contain only carbon and hydrogen (hydrocarbons) are weakly attracted to each other by London forces. Though weak, these attractive forces increase as molecular size increases. Thus, the larger the molecule, the greater the attractive force for neighboring molecules and the greater the energy required to get two molecules to move apart. This trend can be seen if we compare

CH_4

Methane
mp $-182°C$
bp $-162°C$
Gas at room temp.

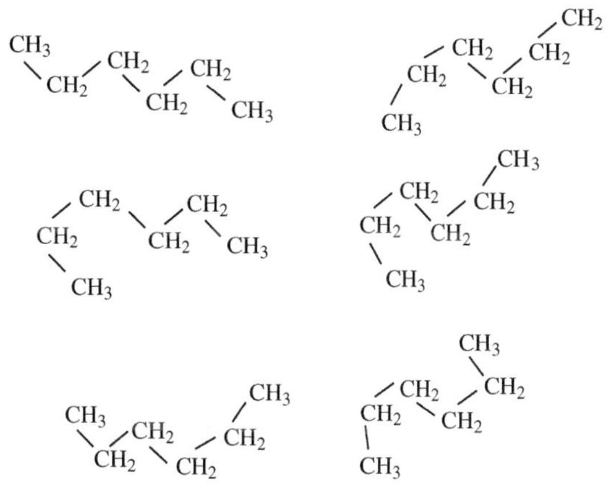

Hexane, C_6H_{14}
mp $-95°C$
bp $+69°C$
Liquid at room temp.

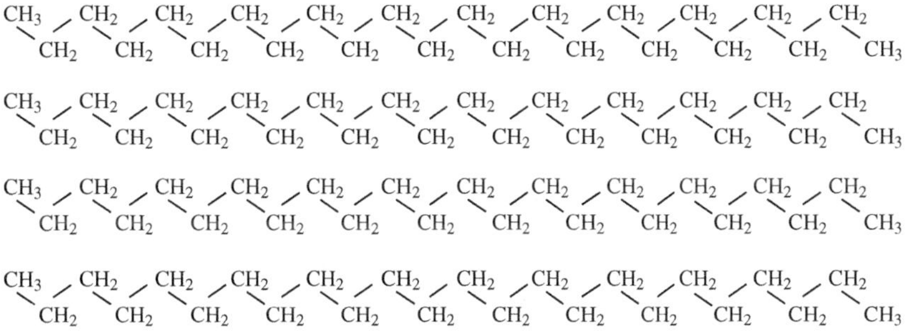

Tetracosane, $C_{24}H_{50}$
mp $+51°C$
bp $+391°C$
Solid at room temp.

the melting points and boiling points of three hydrocarbons of different size: methane, hexane, and tetracosane.

We know that methane is called natural gas because methane's physical state at room temperature (20°C = 68°F) is a gas. Its London forces are so weak that methane must be cooled to −162°C at 1 atm of pressure before the molecules will stick together enough to form a liquid. Hexane is a very common liquid solvent found on most organic laboratory shelves. The intermolecular forces between its molecules are strong enough to keep them from flying apart, but the molecules are still able to flex and slide by each other to form a fluid. Hexane must be heated above 69°C, which is 231°C hotter than methane, to convert all its molecules to a gas. Tetracosane, a C_{24} solid hydrocarbon, is four times larger than hexane, and its London forces are strong enough to hold the molecules rigidly in place at room temperature. Tetracosane is one of the many long-chain hydrocarbons found in candle wax, which must be heated in order to disrupt the intermolecular forces and melt the wax into a liquid. A lot of energy is required to convert liquid tetracosane to a gas, as evidenced by its extremely high boiling point (391°C).

2. Dipole-Dipole Attractive Forces

The attractive forces between molecules increases when functional groups containing electronegative atoms such as chlorine, oxygen, and nitrogen are present because these atoms are more electronegative than carbon. These atoms pull electrons toward themselves, making their end of the bond slightly negatively charged (δ^-) and leaving the carbon slightly positively charged (δ^+), as shown for isopropyl chloride and acetone.

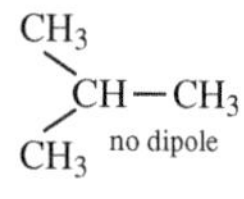

Isobutane	Isopropyl chloride	Acetone
mp −137°C	mp −117°C	mp −94°C
bp 0°C	bp +35°C	bp +56°C
Gas at room temp.	Liquid at room temp.	Liquid at room temp.

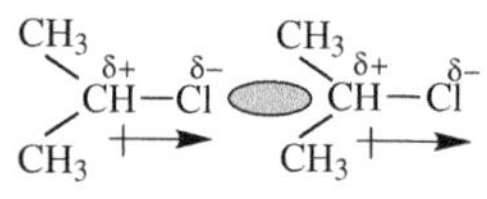

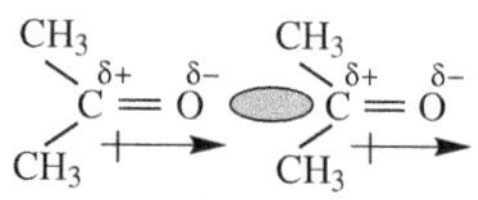

A bond with a slight charge separation is termed a *polar* bond, and polar bonds often give a molecule a *dipole*: slightly positive and negative ends symbolized by an arrow in the direction of the negative charge (+→). Attraction of the positive end of one molecule's dipole to the negative end of another's dipole occurs between polar molecules, which increases the intermolecular attractive force. Dipole-dipole attractive forces are stronger than London forces, as demonstrated in the previous examples that show an increase in melting point and boiling point when a methyl group of isobutane is replaced by chlorine or oxygen.

3. Hydrogen Bonding

Hydrogen bonding is an even stronger intermolecular attractive force, as evidenced by the large increase in the melting point and boiling points of the alcohol methanol (MW = 32) relative to those of the hydrocarbon ethane (MW = 30), both of comparable molecular weight. Hydrogen bonding occurs with organic molecules containing O—H groups (for example, alcohols and carboxylic acids) or N—H groups (for example, amines or amides). The hydrogen in these groups is attracted to the unshared pair of electrons on the O or N of another molecule, forming a hydrogen bond, often symbolized by a dashed line, which is shown for methanol.

CH_3—CH_3

Ethane
mp –172°C
bp –88°C
Gas at room temp.

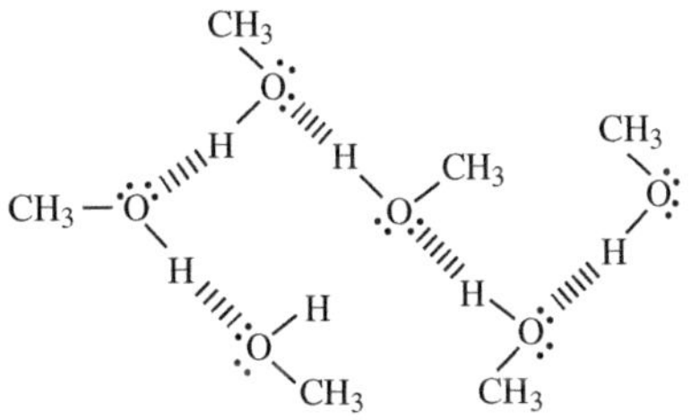

Methanol
mp –97°C
bp +65°C
Liquid at room temp.

As this example indicates, the hydrogen bonds extend throughout the liquid.

One can think of these hydrogen bonds as molecular Velcro that can be pulled apart if there is sufficient energy. Hydrogen bonding plays a major role in the special physical behavior of water and is a major determinant of the chemistry of proteins and nucleic acids in living systems.

4. Ionic Attractive Forces

Recall that ionic substances, such as table salt (NaCl), are usually solids with high melting points (>300°C) due to the strong attractive forces between positive and negative ions. Most organic molecules contain only covalent bonds and have no ionic attractive forces between them. However, there are three important exceptions involving acidic or basic functional groups that can form ionic structures as the pH is raised or lowered.

1. The hydrogen on the —OH of the carboxyl group in carboxylic acids, such as acetic acid, is acidic (H^+ donating) and reacts with bases such as potassium hydroxide (KOH) and sodium bicarbonate ($NaHCO_3$) to form salts. The process is reversed by lowering the pH by adding an acid.

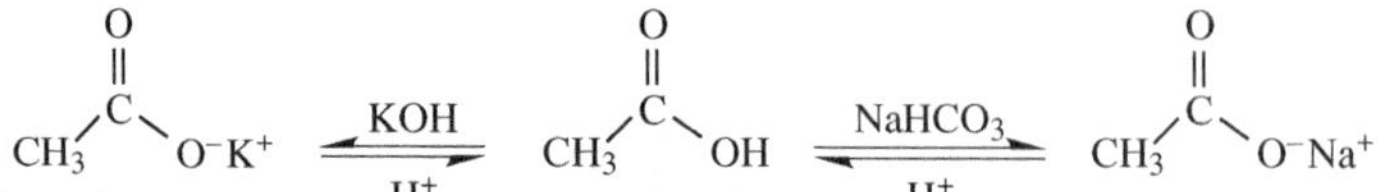

Potassium acetate
mp 306°C

Acetic acid
mp 17°C

Sodium acetate
mp >300°C

The dry salts are ionic and have very high melting points, which is expected for ionic substances. Note that this acidity is *not* observed for alcohols where the —OH group is attached to a singly bonded (sp^3 or saturated) carbon.

2. The hydrogen on an —OH group that is attached to an aromatic ring is weakly acidic and reacts with strong bases such as sodium hydroxide (NaOH) to form high melting ionic salts, as evidenced by the reaction of phenol to sodium phenolate. Again, the reaction is reversed by the addition of an acid.

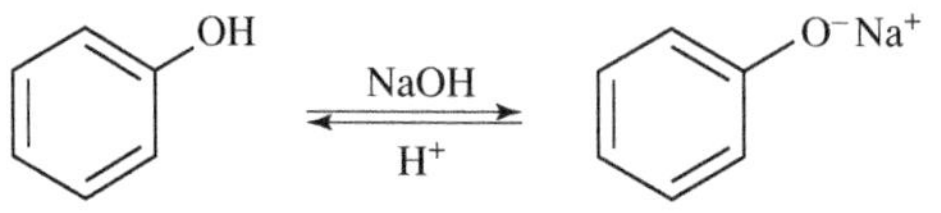

Phenol
mp 41°C

Sodium phenolate
mp 382°C

Note again that this acidity is *not* observed for alcohols where the —OH group is attached to a singly bonded (sp^3 or saturated) carbon.

3. Amines (but not amides) are basic (H^+ accepting) and will react with acid to form ionic amine salts with elevated melting points, as shown, for example, for isopropyl amine.

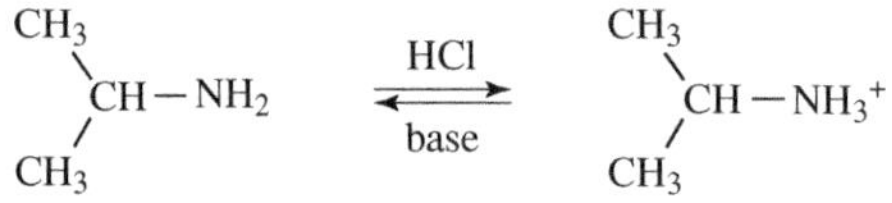

Isopropyl amine
mp –95°C

Isopropyl amine hydrochloride
mp 162°C

Amine salts can be converted back to amines by raising the pH by adding a base.

5. Competing Intermolecular Forces and Solubility

For pure compounds containing identical molecules, the total attractive force between molecules is the sum of all the attractive forces listed previously, both weak and strong. These forces tend to work together to raise melting and boiling points as the size of the molecule's hydrocarbon skeleton increases and as polar, hydrogen bonded, or ionic functional groups are incorporated into the molecule.

However, solubility involves the interaction of two different molecules, which may have different types of attractive forces. When we try to dissolve one substance in another, we have to disrupt the attractive forces in both substances to get the molecules of the two substances to intermingle. For example, to get water (polar with hydrogen bonding) and the hydrocarbons (nonpolar and no hydrogen bonding) in motor oil to mix and dissolve in one another, we would have to disrupt the London attractive forces between the oil molecules and the hydrogen bonds between the water molecules. Because London forces are weak, separating the oil molecules does not require much energy. However, breaking apart the much stronger hydrogen bonds by inserting oil molecules between the water molecules requires considerable energy and is unfavorable. Therefore, oil or even the simplest hydrocarbon, methane, is insoluble in water. This is the molecular basis of the old adage "Oil and water don't mix."

In addition to carbon and hydrogen, the majority of organic molecules contain other elements, such as nitrogen and oxygen, in functional groups that can be polar, exhibit hydrogen bonding, show ionic tendencies, or have any combination thereof. Can we predict the water solubility of these organic substances? Let's look at some examples and see.

Figure 3.1 shows a collection of small organic molecules of about the same size and molecular weight listed in order of increasing boiling point or melting point, which is consistent with the types of intermolecular forces discussed. With the exception of the hydrocarbon butane, all of these substances are very soluble in water—at least 100 g will dissolve in 1 L of water. It appears that the intermolecular attractive forces between these polar, hydrogen bonded, or ionic molecules and water compensates for the disruption of hydrogen bonding between water molecules so the organic molecules can move into and intermingle with the water molecules—in other words, dissolve.

Figure 3.2 shows a collection of larger organic molecules than those in Figure 3.1, again listed in order of increasing melting and boiling point, which is consistent with our knowledge of the strength of intermolecular attractive forces. The important difference for this group is that the hydrocarbon portion of each molecule is four carbons larger than for those in Figure 3.1. We might predict that the larger hydrocarbon portion makes them behave more like the water-insoluble hydrocarbon octane. Indeed, the larger hydrocarbon portion of these molecules greatly reduces their solubility in water to less than 5 g/L of water except for the two ionic compounds. These two, the amine hydrochloride and the sodium carboxylate compounds, have higher solubility—tens of grams per liter of water—proof that ionic charges can interact strongly with water molecules.

This trend in water solubility based on the size of the hydrocarbon portion continues for the set of even larger organic molecules shown in Figure 3.3 containing C_{16} to C_{18} hydrocarbon chains. Most are virtually insoluble in water; even the ionic forms have solubilities of less than 1 g/L of water.

In addition to water, many other liquid solvents are used in the organic laboratory to dissolve substances, including acetone, dichloromethane (CH_2Cl_2), ethanol (CH_3CH_2OH), diethyl ether ($CH_3CH_2OCH_2CH_3$), hexane, methanol, and toluene ($C_6H_5CH_3$). Predicting the solubility of different organic compounds,

■ **FIG. 3.1**

Some small organic molecules containing 4 atoms (carbon, oxygen, and nitrogen) in order of increasing melting or boiling points.

C₄ Hydrocarbon, butane
mp –138°C; bp 0°C
Insoluble in H₂O

C₃ Amine
mp –43°C; bp 48°C
Soluble in H₂O; Soluble in organic solvents

C₃ Ketone, acetone
mp –94°C; bp 56°C
Soluble in H₂O; Soluble in organic solvents

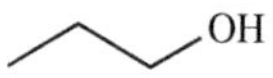

C₃ Alcohol
mp –127°C; bp 97°C
Soluble in H₂O; Soluble in organic solvents

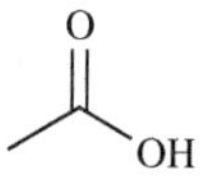

C₂ Carboxylic acid. acetic acid
mp 17°C; bp 117°C
Soluble in H₂O; Soluble in organic solvents

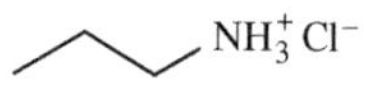

C₃ Amine hydrochloride
mp 161°C
Soluble in H₂O; Insoluble in organic solvents

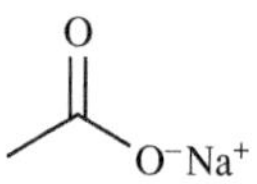

Sodium salt of C₂ carboxylic acid
mp 324°C
Soluble in H₂O; Insoluble in organic solvents

■ **FIG. 3.2**

Some organic molecules containing 8 atoms (carbon, oxygen, and nitrogen) in order of increasing melting or boiling points.

C₈ Hydrocarbon, octane
mp –57°C; bp 126°C
Insoluble in H₂O

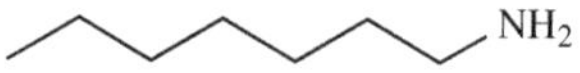

C₇ Amine
mp –18°C; bp 154°C
Insoluble in H₂O; Soluble in organic solvents

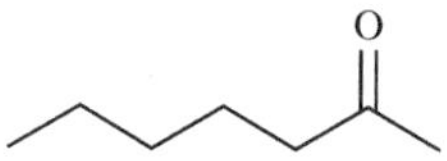

C₇ Ketone
mp –35°C; bp 150°C
Insoluble in H₂O; Soluble in organic solvents

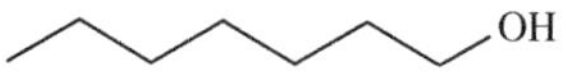

C₇ Alcohol
mp –34°C; bp 175°C
Insoluble in H₂O; Soluble in organic solvents

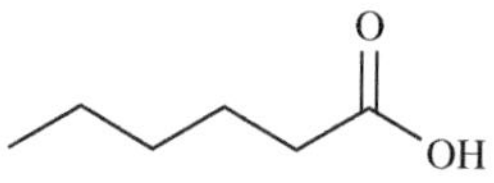

C₆ Carboxylic acid
mp –2°C; bp 200°C
Insoluble in H₂O; Soluble in organic solvents

C₇ Amine hydrochloride
mp 242°C
Soluble in H₂O; Insoluble in organic solvents

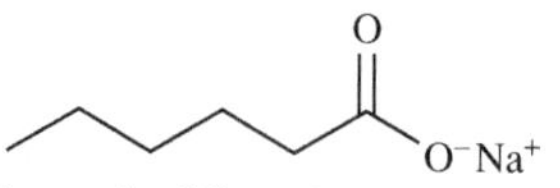

Sodium salt of C₆ carboxylic acid
mp 245°C
Soluble in H₂O; Insoluble in organic solvents

such as those shown in Figures 3.1, 3.2, and 3.3, in these solvents can be done using the intermolecular attractive force concepts discussed here. For example, because the molecules in Figure 3.3 have long hydrocarbon skeletons, you would predict that these would probably be soluble in the hydrocarbon hexane. This predictive rule can be summed up as "Like dissolves like." You might also predict

■ FIG. 3.3

Some organic molecules containing 18 atoms (carbon, oxygen, and nitrogen) in order of increasing melting or boiling points.

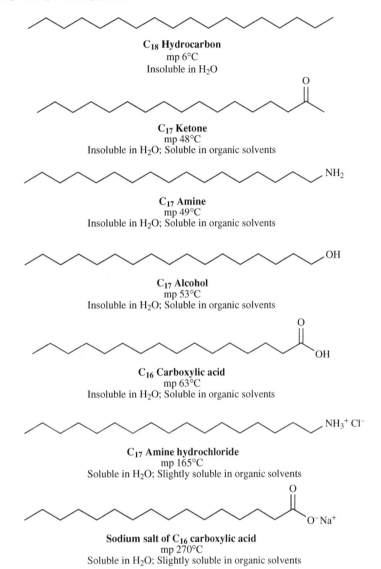

C$_{18}$ Hydrocarbon
mp 6°C
Insoluble in H$_2$O

C$_{17}$ Ketone
mp 48°C
Insoluble in H$_2$O; Soluble in organic solvents

C$_{17}$ Amine
mp 49°C
Insoluble in H$_2$O; Soluble in organic solvents

C$_{17}$ Alcohol
mp 53°C
Insoluble in H$_2$O; Soluble in organic solvents

C$_{16}$ Carboxylic acid
mp 63°C
Insoluble in H$_2$O; Soluble in organic solvents

C$_{17}$ Amine hydrochloride
mp 165°C
Soluble in H$_2$O; Slightly soluble in organic solvents

Sodium salt of C$_{16}$ carboxylic acid
mp 270°C
Soluble in H$_2$O; Slightly soluble in organic solvents

that the two ionic materials would be the least soluble in hexane because of the strong intermolecular forces in these solids, as evidenced by their high melting points, which are so unlike the weak London forces in hexane. The solubility of organic compounds in organic solvents and water at low, neutral, and high pH will be considered in more detail when you learn about recrystallization in Chapter 4 and extraction in Chapter 7.

PART 2: Melting Points
A. Thermometers

There are a few types of thermometers that can be used to read melting point (and boiling point) temperatures: mercury-in-glass thermometers, nonmercury thermometers, and digital thermometers. Mercury-in-glass thermometers provide highly accurate readings and are ideal for use at high temperatures (260°C–400°C). Care should be taken not to break the thermometer, which will release the toxic mercury. Teflon-coated mercury thermometers are usable up to 260°C and are less likely to spill mercury if broken. If breakage does happen, inform your instructor immediately because special equipment is required to clean up mercury spills. A digital thermometer (Fig. 3.4) has a low heat capacity and a fast response time. It is more robust than a glass thermometer and does not, of course, contain mercury. Nonmercury thermometers may be filled with isoamyl benzoate (a biodegradable liquid) or with a custom organic red-spirit liquid instead of mercury. These thermometers give reasonably accurate readings at temperatures up to 150°C, but above this temperature they need to be carefully calibrated. These thermometers should be stored vertically to prevent thread separation.

CAUTION: Mercury is toxic. Immediately report any broken thermometers to your instructor.

B. Melting Points

The melting point of a pure solid organic compound is one of its characteristic physical properties, along with molecular weight, boiling point, refractive index, and density. A pure solid will melt reproducibly over a narrow range of temperatures, typically less than 1°C. The process of determining this melting point is done on a truly micro scale using less than 1 mg of material. The apparatus is simple, consisting of a thermometer, a capillary tube to hold the sample, and a heating bath.

Melting points are determined for three reasons: (1) If the compound is a known one, the melting point will help to characterize the sample. (2) If the compound is new, then the melting point is recorded to allow future characterization by others. (3) The range of the melting point is indicative of the purity of the compound; an impure compound will melt over a wide range of temperatures. Recrystallization of the compound will purify it and decrease the melting point range. In addition, the entire range will be displaced upward. For example, an impure sample might melt from 120°C to 124°C and after recrystallization melt at 125°C–125.5°C. A solid is considered pure if the melting point does not rise after recrystallization.

A crystal is an orderly arrangement of molecules in a solid. As heat is added to the solid, the molecules will vibrate and perhaps rotate but still remain a solid. At a characteristic temperature the crystal will suddenly acquire the necessary energy to overcome the forces that attract one molecule to another and will undergo translational motion—in other words, it will become a liquid.

The forces by which one molecule is attracted to another include ionic attractions, London forces, hydrogen bonds, and dipole-dipole attractions. Most, but by no means all, organic molecules are covalent in nature and melt at temperatures below 300°C. Typical inorganic compounds are ionic and have much higher

Melting points—a micro technique

Characterization

An indication of purity

■ FIG. 3.4
A digital thermometer.

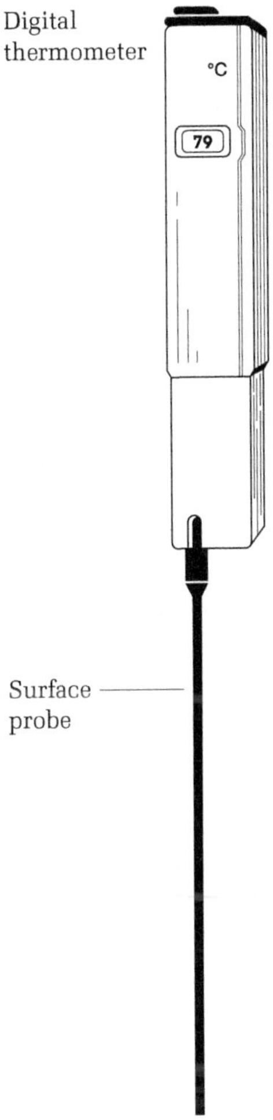

Digital thermometer

°C

79

Surface probe

■ FIG. 3.5

A melting point–composition diagram for mixtures of the solids X and Y.

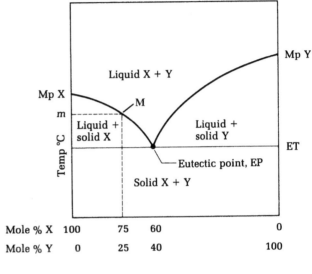

| Mole % X | 100 | 75 | 60 | | 0 |
| Mole % Y | 0 | 25 | 40 | | 100 |

melting points (e.g., sodium chloride melts at 800°C). Ionic organic molecules often decompose before melting, as do compounds having many strong hydrogen bonds, such as sucrose.

Melting point generalizations

Other factors being equal, larger molecules melt at higher temperatures than smaller molecules. Among structural isomers, the more symmetrical isomers will have the higher melting point. Among optical isomers, the *R* and *S* enantiomers will have the same melting points; but the racemate (a mixture of equal parts of *R* and *S*) will usually possess a different melting point. Diastereomers, another type of stereoisomer, will have different melting points. Molecules that can form hydrogen bonds will usually possess higher melting points than their counterparts of similar molecular weight.

A phase diagram

The melting point behavior of impure compounds is best understood by considering a simple binary mixture of compounds X and Y (Fig. 3.5). This melting point–composition diagram shows melting point behavior as a function of composition. The melting point of a pure compound is the temperature at which the vapor pressures of the solid and liquid are equal. But in dealing with a mixture, the situation is different. Consider the case of a mixture of 75% X and 25% Y. At a temperature below the eutectic temperature (ET), the mixture is solid Y and solid X. At ET, the solid begins to melt. The melt is a solution of Y dissolved in liquid X. The vapor pressure of the solution of X and Y together is less than that of pure X at the melting point; therefore, the temperature at which X will melt is lower when mixed with Y. This is an application of Raoult's law (*see* Chapter 5). As the temperature is increased, more and more of solid X melts until it is all gone at point **M** (temperature *m*). The melting point range is thus from ET to *m*. In practice it is very difficult to detect the ET when a melting point is determined in a capillary because it represents the point at which an infinitesimal amount of the solid mixture has begun to melt.

In this hypothetical example, the liquid solution becomes saturated with Y at the eutectic point (EP). This is the point at which X and Y and their liquid solutions are in equilibrium. A mixture of X and Y containing 60% X will appear to have a sharp melting point at the ET.

The melting point range of a mixture of compounds is generally broad, and the breadth of the range is an indication of purity. The chances of accidentally coming on the eutectic composition are small. Recrystallization will enrich the predominant compound while excluding the impurity and will, therefore, decrease the melting point range.

It should be apparent that the impurity must be soluble in the compound, so an insoluble impurity such as sand or charcoal will not depress the melting point. The impurity does not need to be a solid. It can be a liquid such as water (if it is soluble) or an organic solvent, such as the one used to recrystallize the compound; this advocates the necessity for drying the compound before determining its melting point.

Advantage is taken of the depression of melting points of mixtures to prove whether or not two compounds having the same melting points are identical. If X and Y are identical, then a mixture of the two will have the same melting point; if X and Y are not identical, then a small amount of X in Y or of Y in X will reduce the melting point.

Melting point depression

The eutectic point

Mixed melting points

Apparatus

Melting Point Capillaries

Before using a melting point apparatus, the sample needs to be prepared for analysis. Most melting point determinations require that the sample be placed in a capillary tube. The experiments in this book require capillary tubes for sample preparation. Capillaries may be obtained commercially or may be produced by drawing out 12-mm soft-glass tubing. The tubing is rotated in the hottest part of a Bunsen burner flame until it is very soft and begins to sag. It should not be drawn out during heating; it is removed from the flame and after a slight hesitation drawn steadily and not too rapidly to arm's length. With some practice it is possible to

■ FIG. 3.6
Sealing a melting point
capillary tube.

Samples that sublime

A small rubber band can be
made by cutting off a very
short piece of ¼" gum rubber
tubing.

Thomas–Hoover Uni-Melt

produce 10–15 good tubes in a single drawing. The long capillary tube can be cut into 100-mm lengths with a glass scorer. Each tube is sealed by rotating the end in the edge of a small flame, as seen in Figure 3.6.

Filling Melting Point Capillaries. The dry sample is ground to a fine powder on a watch glass or a piece of glassine paper on a hard surface using the flat portion of a spatula. It is formed into a small pile, and the open end of the melting point capillary is forced down into the pile. The sample is shaken into the closed end of the capillary by rapping sharply on a hard surface or by dropping it down a 2-ft length of glass tubing onto a hard surface. The height of the sample should be no more than 2–3 mm.

Sealed Capillaries. Some samples sublime (go from a solid state directly to the vapor phase without appearing to melt) or undergo rapid air oxidation and decompose at the melting point. These samples should be sealed under vacuum. This can be accomplished by forcing a capillary through a hole previously made in a rubber septum and evacuating the capillary using a water aspirator or a mechanical vacuum pump (Fig. 3.7). Using the flame from a small micro burner, the tube is gently heated about 15 mm above the tightly packed sample. This will cause any material in this region to sublime away. It is then heated more strongly in the same place to collapse the tube, taking care that the tube is straight when it cools. It is also possible to seal the end of a Pasteur pipette, add the sample, pack it down, and seal off a sample under vacuum in the same way.

Melting Point Devices. The apparatus required for determining an accurate melting point need not be elaborate; the same results are obtained on the simplest as on the most complex devices. The simplest setup involves attaching the sample-filled, melting point capillary to a thermometer using a rubber band and immersing them into a silicone oil bath (Fig. 3.8). This rubber band must be above the level of the oil bath; otherwise, it will break in the hot oil. The sample should be close to and on a level with the center of the thermometer bulb. This method can analyze compounds whose melting points go up to ~350°C. If determinations are to be done on two or three samples that differ in melting point by as much as 10°C, two or three capillaries can be secured to the thermometer together; the melting points can be observed in succession without removing the thermometer from the bath. As a precaution against the interchange of tubes while they are being attached, use some system of identification, such as one, two, and three dots made with a marking pencil.

More sophisticated melting point devices, some of which can attain temperatures of 500°C, will now be described.

The Thomas–Hoover Uni-Melt apparatus (Fig. 3.9) will accommodate up to seven capillaries in a small, magnified, lighted beaker of high-boiling silicone oil that is stirred and heated electrically. The heating rate is controlled with a variable transformer that is part of the apparatus. The rising mercury column of the thermometer can be observed with an optional traveling periscope device so the eye need not move away from the capillary. For industrial, analytical, and control

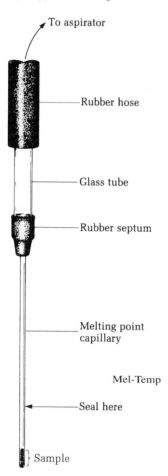

■ **FIG 3.7**
Evacuation of a melting point
capillary prior to sealing.

To aspirator

Rubber hose

Glass tube

Rubber septum

Melting point
capillary

Mel-Temp

Seal here

} Sample

Fisher-Johns

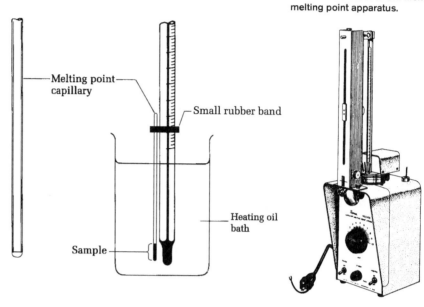

■ **FIG 3.8**
A simple melting point apparatus.

Melting point
capillary

Small rubber band

Heating oil
bath

Sample

■ **FIG. 3.9**
The Thomas–Hoover Uni-Melt
melting point apparatus.

work the Mettler apparatus determines the melting point automatically and displays the result in digital form.

The Mel-Temp apparatus (Fig. 3.10) consists of an electrically heated aluminum block that accommodates three capillaries. The sample is illuminated through the lower port and observed with a six-power lens through the upper port. The heating rate can be controlled and, with a special thermometer, the apparatus can be used up to 500°C—far above the useful limit of silicone oil (which is about 350°C). For this melting point apparatus it is advisable to use a digital thermometer rather than the mercury-in-glass thermometer.

The Fisher-Johns melting point apparatus (Fig. 3.11) is used to determine the melting point of a single sample. Instead of a capillary tube, the sample is placed between two thin glass disks that are placed on an aluminum heating stage. Heating is controlled by a variable transformer, and melting is observed through a magnifier; the melting temperature is read from a mercury-in-glass thermometer. This apparatus can be used for compounds that melt between 20°C and 300°C.

Determining the Melting Point

The accuracy of the melting point depends on the accuracy of the thermometer, so the first exercise in the following experiments will be to calibrate the thermometer. Melting points of pure, known compounds will be determined and deviations recorded so a correction can be applied to future melting points. Be forewarned, however, that thermometers are usually fairly accurate.

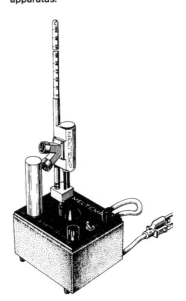

■ FIG. 3.10
The Mel-Temp melting point apparatus.

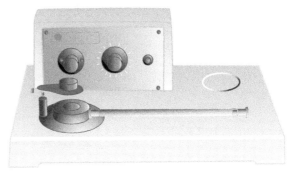

■ FIG 3.11
The Fisher-Johns melting point apparatus.

The rate of heating is the most important factor in obtaining accurate melting points. Heat no faster than 1°C per minute.

The most critical factor in determining an accurate melting point is the rate of heating. At the melting point the temperature increase should not be greater than 1°C per minute. This may seem extraordinarily slow, but it is necessary in order for heat from the oil bath or the heating block to be transferred equally to the sample and to the glass and mercury of the thermometer.

From your own experience you know the rate at which ice melts. Consider performing a melting point experiment on an ice cube. Because water melts at 0°C, you would need to have a melting point bath a few degrees below zero. To observe the true melting point of the ice cube, you would need to raise the temperature extraordinarily slowly. The ice cube would appear to begin to melt at 0°C and, if you waited for temperature equilibrium to be established, it would all be melted at 0.5°C. If you were impatient and raised the temperature too rapidly, the ice might appear to melt over a range of 0°C to 20°C. Similarly, melting points determined in capillaries will not be accurate if the rate of heating is too fast.

EXPERIMENTS

1. Calibration of the Thermometer

Determine the melting point of standard substances (Table 3.1) over the temperature range of interest. The difference between the values found and those expected constitutes the correction that must be applied to future temperature readings. If the thermometer has been calibrated previously, then determine one or more melting points of known substances to familiarize yourself with the technique. If the

TABLE 3.1 • Melting Point Standards		
Compound	**Structure**	**Melting Point (°C)**
Naphthalene	(a)	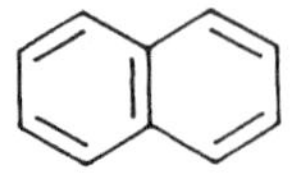 80–82
Urea	(b) H₂NCNH₂	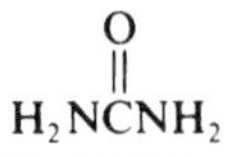 132.5–133
Sulfanilamide	(c)	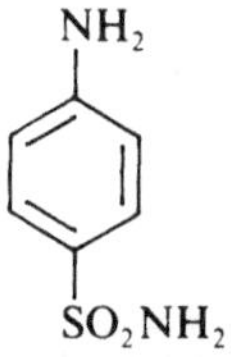 164–165
4-Toluic acid	(d)	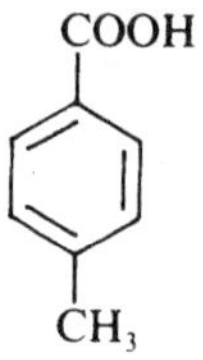 180–182
Anthracene	(e)	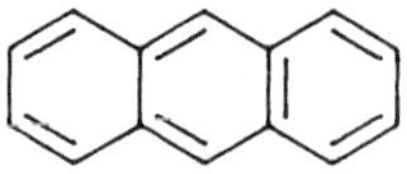 214–217
Caffeine (evacuated capillary)	(f)	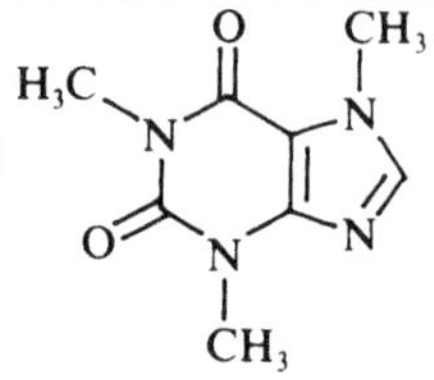 234–236.5

determinations do not agree within 1°C, then repeat the process. Both mercury-in-glass and digital thermometers will need to be calibrated; nonmercury thermometers are not typically used for melting point determination.

2. Melting Points of Pure Urea and Cinnamic Acid

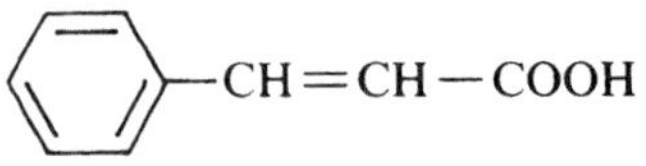

$-CH=CH-COOH$

Cinnamic acid

Using a metal spatula, crush the sample to a fine powder on a hard surface such as a watch glass. Push the open end of a melting point capillary into the powder and force the powder down in the capillary by tapping the capillary or by dropping it through a long glass tube held vertically and resting on a hard surface. The column of solid should be no more than 2–3 mm in height and should be tightly packed.

Heat rapidly to within 20°C of
the melting point.

Heat slowly (<1°C/min) near the
melting temperature.

If the approximate melting temperature is known, the bath can be heated rapidly until the temperature is about 20°C below this point; the heating during the last 15°C–20°C should slow down considerably so *the rate of heating at the melting point is no more than 1°C per minute* while the sample is melting. As the melting point is approached, the sample may shrink because of crystalline structure changes. However, the melting process begins when the first drops of liquid are seen in the capillary and ends when the last trace of solid disappears. For a pure compound this whole process may occur over a range of only 0.5°C; hence, it is necessary to slowly increase the temperature during the determination.

Determine the melting point (mp) of either urea (mp 132.5°C–133°C) or cinnamic acid (mp 132.5°C–133°C). Repeat the determination; if the two determinations do not check within 1°C, repeat a third time.

3. Melting Points of Urea-Cinnamic Acid Mixtures

> **IN THIS EXPERIMENT** you will see dramatic evidence of the phenomenon of melting point depression, which will allow you to prepare a phase diagram like that shown in Figure 3.5 (on page 47).

Make mixtures of urea and cinnamic acid in the approximate proportions 1:4, 1:1, and 4:1 by putting side by side the correct number of equal-sized small piles of the two substances and then mixing them. Grind the mixture thoroughly for at least a minute on a watch glass using a metal spatula. Note the ranges of melting of the three mixtures and use the temperatures of complete liquefaction to construct a rough diagram of melting point versus composition.

4. Unknowns

Determine the melting point of one or more of the unknowns selected by your instructor and identify the substance based on its melting point (Table 3.2). Prepare two capillaries of each unknown. Run a very fast determination on the first sample to ascertain the approximate melting point. Cool the melting point bath to just below the melting point and make a slow, careful determination using the other capillary.

5. An Investigation: Determination of Molecular Weight Using Melting Point Depression

Before the mass spectrometer came into common usage, the molal freezing point depression of camphor was used to determine molecular weights. Whereas a 1% solid solution of urea in cinnamic acid will cause a relatively small melting point depression, a 1% by weight solid solution in camphor of any organic compound with a molecular weight of 100 will cause a 4.0°C depression in the melting point of the camphor. Quantitative use of this relationship can be used to determine the

TABLE 3.2 • Melting Point Unknowns	
Compound	**Melting Point (°C)**
Benzophenone	49–51
Maleic anhydride	52–54
4-Nitrotoluene	54–56
Naphthalene	80–82
Acetanilide	113.5–114
Benzoic acid	121.5–122
Urea	132.5–133
Salicylic acid	158.5–159
Sulfanilamide	165–166
Succinic acid	184.5–185
3,5-Dinitrobenzoic acid	205–207
p-Terphenyl	210–211

molecular weight of an unknown. You can learn more about the details of this technique in very old editions of *Organic Experiments*[1] or by searching the Web for "colligative properties, molal freezing point depression." Visit this book's website for more information.

PART 3: Boiling Points

The boiling point of a pure organic liquid is one of its characteristic physical properties, just like its density, molecular weight, and refractive index, and the melting point of its solid state. The boiling point is used to characterize a new organic liquid, and knowledge of the boiling point helps to compare one organic liquid to another, as in the process of identifying an unknown organic substance.

A comparison of boiling points with melting points is instructive. The process of determining the boiling point is more complex than that for the melting point: It requires more material, and because it is less affected by impurities, it is not as good an indication of purity. Boiling points can be determined on a few microliters of a liquid, but on a small scale it is difficult to determine the boiling point *range*. This requires enough material to distill—about 1 to 2 mL. Like the melting point, the boiling point of a liquid is affected by the forces that attract one molecule to another—ionic attraction, London forces, dipole-dipole interactions, and hydrogen bonding, as discussed in Part 1 of this chapter.

Structure and Boiling Point

In a homologous series of molecules the boiling point increases in a perfectly regular manner. The normal saturated hydrocarbons have boiling points ranging from −162°C for methane to 330°C for n-$C_{19}H_{40}$, an increase of about 27°C for each

1. Fieser L. F. *Organic Experiments*, 2nd ed.; D.C. Heath: Lexington, MA, 1968; 38–42.

CH$_2$ group. It is convenient to remember that *n*-heptane with a molecular weight of 100 has a boiling point near 100°C (98.4°C). A spherical molecule such as 2,2-dimethylpropane has a lower boiling point than *n*-pentane because it does not have as many points of attraction to adjacent molecules. For molecules of the same molecular weight, those with dipoles, such as carbonyl groups, will have higher boiling points than those without, and molecules that can form hydrogen bonds will boil even higher. The boiling point of such molecules depends on the number of hydrogen bonds that can be formed. An alcohol with one hydroxyl group will boil at a lower temperature than an alcohol with two hydroxyl groups if they both have the same molecular weight. A number of other generalizations can be made about boiling point behavior as a function of structure; you will learn about these throughout your study of organic chemistry.

Boiling Point as a Function of Pressure

Because the boiling point of a pure liquid is defined as the temperature at which the vapor pressure of the liquid exactly equals the pressure exerted on it, the boiling point will be a function of atmospheric pressure. At an altitude of 14,000 ft the boiling point of water is 81°C. At pressures near that of the atmosphere at sea level (760 mm), the boiling point of most liquids decreases about 0.5°C for each 10-mm decrease in atmospheric pressure. This generalization does not hold for greatly reduced pressures because the boiling point decreases as a nonlinear function of pressure (*see* Fig. 5.2 on page 91). Under these conditions a nomograph relating the observed boiling point, the boiling point at 760 mm, and the pressure in millimeters should be consulted (*see* Fig. 6.19 on page 128). This nomograph is not highly accurate; the change in boiling point as a function of pressure also depends on the type of compound (polar, nonpolar, hydrogen bonding, etc.). Consult the *CRC Handbook of Chemistry and Physics*[2] for the correction of boiling points to standard pressure.

Most mercury-in-glass laboratory thermometers have a mark around the stem that is 3 in. (76 mm) from the bottom of the bulb. This is the immersion line; the thermometer will record accurate temperatures if immersed to this line. If you break a mercury thermometer, immediately inform your instructor, who will use special apparatus to clean up the mercury. Mercury vapor is very toxic.

Calibrating the Thermometer

If you have not previously carried out a calibration, test the 0°C point of your thermometer with a well-stirred mixture of crushed ice and distilled water. To check the 100°C point, put 2 mL of water in a test tube with a boiling chip to prevent bumping and boil the water gently over a hot sand bath with the thermometer in the vapor from the boiling water. Make sure that the thermometer does not touch the side of the test tube. Then immerse the bulb of the thermometer into the liquid and see if you can observe superheating. Check the atmospheric pressure to determine the true boiling point of the water.

Boiling points decrease about 0.5°C for each 10-mm decrease in atmospheric pressure.

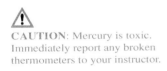

CAUTION: Mercury is toxic. Immediately report any broken thermometers to your instructor.

---

2. Lide, D. R., ed. *CRC Handbook of Chemistry and Physics,* 86th ed.; CRC Press: Boca Raton, FL, 2005.

Distillation Considerations

Prevention of Superheating—Boiling Sticks and Boiling Stones

Superheating occurs when a very clean liquid in a very clean vessel is heated to a temperature above its boiling point without ever actually boiling. That is, if a thermometer is placed in the superheated liquid, the thermometer will register a temperature higher than the boiling point of the liquid. If boiling does occur under these conditions, it occurs with explosive violence. To avoid this problem, boiling stones or boiling sticks are always added to liquids before heating them to boiling—whether to determine a boiling point or to carry out a reaction or distillation. These boiling stones or sticks provide the nuclei on which the bubble of vapor indicative of a boiling liquid can form; be careful not to confuse the bubbling for boiling. Some boiling stones, also called boiling chips, are composed of porous unglazed porcelain. This material is filled with air in numerous fine capillaries. With heating, this air expands to form the fine bubbles on which even boiling can take place. Once the liquid cools, it will fill these capillaries and the boiling chip will become ineffectual, so another chip must be added each time the liquid is reheated to boiling. Wooden boiling sticks about 1.5 mm in diameter—often called applicator sticks—also promote even boiling and, unlike stones, are removed easily from the solution. None of these work well for vacuum distillation (*see* Chapter 6).

Closed Systems

Distillations that are run at atmospheric pressure need to be open to the atmosphere to avoid pressure buildup, which could lead to an explosion. Therefore, always make sure that a distillation setup is not a closed system—unless, of course, you are running a vacuum distillation.

Boiling Point Determination: Apparatus and Technique

Boiling Point Determination by Distillation

All procedures involving volatile and/or flammable solvents should be conducted in a fume hood.

When enough material is available (at least 3 mL), the best method for determining the boiling point of a liquid is to distill it (*see* Chapter 5). Distillation allows the boiling range to be determined and thus gives an indication of purity. Bear in mind, however, that a constant boiling point is not a guarantee of homogeneity and thus purity. Constant-boiling azeotropes such as 95% ethanol are common.

Boiling Point Determination Using a Digital Thermometer and a Reaction Tube

Online Study Center

Photo: Boiling Point
Determination with a Digital
Thermometer

Boiling points can be measured rapidly and accurately using an electronic digital thermometer, as depicted in Figure 3.12. Although digital thermometers are currently too expensive for each student to have, several of these in the laboratory can greatly speed up the determination of boiling points. Digital thermometers are

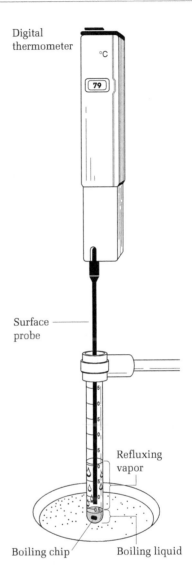

■ FIG. 3.12
Using a digital thermometer
for determining boiling points.

Digital thermometer

Surface probe

Refluxing vapor

Boiling chip   Boiling liquid

much safer to use because there is no danger from toxic mercury vapor if the thermometer is accidentally dropped.

The surface probe of the digital thermometer is the active element. Unlike the bulb of mercury at the end of a thermometer, this element has a very low heat capacity and a very fast response time, so boiling points can be determined very quickly in this apparatus. About 0.2 mL to 0.3 mL of the liquid and a boiling chip are heated on a sand bath until the liquid refluxes about 3 cm up the tube. The probe should not touch the side of the reaction tube but should be placed approximately 5 mm above the liquid. The boiling point is the highest temperature recorded by the thermometer and is maintained for about 1 min. The application

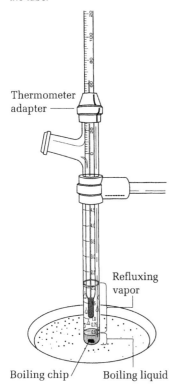

Thermometer
adapter

Refluxing
vapor

Boiling chip   Boiling liquid

Smaller-scale boiling point apparatus

of heat will drive tiny bubbles of air from the boiling chip; do not mistake these tiny bubbles for true boiling; this mistake can readily happen if the unknown has a very high boiling point.

Boiling Point Determination in a Reaction Tube

If a digital thermometer is not available, use the apparatus shown in Figure 3.13. Using a distilling adapter at the top of a reaction tube allows access to the atmosphere. Place 0.3 mL of the liquid along with a boiling stone in a 10×100-mm reaction tube, clamp a thermometer so that the bulb is just above the level of the liquid, and then heat the liquid with a sand bath. It is *very important* that no part of the thermometer touch the reaction tube. Heating is regulated so that the boiling liquid refluxes about 3 cm up the thermometer but does not boil out of the apparatus. If you cannot see the refluxing liquid, carefully run your finger down the side of the reaction tube until you feel heat. This indicates where the liquid is refluxing. Droplets of liquid must drip from the thermometer bulb to thoroughly heat the mercury. The boiling point is the highest temperature recorded by the thermometer and is maintained over about a 1-min time interval.

The application of heat will drive out tiny air bubbles from the boiling chip. Do not mistake these tiny bubbles for true boiling. This can occur if the unknown has a very high boiling point. It may take several minutes to heat up the mercury in the thermometer bulb. True boiling is indicated by drops dripping from the thermometer, with a constant temperature recorded on the thermometer. If the temperature is not constant, then you are probably not observing true boiling.

Boiling Point Determination Using a 3-mm to 5-mm Tube

For smaller quantities, a 3-mm to 5-mm diameter tube is attached to the side of the thermometer by a rubber band (Fig. 3.14) and heated with a liquid bath. The tube, which can be made from tubing 3 mm to 5 mm in diameter, contains a small inverted capillary. This is made by cutting a 6-mm piece from the sealed end of a melting point capillary, inverting it, and sealing the closed end to the capillary. A centimeter ruler is printed on the inside back cover of this book.

When the sample is heated in this device, the air in the inverted capillary will expand, and an occasional bubble will escape. At the true boiling point a continuous and rapid stream of bubbles will emerge from the inverted capillary. When this occurs, the heating is stopped, and the bath is allowed to cool. A time will come when bubbling ceases and the liquid just begins to rise in the inverted capillary. The temperature at which this happens is recorded. The liquid is allowed to partially fill the small capillary, and the heat is applied carefully until the first bubble emerges from the capillary. The temperature is recorded at that point. The two temperatures approximate the boiling point range for the liquid.

As the liquid was being heated, the air expanded in the inverted capillary and was replaced by vapor of the liquid. The liquid was actually slightly superheated when rapid bubbles emerged from the capillary, but on cooling a point was reached at which the pressure on the inside of the capillary matched the outside (atmospheric) pressure. This is, by definition, the boiling point.

■ **FIG. 3.14**
A smaller-scale boiling point apparatus.

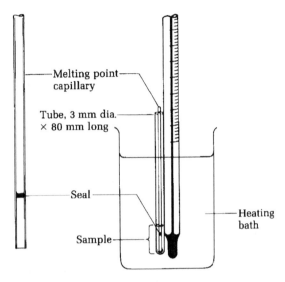

Melting point capillary

Tube, 3 mm dia. × 80 mm long

Seal

Sample

Heating bath

Cleaning Up. Place the boiling point sample in either the halogenated or nonhalogenated waste container. Do not pour it down the sink.

QUESTIONS

1. What effect would poor circulation of the melting point bath liquid have on the observed melting point?
2. What is the effect of an insoluble impurity, such as sodium sulfate, on the observed melting point of a compound?
3. Three test tubes, labeled A, B, and C, contain substances with approximately the same melting points. How could you prove the test tubes contain three different chemical compounds?
4. One of the most common causes of inaccurate melting points is too rapid heating of the melting point bath. Under these circumstances, how will the observed melting point compare to the true melting point?
5. Strictly speaking, why is it incorrect to speak of a melting *point*?
6. What effect would the incomplete drying of a sample (for example, the incomplete removal of a recrystallization solvent) have on the melting point?
7. Why should the melting point sample be finely powdered?
8. You suspect that an unknown is acetanilide (mp 113.5°C–114°C). Give a qualitative estimation of the melting point when the acetanilide is mixed with 10% by weight of naphthalene.
9. You have an unknown with an observed melting point of 90°C–93°C. Is your unknown compound A with a reported melting point of 95.5°C–96°C or compound B with a reported melting point of 90.5°C–91°C? Explain.

10. Why is it important to heat the melting point bath or block slowly and steadily when the temperature gets close to the melting point?

11. Why is it important to pack the sample tightly in the melting point capillary?

12. An unknown compound is suspected to be acetanilide (mp 113.5°C–114°C). What would happen to the melting point if this unknown were mixed with (a) an equal quantity of pure acetanilide? (b) an equal quantity of benzoic acid?

Online Study Center

General Resources
Web Links

13. Which would be expected to have the higher boiling point, *t*-butyl alcohol (2-methyl-2-propanol) or *n*-butyl alcohol (1-butanol)? Explain.

14. What is the purpose of the side arm of the thermometer adapter in Figure 3.13?

REFERENCE

Weissberger, Arnold, and Bryant W. Rossiter (eds.). *Physical Methods of Chemistry*, Vol. 1, Part V. New York: Wiley-Interscience, 1971.

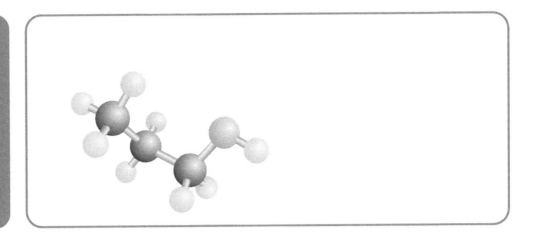

CHAPTER 4

Recrystallization

Online Study Center

This icon will direct you to techniques, equipment setups, and online resources at http://college.hmco.com/PIC/williamsonMMOE5e.

Recrystallization: the most important purification method for solids, especially for small-scale experiments.

PRELAB EXERCISE: Write an expanded outline for the seven-step process of recrystallization.

Recrystallization is the most important method for purifying solid organic compounds. It is also a very powerful, convenient, and efficient method of purification, and it is an important industrial technique that is still relevant in the chemical world today. For instance, the commercial purification of sugar is done by recrystallization on an enormous scale.

A pure, crystalline organic substance is composed of a three-dimensional array of molecules held together primarily by London forces. These attractive forces are fairly weak; most organic solids melt in the range between 22°C and 250°C. An impure organic solid will not have a well-defined crystal lattice because impurities do not allow the crystalline structure to form. The goal of recrystallization is to remove impurities from a solid to allow a perfect crystal lattice to grow.

There are four important concepts to consider when discussing the process of recrystallization:

1. Solubility
2. Saturation level
3. Exclusion
4. Nucleation

Recrystallization involves dissolving the material to be purified (the solute) in an appropriate hot solvent to yield a solution (*solubility*). As the solvent cools, the solution becomes saturated with respect to the solute (*saturation level*), which then recrystallizes. As the perfectly regular array of a crystal is formed, impurities are excluded (*exclusion*), and the crystal is thus a single pure substance. Soluble impurities remain in solution because they are not concentrated enough to saturate the solution. Recrystallization is initiated at a point of *nucleation*—a seed crystal, a speck of dust, or a scratch on the wall of the test tube if the solution is super-saturated with the solute.

In this chapter you will carry out the recrystallization process, one of the most important laboratory operations of organic chemistry, by following its seven steps. Then you will perform several actual recrystallization experiments.

The Seven Steps of Recrystallization

The process of recrystallization can be broken into seven discrete steps: (1) choosing the solvent and solvent pairs; (2) dissolving the solute; (3) decolorizing the solution with pelletized Norit; (4) filtering suspended solids; (5) recrystallizing the solute; (6) collecting and washing the crystals; and (7) drying the crystals. A detailed description of each of these steps is given in the following sections.

Step 1. Choosing the Solvent and Solvent Pairs

Similia similibus solvuntur.

Online Study Center
Video: The Reaction Tube in Use

The ideal solvent

Miscible: capable of being mixed

In choosing the solvent, the chemist is guided by the dictum "like dissolves like." Even the nonchemist knows that oil and water do not mix and that sugar and salt dissolve in water but not in oil. Hydrocarbon solvents such as hexane will dissolve hydrocarbons and other nonpolar compounds, and hydroxylic solvents such as water and ethanol will dissolve polar compounds. It is often difficult to decide, simply by looking at the structure of a molecule, just how polar or nonpolar it is and which solvent would be best. Therefore, the solvent is often chosen by experimentation. If an appropriate single solvent cannot be found for a given substance, a solvent pair system may be used. The requirement for this solvent pair is miscibility; both solvents should dissolve in each other for use as a recrystallization solvent system.

The best recrystallization solvent (and none is ideal) will dissolve the solute when the solution is hot but not when the solution is cold; it will either not dissolve the impurities at all or it will dissolve them very well (so they do not recrystallize out along with the solute); it will not react with the solute; and it will be nonflammable, nontoxic, inexpensive, and very volatile (so it can be removed from the crystals).

Some common solvents and their properties are presented in Table 4.1 in order of decreasing polarity of the solvent. Solvents adjacent to each other in the list will dissolve in each other; that is, they are miscible with each other, and each solvent will, in general, dissolve substances that are similar to it in chemical structure. These solvents are used both for recrystallization and as solvents in which reactions are carried out.

Procedure

Choosing a Solvent

Online Study Center
Video: Picking a Solvent

To choose a solvent for recrystallization, place a few crystals of the impure solute in a small test tube or centrifuge tube and add a very small drop of the solvent. Allow the drop to flow down the side of the tube and onto the crystals. If the crystals

TABLE 4.1 • Recrystallization Solvents		
Solvent	Boiling Point (°C)	Remarks
Water (H_2O)	100	It is the solvent of choice because it is cheap, nonflammable, and nontoxic and will dissolve a large variety of polar organic molecules. Its high boiling point and high heat of vaporization make it difficult to remove from crystals.
Acetic acid (CH_3COOH)	118	It will react with alcohols and amines, and it is difficult to remove from crystals. It is not a common solvent for recrystallizations, although it is used as a solvent when carrying out oxidation reactions.
Dimethyl sulfoxide (DMSO; CH_3SOCH_3)	189	It is not a commonly used solvent for recrystallization, but it is used for reactions.
Methanol (CH_3OH)	64	It is a very good solvent that is often used for recrystallization. It will dissolve molecules of higher polarity than other alcohols.
95% Ethanol (CH_3CH_2OH)	78	It is one of the most commonly used recrystallization solvents. Its high boiling point makes it a better solvent for less polar molecules than methanol. It evaporates readily from crystals. Esters may undergo an interchange of alcohol groups on recrystallization.
Acetone (CH_3COCH_3)	56	It is an excellent solvent, but its low boiling point means there is not much difference in the solubility of a compound at its boiling point compared to about 22°C.
2-Butanone; also known as methyl ethyl ketone (MEK; $CH_3COCH_2CH_3$)	80	It is an excellent solvent that has many of the most desirable properties of a good recrystallization solvent.
Ethyl acetate ($CH_3COOC_2H_5$)	78	It is an excellent solvent that has about the right combination of moderately high boiling point and the volatility needed to remove it from crystals.
Dichloromethane; also known as methylene chloride (CH_2Cl_2)	40	Although a common extraction solvent, dichloromethane boils too to make it a good recrystallization solvent. It is useful in a solvent pair with ligroin.
Diethyl ether; also known as ether ($CH_3CH_2OCH_2CH_3$)	35	Its boiling point is too low for recrystallization, although it is an extremely good solvent and fairly inert. It is used in a solvent pair with ligroin.
Methyl t-butyl ether ($CH_3OC(CH_3)_3$)	52	It is a relatively new and inexpensive solvent because of its large-scale use as an antiknock agent and oxygenate in gasoline. It does not easily form peroxides; it is less volatile than diethyl ether, but it has the same solvent characteristics. (See also Chapter 7.)
Dioxane ($C_4H_8O_2$)	101	It is a very good solvent that is not too difficult to remove from crystals; it is a mild carcinogen, and it forms peroxides.
Toluene ($C_6H_5CH_3$)	111	It is an excellent solvent that has replaced the formerly widely used benzene (a weak carcinogen) for the recrystallization of aryl compounds. Because of its boiling point, it is not easily removed from crystals.

(continued)

TABLE 4.1 • *(continued)*

Solvent	Boiling Point (°C)	Remarks
Pentane (C_5H_{12})	36	It is a widely used solvent for nonpolar substances. It is not often used alone for recrystallization, but it is good in combination with several other solvents as part of a solvent pair.
Hexane (C_6H_{14})	69	It is frequently used to recrystallize nonpolar substances. It is inert and has the correct balance between boiling point and volatility. It is often used as part of a solvent pair. (*See also* ligroin.)
Cyclohexane (C_6H_{12})	81	It is similar in all respects to hexane. (*See also* ligroin.)
Petroleum ether	30–60	It is a mixture of hydrocarbons, of which pentane is the chief component. It is used interchangeably with pentane because it is cheap. Unlike diethyl ether or *t*-butyl methyl ether, it is not an ether in the modern chemical sense.
Ligroin	60–90	It is a mixture of hydrocarbons with the properties of hexane and cyclohexane. It is a very commonly used recrystallization solvent. It is also sold as "hexanes."

Note: The solvents in this table are listed in decreasing order of polarity. Adjacent solvents in the list are, in general, miscible with each other.

dissolve instantly at about 22°C, that solvent cannot be used for recrystallization because too much of the solute will remain in solution at low temperatures. If the crystals do not dissolve at about 22°C, warm the tube on a hot sand bath and observe the crystals. If they do not go into solution, add 1 more drop of solvent. If the crystals go into solution at the boiling point of the solvent and then recrystallize when the tube is cooled, you have found a good recrystallization solvent. If not, remove the solvent by evaporation and try a different solvent. In this trial-and-error process it is easiest to try low-boiling solvents first because they can be easily removed. Occasionally, no single satisfactory solvent can be found, so mixed solvents, or *solvent pairs*, are used.

Solvent Pairs

To use a mixed solvent pair, dissolve the crystals in the better solvent (more solubilizing) and add the poorer solvent (less solubilizing) to the hot solution until it becomes cloudy, and the solution is saturated with the solute. The two solvents must, of course, be miscible with each other. Some useful solvent pairs are given in Table 4.2.

TABLE 4.2 • **Solvent Pairs**

Acetic acid–water	Ethyl acetate–cyclohexane
Ethanol-water	Acetone-ligroin
Acetone-water	Ethyl acetate–ligroin
Dioxane-water	*t*-Butyl methyl ether–ligroin
Acetone-ethanol	Dichloromethane-ligroin
Ethanol–*t*-butyl methyl ether	Toluene-ligroin

Step 2. Dissolving the Solute

Microscale Procedure

Once a recrystallization solvent has been found, the impure crystals are placed in a reaction tube, the solvent is added dropwise, the crystals are stirred with a microspatula or a small glass rod, and the tube is warmed on a steam bath or a sand bath until the crystals dissolve. Care must be exercised to use the minimum amount of solvent at or near boiling. Observe the mixture carefully as solvent is being added. Allow sufficient time for the boiling solvent to dissolve the solute and note the rate at which most of the material dissolves. When you believe most of the material has been dissolved, stop adding solvent. There is a possibility that your sample is contaminated with a small quantity of an insoluble impurity that never will dissolve. To hasten the solution process, crush large crystals with a stirring rod, taking care not to break the reaction tube.

If the solution contains no undissolved impurities and is not colored from impurities, you can simply let it cool, allowing the solute to recrystallize (step 5), and then collect the crystals (step 6). On the other hand, if the solution is colored, it must be treated with activated (decolorizing) charcoal and then filtered before recrystallization (step 3). If it contains solid impurities, it must be filtered before recrystallization takes place (step 4).

On a microscale, there is a tendency to use too much solvent so that on cooling the hot solution little or no material recrystallizes. This is not a hopeless situation. The remedy is to evaporate some of the solvent (by careful boiling) and repeat the cooling process. Inspect the hot solution.

A solution (solute dissolved in solvent) can become *superheated*; that is, heated above its boiling point without actually boiling. When boiling does suddenly occur, it can happen with almost explosive violence, a process called *bumping*. To prevent this from happening, a *wood applicator* stick can be added to the solution (Fig. 4.1). Air trapped in the wood comes out of the stick and forms the nuclei on which even boiling can occur. Porous porcelain *boiling chips* work in the same way. Never add a boiling chip or a boiling stick to a hot solution because the hot solutuion may be superheated and boil over or bump.

Macroscale Procedure

Place the substance to be recrystallized in an Erlenmeyer flask (never use a beaker), add enough solvent to cover the crystals, and then heat the flask on a steam bath (if the solvent boils below 90°C) or a hot plate until the solvent boils. (Note: Adding a boiling stick or a boiling chip to the solution will promote even boiling. It is easy to superheat the solution; that is, heat it above the boiling point with no boiling taking place. Once the solution does boil, it does so with explosive violence; it bumps.) Never add a boiling chip or boiling stick to a hot solution.

Stir the mixture or, better, swirl it (Fig. 4.2) to promote dissolution. Add solvent gradually, keeping it at the boiling point, until all of the solute dissolves. A glass rod with a flattened end can sometimes be useful in crushing large particles of solute to speed up the dissolving process. Be sure no flames are nearby when working with flammable solvents.

Do not use too much solvent.

Online Study Center

Video: Recrystallization

Prevention of bumping

Do not use wood applicator sticks (boiling sticks) in place of boiling chips in a reaction. Use them only for recrystallization.

Online Study Center

Video: Macroscale Crystallization

All procedures involving volatile and/or flammable solvents should be conducted in a fume hood.

■ **FIG. 4.1**
A reaction tube being used for recrystallization. The wood applicator stick ("boiling stick") promotes even boiling and is easier to remove than a boiling chip. The Thermowell sand is cool on top and hotter deeper down, so it provides a range of temperatures. The reaction tube is long and narrow; it can be held in the hand while the solvent refluxes. Do not use a boiling stick in place of a boiling chip in a reaction.

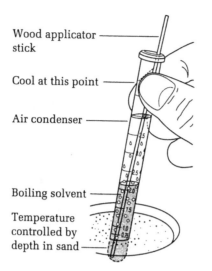

Wood applicator stick

Cool at this point

Air condenser

Boiling solvent

Temperature controlled by depth in sand

■ **FIG. 4.2**
Swirling of a solution to mix contents and help dissolve material to be recrystallized.

Be careful not to add too much solvent. Note how rapidly most of the material dissolves; stop adding solvent when you suspect that almost all of the desired material has dissolved. It is best to err on the side of too little solvent rather than too much. Undissolved material noted at this point could be an insoluble impurity that never will dissolve. Allow the solvent to boil, and if no further material dissolves, proceed to step 4 to remove suspended solids from the solution by filtration or if the solution is colored, go to step 3 to carry out the decolorization process. If the solution is clear, proceed to step 5.

Step 3. Decolorizing the Solution with Pelletized Norit

Online Study Center

Video: Decolorization of a Solution with Norit

Activated charcoal = decolorizing carbon = Norit

The vast majority of pure organic chemicals are colorless or a light shade of yellow; consequently, this step is not usually required. Occasionally, a chemical reaction will produce high molecular weight byproducts that are highly colored. The impurities can be adsorbed onto the surface of activated charcoal by simply boiling the solution with charcoal. Activated charcoal has an extremely large surface area per gram (several hundred square meters) and can bind a large number of molecules to this surface. On a commercial scale, the impurities in brown sugar are adsorbed onto charcoal in the process of refining sugar.

Add a small amount (0.1% of the solute weight is sufficient) of pelletized Norit to the colored solution and then boil the solution for a few minutes. Be careful not to add the charcoal pieces to a superheated solution; the charcoal functions like hundreds of boiling chips and will cause the solution to boil over. Remove the Norit by filtration as described in step 4.

Step 4. Filtering Suspended Solids

The filtration of a hot, saturated solution to remove solid impurities or charcoal can be performed in a number of ways. These processes include gravity filtration, pressure filtration, decantation, or removing the solvent with a Pasteur pipette. Vacuum filtration is not used because the hot solvent will cool during the process, and the product will recrystallize in the filter. Filtration can be one of the most vexing operations in the laboratory if the desired compound recrystallizes during filtration. Test the solution or a small portion of it before filtration to ensure that no crystals form at about 22°C. Like decolorization with charcoal, the removal of solid impurities by filtration is usually not necessary.

Microscale Procedure

(A) Removal of Solution with a Pasteur Pipette

If the solid impurities are large in size, they can be removed by filtration of the liquid through the small space between the square end of a Pasteur pipette and the bottom of a reaction tube (Fig. 4.3). Expel air from the pipette by squeezing the pipette bulb as the pipette is being pushed to the bottom of the tube. Use a small additional quantity of solvent to rinse the tube and pipette. Anhydrous calcium chloride, a drying agent, is easily removed in this way. The removal of very fine material, such as traces of charcoal, is facilitated by filtration of the solution through a small piece of filter paper (3 mm^2) placed in a reaction tube. This process is even easier if the filter paper is the thick variety, such as that from which Soxhlet extraction thimbles are made.[1]

(B) Filtration in a Pasteur Pipette

To filter 0.1 mL to 2 mL of a solution, dilute the solution with enough solvent so that the solute will not recrystallize at about 22°C. Prepare a filter pipette by pushing a tiny bit of cotton into a Pasteur pipette, put the solution to be filtered into this filter pipette using another Pasteur pipette, and then force the liquid through the filter using air pressure from a pipette bulb (Fig. 4.4). Fresh solvent should be added to rinse the pipette and cotton. The filtered solution is then concentrated by evaporation. One problem encountered with this method is using too much cotton packed too tightly in the pipette so that the solution cannot be forced through it. To remove very fine impurities, such as traces of decolorizing charcoal, a 3-mm to 4-mm layer of Celite filter aid can be added to the top of the cotton.

(C) Removal of Fine Impurities by Centrifugation

To remove fine solid impurities from up to 4 mL of solution, dilute the solution with enough solvent so that the solute will not recrystallize at about 22°C. Counterbalance the reaction tube and centrifuge for about 2 min at high speed in a laboratory centrifuge. The clear supernatant can be decanted (poured off) from the solid on the bottom of the tube. Alternatively, with care, the solution can be removed with a Pasteur pipette, leaving the solid behind.

---

1. Belletire, J. L.; Mahmoodi, N. O. *J. Chem. Educ.* **1989**, *66*, 964.

Online Study Center Video: Filtration of Crystals Using the Pasteur Pipette

Online Study Center Photo: Preparation of a Filter Pipette

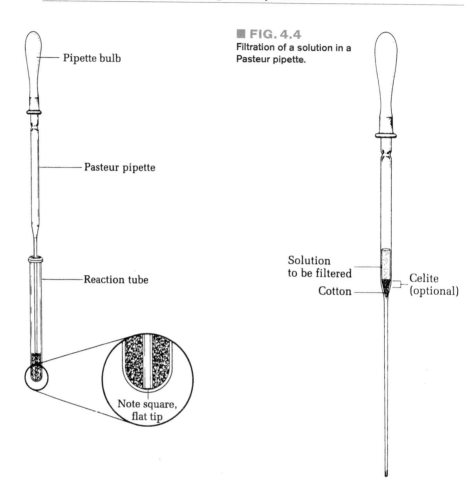

■ **FIG. 4.3**
Filtration using a Pasteur
pipette and a reaction tube.

Pipette bulb

Pasteur pipette

Reaction tube

Note square,
flat tip

■ **FIG. 4.4**
Filtration of a solution in a
Pasteur pipette.

Solution
to be filtered

Cotton

Celite
(optional)

Online Study Center

Video: Microscale
Crystallization

Use filter paper on top
of frit.

Using the chromatography
column for pressure filtration

(D) Pressure Filtration with a Micro Büchner Funnel

The technique applicable to volumes from 0.1 mL to 5 mL is to use a micro Büchner funnel. It is made of polyethylene and is fitted with a porous polyethylene frit that is 6 mm in diameter. This funnel fits in the bottom of an inexpensive disposable polyethylene pipette in which a hole is cut (Fig. 4.5). The solution to be filtered is placed in the pipette using a Pasteur pipette. The thumb covers the hole in the plastic pipette and pressure is applied to filter the solution. It is good practice to place a 6-mm-diameter piece of filter paper over the frit, which would otherwise become clogged with insoluble material.

The glass chromatography column can be used in the same way. A piece of filter paper is placed over the frit. The solution to be filtered is placed in the chromatography column, and pressure is applied to the solution using a pipette bulb. In both procedures, dilute the solution to be filtered so that it does not recrystallize in the apparatus and use a small amount of clean solvent to rinse the apparatus. The filtered solution is then concentrated by evaporation.

FIG. 4.5
A pressure filtration apparatus.
The solution to be filtered is
added through the aperture, which
is closed by a finger as pressure
is applied.

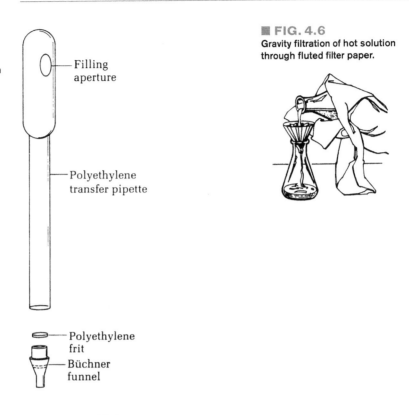

Filling
aperture

Polyethylene
transfer pipette

Polyethylene
frit
Büchner
funnel

FIG. 4.6
Gravity filtration of hot solution
through fluted filter paper.

Macroscale Procedure

(A) Decantation

Decant: to pour off. A fast, easy
separation procedure

On a large scale, it is often possible to pour off (decant) the hot solution, leaving the insoluble material behind. This is especially easy if the solid is granular like sodium sulfate. The solid remaining in the flask and the inside of the flask should be rinsed with a few milliliters of the solvent in order to recover as much of the product as possible.

(B) Gravity Filtration

The most common method for the removal of insoluble solid material is gravity filtration through fluted filter paper (Fig. 4.6). This is the method of choice for removing finely divided charcoal, dust, lint, and so on. The following equipment is needed for this process: three labeled Erlenmeyer flasks on a steam bath or a hot plate (flask A contains the solution to be filtered, flask B contains a few milliliters of solvent and a stemless funnel, and flask C contains several milliliters of the crystallizing solvent to be used for rinsing purposes), a fluted piece of filter paper, a towel for holding the hot flask and drying out the stemless funnel, and boiling chips for all solutions.

A piece of filter paper is fluted as shown in Figure 4.7 and is then placed in a stemless funnel. Appropriate sizes of Erlenmeyer flasks, stemless funnels, and

■ FIG. 4.7
Fluting a piece of filter paper.

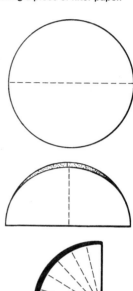

■ FIG. 4.8
Assemblies for gravity filtration. Stemless funnels have diameters of 2.5, 4.2, 5.0, and 6.0 cm.

 Be aware that the vapors of low-boiling solvents can ignite on an electric hot plate.

filter paper are shown in Figure 4.8. The funnel is stemless so that the saturated solution being filtered will not have a chance to cool and clog the stem with crystals. The filter paper should fit entirely inside the rim of the funnel; it is fluted to allow rapid filtration. Test to see that the funnel is stable in the neck of flask B. If not, support it with a ring attached to a ring stand. A few milliliters of solvent and a boiling chip should be placed in flask B into which the solution is to be filtered. This solvent is brought to a boil on the steam bath or hot plate along with the solution to be filtered.

The solution to be filtered (in flask A) should be saturated with the solute at the boiling point. Note the volume and then add 10% more solvent (from flask C). The resulting slightly dilute solution is not as likely to recrystallize in the funnel during filtration. Bring the solution to be filtered to a boil, grasp flask A in a towel, and pour the solution into the filter paper in the stemless funnel equipped in flask B (Fig. 4.6). The funnel should be warm to prevent recrystallization from occurring in the funnel. This can be accomplished in two ways: (1) Invert the funnel over a steam bath for a few seconds, pick up the funnel with a towel, wipe it perfectly dry, place it on top of flask B, and then add the fluted filter paper; or (2) place the stemless funnel in the neck of flask B and allow the solvent to reflux into the funnel, thereby warming it.

Pour the solution to be filtered (in flask A) at a steady rate into the fluted filter paper (equipped in flask B). Check to see whether recrystallization is occurring in the filter. If it does, add boiling solvent (from flask C heated on a steam bath or a hot plate) until the crystals dissolve, dilute the solution being filtered, and carry on. Rinse flask A with a few milliliters of boiling solvent (from flask C) and rinse the fluted filter paper with this same solvent.

Because the filtrate has been diluted to prevent it from recrystallizing during the filtration process, the excess solvent must now be removed by boiling the solution. This process can be speeded up somewhat by blowing a slow current of air into the flask in the hood or using an aspirator tube to pull vapors into the aspirator (Fig. 4.9 and Fig. 4.10). However, the fastest method is to heat the solvent in the filter flask on a sand bath while the flask is connected to the water aspirator.

■ FIG. 4.9

An aspirator tube in use. A boiling stick may be necessary to promote boiling.

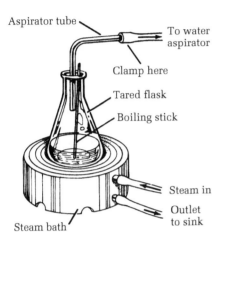

Aspirator tube

To water aspirator

Clamp here

Tared flask

Boiling stick

Steam in

Outlet to sink

Steam bath

■ FIG. 4.10

A tube being used to remove solvent vapors.

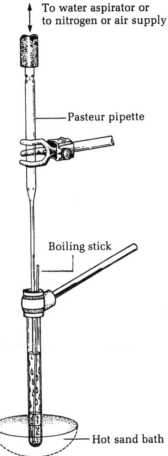

To water aspirator or to nitrogen or air supply

Pasteur pipette

Boiling stick

Hot sand bath

■ FIG. 4.11

Evaporation of a solvent under a vacuum.

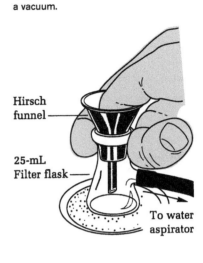

Hirsch funnel

25-mL Filter flask

To water aspirator

Be sure that you are wearing gloves when doing this step!

The vacuum is controlled with the thumb (Fig. 4.11).[2] Be sure that you are wearing gloves when doing this step! If your thumb is not large enough, put a one-holed rubber stopper into the Hirsch funnel or the filter flask and again control the vacuum with your thumb. If the vacuum is not controlled, the solution may boil over and go out the vacuum hose.

Step 5. Recrystallizing the Solute

On both a macroscale and a microscale, the recrystallization process should normally start from a solution that is saturated with the solute at the boiling point. If it has been necessary to remove impurities or charcoal by filtration, the solution

2. See also Mayo, D. W.; Pike, R. M.; Butcher, S. M. *Microscale Organic Laboratory*; Wiley: New York, 1986; 97.

has been diluted. To concentrate the solution, simply boil off the solvent under an aspirator tube as shown in Figure 4.9 (macroscale) or blow off solvent using a gentle stream of air or, better, nitrogen in the hood as shown in Figure 4.10 (microscale). Be sure to have a boiling chip (macroscale) or a boiling stick (microscale) in the solution during this process but make sure you remove it before initiating recrystallization.

A saturated solution

Slow cooling is important.

Once it has been ascertained that the hot solution is saturated with the compound just below the boiling point of the solvent, allow it to cool slowly to about 22°C. Slow cooling is a critical step in recrystallization. If the solution is not allowed to cool slowly, precipitation will occur, resulting in impurities "crashing out" of solution along with the desired solute; thus, no exclusion will occur. On a microscale, it is best to allow the reaction tube to cool in a beaker filled with cotton or paper towels, which acts as insulation, so cooling takes place slowly. Even insulated in this manner, the small reaction tube will cool to about 22°C within a few minutes. Slow cooling will guarantee the formation of large crystals, which are easily separated by filtration and easily washed free of adhering impure solvent. On a small scale, it is difficult to obtain crystals that are too large and occlude impurities. Once the tube has cooled to about 22°C without disturbance, it can be cooled in ice to maximize the amount of product that comes out of solution. On a macroscale, the Erlenmeyer flask is set atop a cork ring or other insulator and allowed to cool gradually to about 22°C. If the flask is moved during recrystallization, many nuclei will form, and the crystals will be small and have a large surface area. They will not be easy to filter and wash clean of the mother liquor. Once recrystallization ceases at about 22°C, the flask should be placed in ice to cool further. Make sure to clamp the flask in the ice bath so that it does not tip over.

Online Study Center

Videos: Recrystallization, Microscale Crystallization

Add a seed crystal or scratch the tube.

With slow cooling, recrystallization should begin immediately. If not, add a seed crystal or scratch the inside of the tube with a glass rod at the liquid-air interface. Recrystallization must start on some nucleation center. A minute crystal of the desired compound saved from the crude material will suffice. If a seed crystal is not available, recrystallization can be started on the rough surface of a fresh scratch on the inside of the container.

Step 6. Collecting and Washing the Crystals

Once recrystallization is complete, the crystals must be separated from the ice-cold mother liquor, washed with ice-cold solvent, and dried.

Microscale Procedure

(A) Filtration Using a Pasteur Pipette

Online Study Center

Video: Filtration of Crystals Using the Pasteur Pipette

The most important filtration technique used in microscale organic experiments employs a Pasteur pipette (Fig. 4.12). About 70% of the crystalline products from the experiments in this text can be isolated in this way. The others will be isolated by filtration on a Hirsch funnel.

The ice-cold crystalline mixture is stirred with a Pasteur pipette and, while air is being expelled from the pipette, forced to the bottom of the reaction tube. The

■ FIG. 4.12
Filtration using a Pasteur pipette
and a reaction tube.

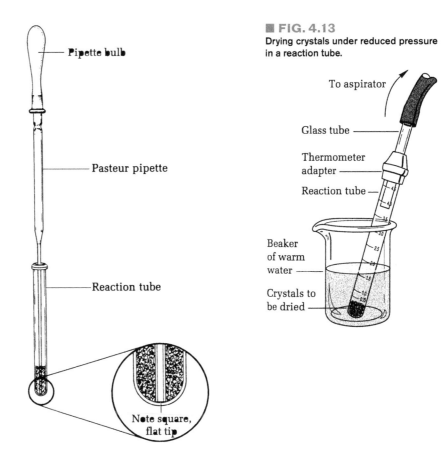

■ FIG. 4.12
Filtration using a Pasteur pipette
and a reaction tube.

Pipette bulb

Pasteur pipette

Reaction tube

Note square,
flat tip

■ FIG. 4.13
Drying crystals under reduced pressure
in a reaction tube.

To aspirator

Glass tube

Thermometer
adapter

Reaction tube

Beaker
of warm
water

Crystals to
be dried

bulb is released, and the solvent is drawn into the pipette through the very small space between the square tip of the pipette and the curved bottom of the reaction tube. When all the solvent has been withdrawn, it is expelled into another reaction tube. It is sometimes useful to rap the tube containing the wet crystals against a hard surface to pack them so that more solvent can be removed. The tube is returned to the ice bath, and a few drops of cold solvent are added to the crystals. The mixture is stirred to wash the crystals, and the solvent is again removed. This process can be repeated as many times as necessary. Volatile solvents can be removed from the damp crystals under vacuum (Fig. 4.13). Alternatively, the last traces of solvent can be removed by centrifugation using a Wilfilter (Fig. 4.14).

(B) Filtration Using a Hirsch Funnel

When the volume of material to be filtered is greater than 1.5 mL, collect the material on a Hirsch funnel. The Hirsch funnel in the Williamson/Kontes kit[3] is unique. It is composed of polypropylene and has an integral molded stopper that fits the 25-mL filter flask. It comes fitted with a 20-μm polyethylene fritted disk, which is not meant to be disposable, although it costs only about twice as much as an 11-cm

Online Study Center

Video: Microscale Filtration
on the Hirsch Funnel

---

3. The microscale kit is available through Kontes (www.kontes.com).

The Wilfilter filtration apparatus. Filtration occurs between the flat face
of the polypropylene Wilfilter and the top of the reaction tube.

The Hirsch funnel being used for vacuum
filtration. This unique design has a
removable and replaceable 20-μm
polyethylene frit. No adapter is needed
because there is a vacuum-tight fit to the
filter flask. Always use a piece of filter paper.

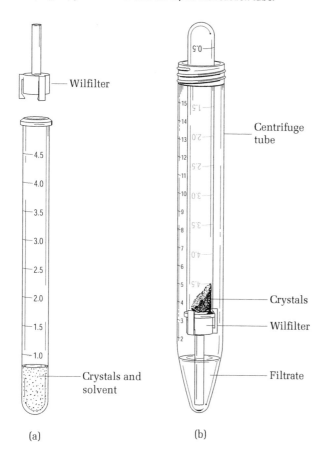

(a)   (b)

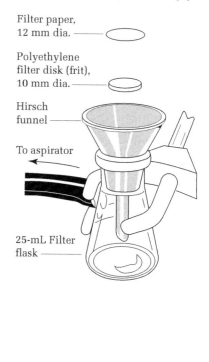

piece of filter paper (Fig. 4.15). Although products can be collected directly on this
disk, it is good practice to place an 11- or 12-mm-diameter piece of no. 1 filter paper
on the disk. In this way the frit will not become clogged with insoluble impurities.
The disk of filter paper can be cut with a cork borer or leather punch. A piece of filter paper *must* be used on the old-style porcelain Hirsch funnels.

Clamp the clean, dry 25-mL filter flask in an ice bath to prevent it from falling
over and place the Hirsch funnel with filter paper in the flask. (The reason for
cooling the filter flask is to keep the mother liquor cold so it will not dissolve the
crystals on the Hirsch funnel when fresh cold solvent is used to wash crystals from
the container onto the funnel.) In a separate flask, cool ~10 mL of solvent in an ice
bath; this solvent is used for washing the recrystallization flask and for washing
the crystals. Wet the filter paper with the solvent used in the recrystallization, turn
on the water aspirator (see "The Water Aspirator and the Trap"), and ascertain that
the filter paper is pulled down onto the frit. Pour and scrape the crystals and

Break the vacuum, add a very small quantity of ice-cold wash solvent, and reapply vacuum.

mother liquor onto the Hirsch funnel and, as soon as the liquid is gone from the crystals, break the vacuum at the filter flask by removing the rubber hose.

Rinse the recrystallization flask with ice-cold fresh solvent. Pour this rinse through the Hirsch funnel and reapply vacuum to the filter flask. As soon as all the liquid has disappeared from the crystals, wash the crystals with a few drops of ice-cold solvent. Repeat this washing process as many times as necessary to remove colored material or other impurities from the crystals. In some cases, only one very small wash will be needed. After the crystals have been washed with ice-cold solvent, the vacuum can be left on to dry the crystals. Sometimes it is useful to press solvent from the crystals by using a cork.

(C) Filtration with a Wilfilter (Replacing a Craig Tube)

The isolation of less than 100 mg of recrystallized material from a reaction tube (or any other container) is not easy. If the amount of solvent is large enough (1 mL or more), the material can be recovered by filtration on a Hirsch funnel. But when the volume of liquid is less than 1 mL, much product is left in the tube during transfer to a Hirsch funnel. The solvent can be removed with a Pasteur pipette pressed against the bottom of the tube, a very effective filtration technique, but scraping the damp crystals from the reaction tube results in major losses. If the solvent is relatively low boiling, it can be evaporated by connecting the tube to a water aspirator (*see* Fig. 4.13 on page 73). Once the crystals are dry, they are easily scraped from the tube with little or no loss. Some solvents—and water is the principal culprit—are not easily removed by evaporation. And even though removal of the solvent under vacuum is not terribly difficult, it takes time.

We have invented a filtration device that circumvents these problems: the Wilfilter. After recrystallization has ceased, most of the solvent is removed from the crystals using a Pasteur pipette in the usual way (*see* Fig. 4.12 on page 73). Then the polypropylene Wilfilter is placed on the top of the reaction tube followed by a 15-mL polypropylene centrifuge tube (*see* Fig. 4.14 on page 74). The assembly is inverted and placed in a centrifuge such as the International Clinical Centrifuge that holds twelve 15-mL tubes. The assembly, properly counterbalanced, is centrifuged for about 1 min at top speed. The centrifuge tube is removed from the centrifuge, and the reaction tube is then removed from the centrifuge tube. The three fingers on the Wilfilter keep it attached to the reaction tube. The filtrate is left in the centrifuge tube.

Filtration with the Wilfilter occurs between the top surface of the reaction tube and the flat surface of the Wilfilter. Liquid will pass through that space during centrifugation; crystals will not. The crystals will be found on top of the Wilfilter and inside the reaction tube. The very large centrifugal forces remove all the liquid, so the crystals will be virtually dry and thus easily removed from the reaction tube by shaking or scraping with the metal spatula.

The Wilfilter replaces an older device known as a Craig tube (Fig. 4.16), which consists of an outer tube of 1-, 2-, or 3-mL capacity with an inner plunger made of Teflon (expensive) or glass (fragile). The material to be recrystallized is transferred to the outer tube and recrystallized in the usual way. The inner plunger

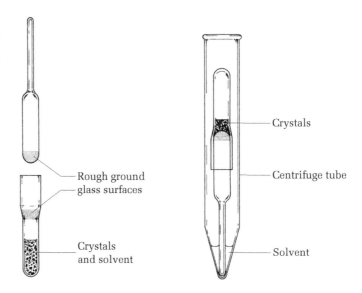

■ FIG. 4.16
The Craig tube filtration apparatus. Filtration occurs between the rough ground glass surfaces when the apparatus is centrifuged.

is added, and a wire hanger is fashioned so that the assembly can be removed from the centrifuge tube without the plunger falling off. Filtration in this device occurs through the rough surface that has been ground into the shoulder of the outer tube.

The Wilfilter possesses several advantages: a special recrystallization device is not needed, no transfers of material are needed, it is not as limited in capacity (which is 4.5 mL), and its cost is one-fifth that of a Craig tube assembly.

(D) Filtration into a Reaction Tube on a Hirsch Funnel

If it is desired to have the filtrate in a reaction tube instead of spread on the bottom of the 25-mL filter flask, then the process described in the previous section can be carried out in the apparatus shown in Figure 4.17. The vacuum hose is connected to the side arm using the thermometer adapter and a short length of glass tubing. Evaporate the filtrate in the reaction tube to collect a second crop of crystals.

(E) Filtration into a Reaction Tube on a Micro Büchner Funnel

If the quantity of material being collected is very small, the bottom of the chromatography column is a micro Büchner funnel, which can be fitted into the top of the thermometer adapter, as shown in Figure 4.18. Again, it is good practice to cover the frit with a piece of 6-mm filter paper (cut with a cork borer).

(F) The Micro Büchner Funnel in an Enclosed Filtration Apparatus

In the apparatus shown in Figure 4.19, recrystallization is carried out in the upper reaction tube in the normal way. The apparatus is then turned upside down, the crystals are shaken down onto a micro Büchner funnel, and a vacuum is applied

Online Study Center

Photo: Vacuum Filtration into Reaction Tube through Hirsch Funnel

Online Study Center

Video: Microscale Crystallization

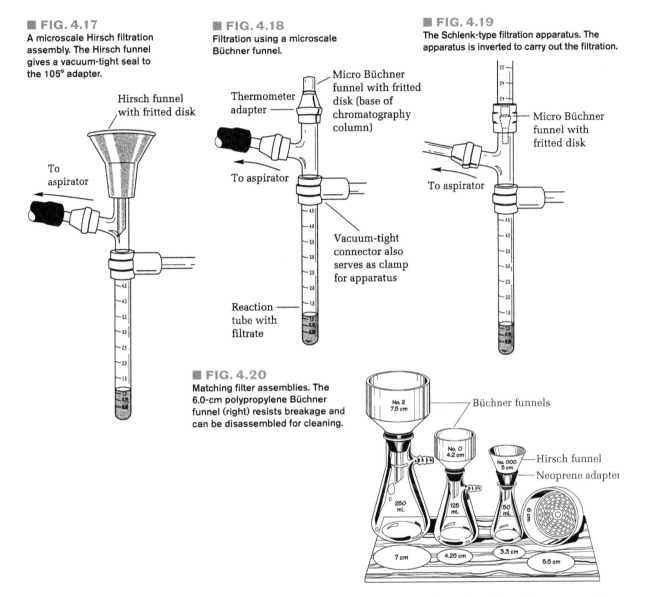

■ FIG. 4.17
A microscale Hirsch filtration assembly. The Hirsch funnel gives a vacuum-tight seal to the 105° adapter.

Hirsch funnel with fritted disk

To aspirator

■ FIG. 4.18
Filtration using a microscale Büchner funnel.

Thermometer adapter

Micro Büchner funnel with fritted disk (base of chromatography column)

To aspirator

Vacuum-tight connector also serves as clamp for apparatus

Reaction tube with filtrate

■ FIG. 4.19
The Schlenk-type filtration apparatus. The apparatus is inverted to carry out the filtration.

Micro Büchner funnel with fritted disk

To aspirator

■ FIG. 4.20
Matching filter assemblies. The 6.0-cm polypropylene Büchner funnel (right) resists breakage and can be disassembled for cleaning.

Büchner funnels

No. 2 7.5 cm

No. 0 4.2 cm

No. 000 5 cm

Hirsch funnel

Neoprene adapter

6 cm

250 mL

125 mL

50 mL

7 cm   4.25 cm   3.3 cm   5.5 cm

through the side arm. In this apparatus, crystals can be collected in an oxygen-free atmosphere (Schlenk conditions).

Macroscale Apparatus

Filtration on a Hirsch Funnel and a Büchner Funnel

If the quantity of material is small (<2 g), a Hirsch funnel can be used in exactly the way described in a previous section. For larger quantities, a Büchner funnel is used. Properly matched Büchner funnels, filter paper, and flasks are shown in Figure 4.20. The Hirsch funnel shown in the figure has a 5-cm bottom plate to accept 3.3-cm paper.

Online Study Center

Video: Microscale Filtration on the Hirsch Funnel

Place a piece of filter paper in the bottom of the Büchner funnel. Wet it with solvent and be sure it lies flat so that crystals cannot escape around the edge and under the filter paper. Then, with the vacuum off, pour the cold slurry of crystals into the center of the filter paper. Apply the vacuum; as soon as the liquid disappears from the crystals, break the vacuum to the flask by disconnecting the hose. Rinse the Erlenmeyer flask with cold solvent. Add this to the crystals and reapply the vacuum just until the liquid disappears from the crystals. Repeat this process as many times as necessary and then leave the vacuum on to dry the crystals.

Clamp the filter flask.

The Water Aspirator and Trap

The most common way to produce a vacuum for filtration in the organic laboratory is by employing a *water aspirator*. Air is entrained efficiently in the water rushing through the aspirator to produce a vacuum roughly equal to the vapor pressure of the water going through it (17 torr at 20°C, 5 torr at 4°C). A check valve is built into the aspirator, but when the water is turned off, it will often back up into the evacuated system. For this reason a *trap* is always installed in the line (Fig. 4.21). *The water passing through the aspirator should always be turned on full force.*

Opening the screw clamp on the trap can open the system to the atmosphere as well as removing the hose from the small filter flask. Open the system and then turn

■ FIG. 4.21
An aspirator, a filter trap, and a funnel. Clamp the small filter flask to prevent it from turning over.

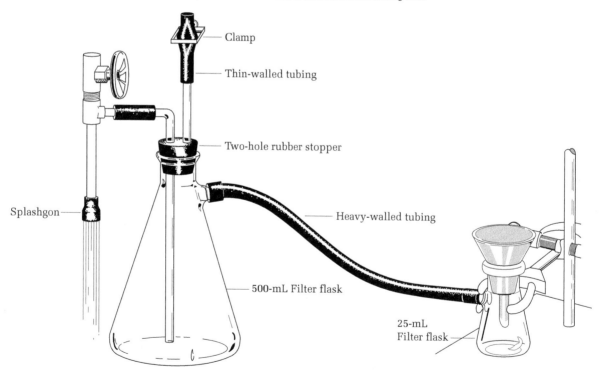

Clamp

Thin-walled tubing

Two-hole rubber stopper

Heavy-walled tubing

Splashgon

500-mL Filter flask

25-mL
Filter flask

off the water to avoid having water sucked back into the filter trap. Thin rubber tubing on the top of the trap will collapse and bend over when a good vacuum is established. You will, in time, learn to hear the differences in the sound of an aspirator when it is pulling a vacuum and when it is working on an open system.

Collecting a Second Crop of Crystals

Regardless of the method used to collect crystals on either a macroscale or a microscale, the filtrate and washings can be combined and evaporated to the point of saturation to obtain a second crop of crystals—hence this advocates having a clean receptacle for the filtrate. This second crop will increase the overall yield, but the crystals will not usually be as pure as the first crop.

Step 7. Drying the Product

Microscale Procedure

If possible, dry the product in the reaction tube after removing the solvent with a Pasteur pipette. Simply connecting the tube to the water aspirator can do this. If the tube is clamped in a beaker of hot water, the solvent will evaporate more rapidly under vacuum but make sure not to melt the product (*see* Fig. 4.13 on page 73). Water, which has a high heat of vaporization, is difficult to remove in this way. Scrape the product onto a watch glass and allow it to dry to constant weight, which will indicate that all the solvent has been removed. If the product is collected on a Hirsch funnel or a Wilfilter, the last traces of solvent can be removed by squeezing the crystals between sheets of filter paper before drying them on the watch glass.

Macroscale Procedure

Once the crystals have been washed on a Hirsch funnel or a Büchner funnel, press them down with a clean cork or other flat object and allow air to pass through them until they are substantially dry. Final drying can be done under reduced pressure (Fig. 4.22). The crystals can then be turned out of the funnel and squeezed between sheets of filter paper to remove the last traces of solvent before final drying on a watch glass.

■ FIG. 4.22
Drying a solid by reduced air pressure.

Experiments

1. Solubility Tests

To test the solubility of a solid, transfer an amount roughly estimated to be about 10 mg (the amount that forms a symmetrical mound on the end of a stainless steel spatula) into a reaction tube and add about 0.25 mL of solvent from a calibrated dropper or pipette. Stir with a fire-polished stirring rod that is 4 mm in diameter, break up any lumps, and determine if the solid is readily soluble at room temperature (about 22°C). If the substance is readily soluble at about 22°C in methanol, ethanol, acetone, or acetic acid, add a few drops of water from a wash bottle to see if a solid precipitates. If so, heat the mixture, adjust the composition of the solvent pair to

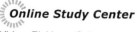
Online Study Center

Video: Picking a Solvent

Test Compounds:

OH

OH

Resorcinol

Anthracene

O   OH
 \\ //
  C

Benzoic acid

SO₃⁻ Na⁺

NH₂

**4-Amino-1-
naphthalenesulfonic acid,
sodium salt**

produce a hot solution saturated at the boiling point, let the solution stand undisturbed, and note the character of the crystals that form.

If the substance fails to dissolve in a given solvent at about 22°C, heat the suspension and observe if a solution occurs. If the solvent is flammable, heat the test tube on a steam bath or in a small beaker of water kept warm on a steam bath or a hot plate. If the solid dissolves completely, it can be declared readily soluble in the hot solvent; if some but not all dissolves, it is said to be moderately soluble, and further small amounts of solvent should then be added until solution is complete.

When a substance has been dissolved in hot solvent, cool the solution by holding the flask under the tap and, if necessary, induce recrystallization by rubbing the walls of the tube with a stirring rod to make sure that the concentration permits recrystallization. Then reheat to dissolve the solid, let the solution stand undisturbed, and inspect the character of the ultimate crystals.

Perform solubility tests on the test compounds shown in the margin in each of the the following solvents: water (hydroxylic and ionic), toluene (an aromatic hydrocarbon), and ligroin (a mixture of aliphatic hydrocarbons). Note the degree of solubility in the solvents—cold and hot—and suggest suitable solvents, solvent pairs, or other expedients for the recrystallization of each substance. Record the crystal form, at least to the extent of distinguishing between needles (pointed crystals), plates (flat and thin), and prisms. How do your observations conform to the generalization that like dissolves like?

Cleaning Up. Place organic solvents and solutions of the compounds in the organic solvents container. Dilute the aqueous solutions with water and flush down the drain. (For this and all other "Cleaning Up" sections, refer to the complete discussion of waste disposal procedures in Chapter 2.)

2. Recrystallization of Pure Phthalic Acid, Naphthalene, and Anthracene

   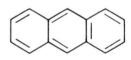

Phthalic acid   **Naphthalene**   **Anthracene**

The process of recrystallization can be readily observed using phthalic acid. In the *CRC Handbook of Chemistry and Physics*, in the table "Physical Constants of Organic Compounds," the entry for phthalic acid gives the following solubility data (in grams of solute per 100 mL of solvent). The superscripts refer to temperature in degrees Celsius:

Water	Alcohol	Ether, etc.
0.54^{14}	11.71^{18}	0.69^{15} eth., i. chl.
18^{99}		

The large difference in solubility in water as a function of temperature suggests that water is the solvent of choice. The solubility in alcohol is high at about 22°C. Ether is difficult to use because it is so volatile; the compound is insoluble in chloroform (i. chl.).

Microscale Procedure for Phthalic Acid

Recrystallize 60 mg (0.060 g) of phthalic acid from the minimum volume of water, using the previous data to calculate the required volume. First, turn on an electrically heated sand bath. Add the solid to a 10×100 mm reaction tube and add water dropwise with a Pasteur pipette. Use the calibration marks found in Chapter 1 (*see* Fig. 1.18 on page 19) to measure the volume of water in the pipette and the reaction tube. Add a boiling stick (a wood applicator stick) to facilitate even boiling and prevent bumping. After a portion of the water has been added, gently heat the solution to boiling on a hot sand bath in the electric heater. The deeper the tube is placed in the sand, the hotter it will be. As soon as boiling begins, continue to add water dropwise until the entire solid just dissolves. Cork the tube, clamp it as it cools, and observe the phenomenon of recrystallization.

After the tube reaches about 22°C, cool it in ice, stir the crystals with a Pasteur pipette, and expel the air from the pipette as the tip is pushed to the bottom of the tube. When the tip is firmly and squarely seated in the bottom of the tube, release the bulb and withdraw the water. Rap the tube sharply on a wood surface to compress the crystals and remove as much of the water as possible with the pipette. Then cool the tube in ice and add a few drops of ice-cold ethanol to the tube to remove water from the crystals. Connect the tube to a water aspirator and warm it in a beaker of hot water (*see* Fig. 4.13 on page 73). Once all the solvent is removed, using the stainless steel spatula, scrape the crystals onto a piece of filter paper, fold the paper over the crystals, and squeeze out excess water before allowing the crystals to dry to constant weight. Weigh the dry crystals and calculate the percent recovery of product.

Microscale Procedure for Naphthalene and Anthracene

Following the previous procedure, recrystallize 40 mg of naphthalene from 80% aqueous methanol or 10 mg of anthracene from ethanol. You may find it more convenient to use a hot water bath to heat these low-boiling alcohols. These are more typical of compounds to be recrystallized in later experiments because they are soluble in organic solvents. It will be much easier to remove these solvents from the crystals under vacuum than it is to remove water from phthalic acid. You will seldom encounter the need to recrystallize less than 30 mg of a solid in these experiments.

Cleaning Up. Dilute the aqueous filtrate with water and flush the solution down the drain. Phthalic acid is not considered toxic to the environment and can be recycled for future recrystallization experiments. Methanol and ethanol filtrates go into the organic solvents container.

Online Study Center

Video: Recrystallization

Set the heater control to about 20% of the maximum.

Alternate procedure: Dry the crystals under vacuum in a steam bath in the reaction tube.

These compounds can also be isolated using a Wilfilter.

Macroscale Procedure

Online Study Center

Video: Macroscale Crystallization

Using the solubility data for phthalic acid to calculate the required volume, recrystallize 1.0 g of phthalic acid from the minimum volume of water. Add the solid to the smallest practical Erlenmeyer flask and then, using a Pasteur pipette, add water dropwise from a full 10-mL graduated cylinder. A boiling stick (a wood applicator stick) facilitates even boiling and will prevent bumping. After a portion of the water has been added, gently heat the solution to boiling on a hot plate. As soon as boiling begins, continue to add water dropwise until the entire solid dissolves. Place the flask on a cork ring or other insulator and allow it to cool undisturbed to about 22°C, during which time the recrystallization process can be observed. Slow cooling favors large crystals. Then cool the flask in an ice bath, decant (pour off) the mother liquor (the liquid remaining with the crystals), and remove the last traces of liquid with a Pasteur pipette. Scrape the crystals onto a filter paper using a stainless steel spatula, squeeze the crystals between sheets of filter paper to remove traces of moisture, and allow the crystals to dry. Alternatively, the crystals can be collected on a Hirsch funnel. Compare the calculated volume of water to the volume of water actually used to dissolve the acid. Calculate the percent recovery of dry, recrystallized phthalic acid.

Cleaning Up. Dilute the filtrate with water and flush the solution down the drain. Phthalic acid is not considered toxic to the environment and can be recycled for future recrystallization experiments.

3. Decolorizing a Solution with Decolorizing Charcoal

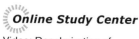
Online Study Center

Video: Decolorization of
a Solution with Norit

Decolorizing using pelletized
Norit

Into a reaction tube place 1.0 mL of a solution of methylene blue dye that has a concentration of 10 mg per 100 mL of water. Add to the tube about 10 or 12 pieces of decolorizing charcoal, shake, and observe the color over a period of 1–2 min. Heat the contents of the tube to boiling (reflux) and observe the color by holding the tube in front of a piece of white paper from time to time. How rapidly is the color removed? If the color is not removed in a minute or so, add more charcoal pellets.

Cleaning Up. Place the Norit in the nonhazardous solid waste container.

4. Decolorization of Brown Sugar (Sucrose, $C_{12}H_{22}O_{11}$)

Raw sugar is refined commercially with the aid of decolorizing charcoal. The clarified solution is seeded generously with small sugar crystals, and excess water is removed under vacuum to facilitate recrystallization. The pure white crystalline product is collected by centrifugation. Brown sugar is partially refined sugar and can be easily decolorized using charcoal.

In a 50-mL Erlenmeyer flask, dissolve 15 g of dark brown sugar in 30 mL of water by heating and stirring. Pour half the solution into another 50-mL flask. Heat one of the solutions nearly to its boiling point, allow it to cool slightly, and add 250 mg (0.25 g) of decolorizing charcoal (Norit pellets) to it. Bring the solution back to near the boiling point for 2 min; then filter the hot solution into an Erlenmeyer flask through a fluted filter paper held in a previously heated funnel. Treat the other half of the sugar solution in exactly the same way but use only 50 mg of decolorizing charcoal. In collaboration with a fellow student, try heating the solutions for only 15 s after adding the charcoal. Compare your results.

Cleaning Up. Decant (pour off) the aqueous layer. Place the Norit in the nonhazardous solid waste container. The sugar solution can be flushed down the drain.

5. Recrystallization of Benzoic Acid from Water and a Solvent Pair

Benzoic acid

Recrystallize 50 mg of benzoic acid from water in the same way phthalic acid was recrystallized. Then in a dry reaction tube dissolve another 50-mg sample of benzoic acid in the minimum volume of hot methanol and add water to the hot solution dropwise. When the hot solution becomes cloudy and recrystallization has begun, allow the tube to cool slowly to about 22°C; then cool it in ice and collect the crystals. Compare recrystallization in water to that in the solvent pair.

Cleaning Up. The methanol-water solution can be disposed in the organic solvents waste container or, if regulations permit, diluted with water and flushed down the drain.

6. Recrystallization of Naphthalene from a Mixed Solvent

Naphthalene

Do not try to grasp Erlenmeyer flasks with a test tube holder.

Support the funnel in a ring stand.

Add 2.0 g of impure naphthalene (a mixture of 100 g of naphthalene, 0.3 g of a dye such as Congo Red, and another substance such as magnesium sulfate, or dust) to a 50-mL Erlenmeyer flask along with 3 mL of methanol and a boiling stick to promote even boiling. Heat the mixture to boiling over a steam bath or a hot plate and then add methanol dropwise until the naphthalene just dissolves when the solvent is boiling. The total volume of methanol should be 4 mL. Remove the flask from the heat and cool it rapidly in an ice bath. Note that the contents of the flask set to a solid mass, which would be impossible to handle. Add enough methanol to bring the total volume to 25 mL, heat the solution to its boiling point, remove the flask from the heat, allow it to cool slightly, and add 30 mg of decolorizing charcoal pellets to remove the colored impurity in the solution. Heat the solution to its boiling point for 2 min; if the color is not gone, add more Norit and boil again, and then filter through a fluted filter paper in a previously warmed stemless funnel into a 50-mL Erlenmeyer flask. Sometimes filtration is slow because the funnel fits so snugly into the mouth of the flask that a back pressure develops. If you note that raising the funnel increases the flow of filtrate, fold a small strip of paper two or three times and insert it between the funnel and flask. Wash the used flask with 2 mL of hot methanol and use this liquid to wash the filter paper, transferring the solvent with a Pasteur pipette in a succession of drops around the upper rim of the filter paper. When the filtration is complete, the volume of methanol should be 15 mL. If it is not, evaporate the excess methanol.

Because the filtrate is far from being saturated with naphthalene at this point, it will not yield crystals on cooling; however, the solubility of naphthalene in methanol can be greatly reduced by the addition of water. Heat the solution to its boiling point and add water dropwise from a 10-mL graduated cylinder using a Pasteur pipette (or a precalibrated pipette). After each addition of water, the solution will become cloudy for an instant. Swirl the contents of the flask and heat to redissolve any precipitated naphthalene. After the addition of 3.5 mL of water, the solution will almost be saturated with naphthalene at the boiling point of the solvent. Remove the flask from the heat and place it on a cork ring or other insulating surface to cool, without being disturbed, to about 22°C.

Immerse the flask in an ice bath along with another flask containing a 30:7 mixture of methanol and water. This cold solvent will be used for washing the crystals. The cold recrystallization mixture is collected by vacuum filtration on a small 50-mm Büchner funnel (Fig. 4.23). The water flowing through the aspirator should always be turned on full force. In collecting the product by suction filtration, use a spatula to dislodge crystals and ease them out of the flask. If crystals still remain in the flask, some filtrate can be poured back into the recrystallization flask as a rinse for washing as often as desired because it is saturated with solute. To free the crystals from contaminating the mother liquor, break the suction, pour a few milliliters of the fresh cold solvent mixture into the Büchner funnel, and immediately reapply suction. Repeat this process until the crystals and the filtrate are free of color. Press the crystals with a clean cork to eliminate excess solvent, pull air through the filter cake for a few minutes, and then put the large, flat, platelike crystals out on a filter paper to dry. The yield of pure white crystalline naphthalene

■ **FIG. 4.23**

A suction filter assembly clamped to provide firm support. The funnel must be pressed down on the Filtervac to establish reduced pressure in the flask.

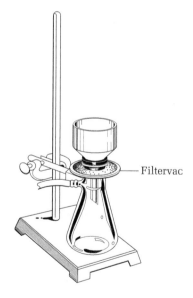

Filtervac

should be about 1.6 g. The mother liquor contains about 0.25 g, and about 0.15 g is retained in the charcoal and on the filter paper.

Cleaning Up. Place the Norit in the nonhazardous solid waste container. The methanol filtrate and washings are placed in the organic solvents container.

7. Purification of an Unknown

Recall the seven-step recrystallization procedure:

1. Choose the solvent.
2. Dissolve the solute.
3. Decolorize the solution (if necessary).
4. Filter suspended solids (if necessary).
5. Recrystallize the solute.
6. Collect and wash the crystals.
7. Dry the product.

You will purify an unknown provided by your instructor, 2.0 g if working on a macroscale and 100 mg if working on a microscale. Conduct tests for solubility and the ability to recrystallize in several organic solvents, solvent pairs, and water. Conserve your unknown by using very small quantities for the solubility tests. If only a drop or two of solvent is used, heating the test tube on a steam bath or a sand bath can evaporate the solvent, and the residue can be used for another test. If decolorization is necessary, dilute the solution before filtration. It is very difficult to filter a hot, saturated solution from decolorizing carbon without recrystallization occurring in the filtration apparatus. Evaporate the decolorized solution to the point of saturation and proceed with the recrystallization. Submit as much pure product as possible with evidence of its purity (i.e., the melting point). From the posted list identify the unknown. If an authentic sample is available, your identification can be verified by a mixed melting point determination (*see* Chapter 3).

Cleaning Up. Place decolorizing charcoal, if used, and filter paper in the nonhazardous solid waste container. Put organic solvents in the organic solvents container and flush aqueous solutions down the drain.

Recrystallization Problems and Their Solutions

Induction of Crystallization

Occasionally, a sample will not crystallize from solution on cooling, even though the solution is saturated with the solute at elevated temperature. The easiest method for inducing crystallization is to add to the supersaturated solution a seed crystal that has been saved from the crude material (if it was crystalline

Seeding

before crystallization was attempted). In a probably apocryphal tale, the great sugar chemist Emil Fischer merely had to wave his beard over a recalcitrant solution, and the appropriate seed crystals would drop out, causing recrystallization to occur. In the absence of seed crystals, scratching the inside of the flask with a stirring rod at the liquid-air interface can often induce recrystallization. One theory holds that part of the freshly scratched glass surface has angles and planes corresponding to the crystal structure, and crystals start growing on these spots. Recrystallization is often very slow to begin. Placing the sample in a refrigerator overnight will bring success. Other expedients are to change the solvent (usually to a less soluble one) and to place the sample in an open container where slow evaporation and dust from the air may help induce recrystallization.

Scratching

Oils and "Oiling Out"

Online Study Center

Video: Formation of an Oil
Instead of Crystals

Crystallize at a lower
temperature

When cooled, some saturated solutions—especially those containing water—deposit not crystals but small droplets referred to as oils. "Oiling out" occurs when the temperature of the solution is above the melting point of the crystals. If these droplets solidify and are collected, they will be found to be quite impure. Similarly, the melting point of the desired compound may be depressed to a point such that a low-melting eutectic mixture of the solute and the solvent comes out of solution. The simplest remedy for this problem is to lower the temperature at which the solution becomes saturated with the solute by simply adding more solvent. In extreme cases it may be necessary to lower this temperature well below 22°C by cooling the solution with dry ice. If oiling out persists, use another solvent.

Recrystallization Summary

Online Study Center

Video: Picking a Solvent

Video: Recrystallization

Video: Decolorization of a
Solution with Norit

Photo: Preparation of a Filter
Pipette; Video: Microscale
Crystallization

1. **Choosing the solvent.** "Like dissolves like." Some common solvents are water, methanol, ethanol, ligroin, and toluene. When you use a solvent pair, dissolve the solute in the better solvent and add the poorer solvent to the hot solution until saturation occurs. Some common solvent pairs are ethanol-water, ether-ligroin, and toluene-ligroin.

2. **Dissolving the solute.** In an Erlenmeyer flask or reaction tube, add solvent to the crushed or ground solute and heat the mixture to boiling. Add more solvent as necessary to obtain a hot, saturated solution.

3. **Decolorizing the solution.** If it is necessary to remove colored impurities, cool the solution to about 22°C and add more solvent to prevent recrystallization from occurring. Add decolorizing charcoal in the form of pelletized Norit to the cooled solution and then heat it to boiling for a few minutes, making sure to swirl the solution to prevent bumping. Remove the Norit by filtration and then concentrate the filtrate.

4. **Filtering suspended solids.** If it is necessary to remove suspended solids, dilute the hot solution slightly to prevent recrystallization from occurring during filtration. Filter the hot solution. Add solvent if recrystallization begins in the funnel. Concentrate the filtrate to obtain a saturated solution.

Photo: Recrystallization; Video: Recrystallization

5. Recrystallizing the solute.   Let the hot saturated solution cool to about 22°C spontaneously. Do not disturb the solution. Then cool it in ice. If recrystallization does not occur, scratch the inside of the container or add seed crystals.

Photos: Use of the Wilfilter, Filtration Using a Pasteur Pipette; Videos: Microscale Filtration on the Hirsch Funnel, Filtration of Crystals Using the Pasteur Pipette

6. Collecting and washing the crystals.   Collect the crystals using the Pasteur pipette method, the Wilfilter, or by vacuum filtration on a Hirsch funnel or a Büchner funnel. If the latter technique is employed, wet the filter paper with solvent, apply vacuum, break vacuum, add crystals and liquid, apply vacuum until solvent just disappears, break vacuum, add cold wash solvent, apply vacuum, and repeat until crystals are clean and filtrate comes through clear.

Photo: Drying Crystals Under Vacuum; Video: Recrystallization

7. Drying the product.   Press the product on the filter to remove solvent. Then remove it from the filter, squeeze it between sheets of filter paper to remove more solvent, and spread it on a watch glass to dry.

QUESTIONS

1. A sample of naphthalene, which should be pure white, was found to have a grayish color after the usual purification procedure. The melting point was correct, and the melting point range was small. Explain the gray color.
2. How many milliliters of boiling water are required to dissolve 25 g of phthalic acid? If the solution were cooled to 14°C, how many grams of phthalic acid would recrystallize out?
3. Why should activated carbon be used during a recrystallization?
4. If a little activated charcoal does a good job removing impurities in a recrystallization, why not use a larger quantity?
5. Under which circumstances is it wise to use a mixture of solvents to carry out a recrystallization?
6. Why is gravity filtration rather than suction filtration used to remove suspended impurities and charcoal from a hot solution?
7. Why is a fluted filter paper used in gravity filtration?
8. Why are stemless funnels used instead of long-stem funnels to filter hot solutions through fluted filter paper?
9. Why is the final product from the recrystallization process isolated by vacuum filtration rather than gravity filtration?

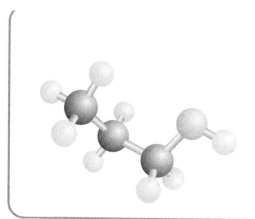

CHAPTER

5

Distillation

Online Study Center

This icon will direct you to techniques, equipment setups, and online resources at http://college.hmco.com/PIC/ williamsonMMOE5e.

Distillation is a common method for purifying liquids and can be used to determine their boiling points.

PRELAB EXERCISE: Predict what a plot of temperature versus the volume of distillate will look like for the simple distillation and the fractional distillation of (a) a cyclohexane-toluene mixture and (b) an ethanol-water mixture.

The origins of distillation are lost in antiquity, when humans in their thirst for more potent beverages found that dilute solutions of fermented alcohol could be separated into alcohol-rich and water-rich portions by heating the solution to boiling and condensing the vapors above the boiling liquid—the process of distillation.

Because ethyl alcohol (ethanol) boils at 78°C and water boils at 100°C, one might naively assume that heating a 50:50 mixture of ethanol and water to 78°C would cause the ethanol molecules to leave the solution as a vapor that could be condensed as pure ethanol. However, in such a mixture of ethanol and water, the water boils at about 87°C, and the vapor above the mixture is not 100% ethanol.

A liquid contains closely packed but mobile molecules of varying energy. When a molecule of the liquid approaches the vapor-liquid boundary and possesses sufficient energy, it may pass from the liquid phase into the gas phase. Some of the molecules present in the vapor phase above the liquid may, as they approach the surface of the liquid, reenter the liquid phase and thus become part of the condensed phase. In so doing, the molecules relinquish some of their kinetic energy (i.e., their motion is slowed). Heating the liquid causes more molecules to enter the vapor phase; cooling the vapor reverses this process.

When a closed system is at equilibrium, many molecules are escaping into the vapor phase from the liquid phase, and an equal number are returning from the vapor phase to the liquid phase. The extent of this equilibrium is measured as the vapor pressure. Even when energy is increased and more molecules in the liquid phase have sufficient energy to escape into the vapor phase, equilibrium is maintained because the number moving from the vapor phase into the liquid phase also increases. However, the number of molecules in the vapor phase increases, which increases the vapor pressure. The number of molecules in the vapor phase depends primarily on the volume of the system, the temperature, the combined pressure of all the gaseous components, and the strength of the intermolecular forces exerted in the liquid phase. Review the introduction to Chapter 3 about the types of intermolecular forces.

Simple Distillation

Simple distillation involves boiling a liquid in a vessel (a distilling flask) and directing the resulting vapors through a condenser, in which the vapors are cooled and converted to a liquid that flows back into a collection vessel (a receiving flask). (*See* Fig. 5.5 on page 95.) Simple distillation is used to purify liquid mixtures by separating one liquid component either from nonvolatile substances or from another liquid that differs in boiling point by at least 75°C. The initial condensate will have essentially the same mole ratio of liquids as the vapor just above the boiling liquid. The closer the boiling points of the components of a liquid mixture, the more difficult they are to completely separate by simple distillation.

Fractional Distillation

Fractional distillation differs from simple distillation in that a fractionating column is placed between the distilling flask and the condenser. This fractionating column allows for successive condensations and distillations and produces a much better separation between liquids with boiling points closer than 75°C. The column is packed with material that provides a large surface area for heat exchange between the ascending vapor and the descending liquid. As a result, multiple condensations and vaporizations occur as the vapors ascend the column. Condensing of the higher-boiling vapor releases heat, which causes vaporization of the lower-boiling liquid on the packing so that the lower-boiling component moves up while the higher-boiling component moves down. Some of the lower-boiling component will run back into the distilling flask. Each successive condensation-vaporization cycle, also called a *theoretical plate*, produces a vapor that is richer in the more volatile fraction. As the temperature of the liquid mixture is increased, the lower-boiling fractions become enriched in the vapor.

A large surface area for the packing material is desirable, but the packing cannot be so dense that pressure changes take place within the column to cause nonequilibrium conditions. Also, if the column packing has a very large surface area, it will absorb (hold up) much of the material being distilled. Several different packings for distilling columns have been tried, including glass beads, glass helices, and carborundum chips. One of the best packings in our experience is a copper or steel sponge (Chore Boy). It is easy to insert into the column; it does not come out of the column as beads do; and it has a large surface area, good heat transfer characteristics, and low holdup. It can be used in both microscale and macroscale apparatus.

The ability of different column packings to separate two materials of differing boiling points is evaluated by calculating the number of theoretical plates. Each theoretical plate corresponds to one condensation-vaporization cycle. Other things being equal, the number of theoretical plates is proportional to the height of

Heat exchange between ascending vapor and descending liquid

Column packing

Holdup: unrecoverable distillate that wets the column packing

Height equivalent to a theoretical plate (HETP)

Equilibration is slow.

Good fractional distillation takes a long time.

■ FIG. 5.1
A bubble plate distilling column.

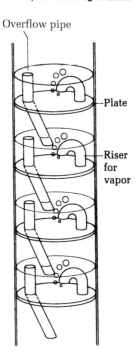

Overflow pipe

Plate

Riser for vapor

the column, so various packings are evaluated according to the *height equivalent to a theoretical plate* (*HETP*); the smaller the HETP, the more plates the column will have and the more efficient it will be. The calculation is made by analyzing the proportion of lower- to higher-boiling material at the top of the column and in the distillation pot.[1]

Although not obvious, the most important variable that contributes to good fractional distillation is the rate at which the distillation is carried out. A series of simple distillations take place within a fractionating column, and it is important that complete equilibrium be attained between the ascending vapors and the descending liquid. This process is not instantaneous. It should be an adiabatic process; that is, heat should be transferred from the ascending vapor to the descending liquid with no gain of heat or net heat loss to the surroundings. Advanced distillation systems use thermally insulated, vacuum-jacketed fractionating columns. They also allow the adjustment of the ratio between the amount of condensate that is directed to the receiving flask and the amount returned to the distillation column. A reflux ratio of 30:1 or 50:1 is not uncommon for a 40-plate column. Although a distillation of this type takes several hours, this is far less time than if one to had to do 40 distillations, one after the other, and yields much better separated compounds.

Perhaps it is easiest to understand the series of redistillations that occur in fractional distillation by examining the bubble plate column used to fractionally distill crude oil (Fig. 5.1). These columns dominate the skyline at oil refineries, with some being 150 ft high and capable of distilling 200,000 barrels of crude oil per day. The crude oil enters the column as a hot vapor. Some of this vapor with high-boiling components condenses on one of the plates. The more volatile substances travel through the bubble cap to the next higher plate, where some of the less-volatile components condense. As high-boiling liquid material accumulates on a plate, it descends through the overflow pipe to the next lower plate, and vapor rises through the bubble cap to the next higher plate. The temperature of the vapor that is rising through a cap is above the boiling point of the liquid on that plate. As bubbling takes place, heat is exchanged, and the less volatile components on that plate vaporize and go on to the next plate. The composition of the liquid on a plate is the same as that of the vapor coming from the plate below. So, on each plate a simple distillation takes place. At equilibrium, vapor containing low-boiling material is ascending through the column, and high-boiling liquid is descending.

As a purification method, distillation, particularly fractional distillation, requires larger amounts of material than recrystallization, liquid/liquid extraction, or chromatography. Performing a fractional distillation on less than 1 g of material is virtually impossible. Fractional distillation can be carried out on a scale of about 3–4 g. As will be seen in Chapters 8, 9, and 10, various types of chromatography are employed for separations of milligram quantities of liquids.

---

1. Weissberger, A., ed. *Techniques of Organic Chemistry*, Vol. IV; Wiley-Interscience: New York, 1951.

Liquid Mixtures

If two different liquid compounds are mixed, the vapor above the mixture will contain some molecules of each component. Let us consider a mixture of cyclohexane and toluene. The vapor pressures, as a function of temperature, are plotted in Figure 5.2. When the vapor pressure of the liquid equals the applied pressure, the liquid boils. Figure 5.2 shows that, at 760 mm Hg (standard atmospheric pressure), these pure liquids boil at about 81°C and 111°C, respectively. If one of these pure liquids were to be distilled, we would find that the boiling point of the liquid would equal the temperature of the vapor and that the temperature of the vapor would remain constant throughout the distillation.

Figure 5.3 is a boiling point–composition diagram for the cyclohexane-toluene system. If a mixture of 75 mole percent toluene and 25 mole percent cyclohexane is heated, we find that it boils at 100°C (point A). Above a binary mixture of cyclohexane and toluene, the vapor pressure has contributions from each component. Raoult's law states that the vapor pressure of the cyclohexane is equal to the product of the vapor pressure of pure cyclohexane and the mole fraction of cyclohexane in the liquid mixture:

Raoult's law of partial pressures

$$P_c = P_c^\circ N_c$$

where P_c is the partial pressure of cyclohexane, P_c° is the vapor pressure of pure cyclohexane at the given temperature, and N_c is the mole fraction of cyclohexane in the mixture. For toluene,

The mole fraction of cyclohexane is equal to the moles of cyclohexane in the mixture divided by the total number of moles (cyclohexane plus toluene) in the mixture.

$$P_t = P_t^\circ N_t$$

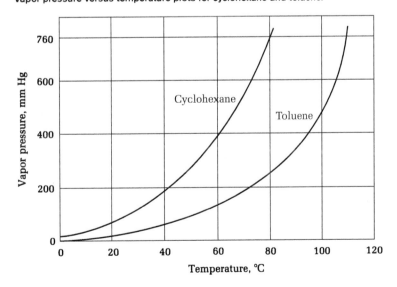

FIG. 5.2
Vapor pressure versus temperature plots for cyclohexane and toluene.

FIG. 5.3 Boiling point–composition curves for a mixture of cyclohexane and toluene.

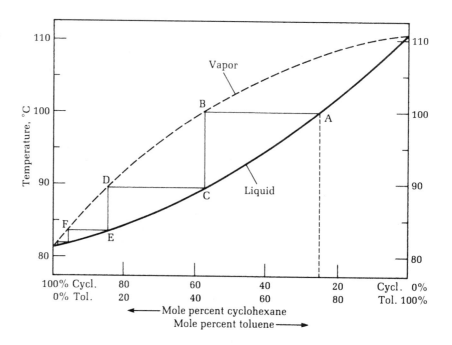

The total vapor pressure above the solution (P_{Tot}) is given by the sum of the partial pressures due to cyclohexane and toluene:

$$P_{Tot} = P_c + P_t$$

Dalton's law states that the mole fraction of cyclohexane (X_c) in the vapor at a given temperature is equal to the partial pressure of the cyclohexane at that temperature divided by the total pressure:

$$X_c = \frac{P_c}{\text{total vapor pressure}}$$

At 100°C cyclohexane has a partial pressure of 433 mm Hg, and toluene has a partial pressure of 327 mm Hg; the sum of the partial pressures is 760 mm Hg, so the liquid boils. If some of the liquid in equilibrium with this boiling mixture were condensed and analyzed, it would be found to be 433/760, or 57 mole percent cyclohexane (point B, Fig. 5.3). This is the best separation that can be achieved on a simple distillation of this mixture. As the simple distillation proceeds, the boiling point of the mixture moves toward 111°C along the line from point A, and the vapor composition becomes richer in toluene as it moves from point B to 110°C. To obtain pure cyclohexane, it would be necessary to condense the liquid at point B and redistill it. When this is done, it is found that the liquid boils at 90°C (point C), and the vapor equilibrium with this liquid is about 85 mole percent cyclohexane (point D). Therefore, to separate a mixture of cyclohexane and toluene, a series of fractions would be collected, and each of these would be partially redistilled. If this fractional distillation were done enough times, the two components could be completely separated.

Azeotropes

Not all liquids form ideal solutions and conform to Raoult's law. Ethanol and water are two such liquids. Because of molecular interaction, a mixture of 95.5% (by weight) of ethanol and 4.5% of water boils *below* the boiling point of pure ethanol (78.15°C versus 78.3°C). Thus, no matter how efficient the distilling apparatus, 100% ethanol cannot be obtained by distillation of a mixture of, say, 75% water and 25% ethanol. A mixture of liquids of a certain definite composition that distills at a constant temperature without a change in composition is called an *azeotrope*; 95% ethanol is such an azeotrope. The boiling point–composition curve for the ethanol-water mixture is seen in Figure 5.4. To prepare 100% ethanol, the water can be removed chemically (by reaction with calcium oxide) or it can be removed as an azeotrope with still another liquid. An azeotropic mixture of 32.4% ethanol and 67.6% benzene (bp 80.1°C) boils at 68.2°C. A ternary azeotrope containing 74.1% benzene, 18.5% ethanol, and 7.4% water boils at 64.9°C. Absolute alcohol (100% ethanol) is made by adding benzene to 95% ethanol followed by removing the water in the volatile ternary azeotrope of benzene, ethanol, and water.

Ethanol and water form a minimum boiling azeotrope. Substances such as formic acid (bp 100.7°C) and water (bp 100°C) form maximum boiling azeotropes. The boiling point of a formic acid–water azeotrope is 107.3°C.

The ethanol-water azeotrope

■ FIG. 5.4
Boiling point–composition curves for a mixture of ethanol and water.

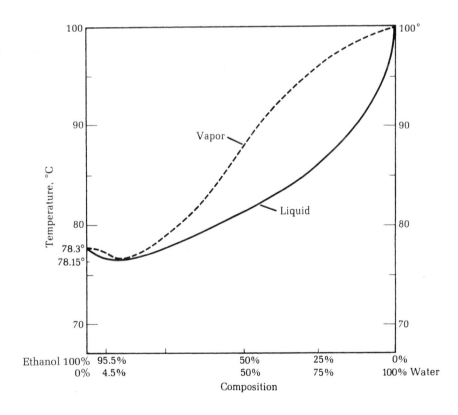

Boiling Points and Distillation

A constant boiling point on distillation does not guarantee that the distillate is a single pure compound.

A pure liquid has a constant boiling point. A change in boiling point during distillation is an indication of impurity. The converse proposition, however, is not always true; that is, constancy of a boiling point does not necessarily mean that the liquid consists of only one compound. For instance, two miscible liquids of similar chemical structure that boil at the same temperature individually will have nearly the same boiling point as a mixture. And, as noted previously, azeotropes have constant boiling points that can be either above or below the boiling points of the individual components.

Distilling a mixture of sugar and water

When a solution of sugar in water is distilled, the boiling point recorded on a thermometer located in the vapor phase is 100°C (at 760 torr) throughout the distillation, whereas the temperature of the boiling sugar solution itself is initially somewhat above 100°C and continues to rise as the concentration of sugar in the remaining solution increases. The vapor pressure of the solution is dependent on the number of water molecules present in a given volume; hence, with increasing concentration of nonvolatile sugar molecules and decreasing concentration of water, the vapor pressure at a given temperature decreases, and a higher temperature is required for boiling. However, sugar molecules do not leave the solution, and the drop clinging to the thermometer is pure water in equilibrium with pure water vapor.

Boiling point changes with pressure.

When a distillation is carried out in a system open to the air (the boiling point is thus dependent on existing air pressure), the prevailing barometric pressure should be noted and allowance made for appreciable deviations from the accepted boiling point temperature (Table 5.1). Distillation can also be done at the lower pressures that can be achieved using an oil pump or an aspirator, which produces a substantial reduction in the boiling point.

EXPERIMENTS

Before beginning any distillation, calibrate the thermometer to ensure accurate readings are made; refer to Part 3 of Chapter 3 for calibration instructions.

TABLE 5.1 • Variation in Boiling Point with Pressure	Boiling Point	
Pressure (mm Hg)	Water (°C)	Benzene (°C)
780	100.7	81.2
770	100.4	80.8
760	100.0	80.3
750	99.6	79.9
740	99.2	79.5
584*	92.8	71.2

*Instituto de Quimica, Mexico City, altitude 7700 ft (2310 m).

1. Simple Distillation

Apparatus for simple distillation

IN THIS EXPERIMENT the two liquids to be separated are placed in a 5-mL round-bottomed, long-necked flask that is fitted to a distilling head (Fig. 5.5). The flask has a larger surface area exposed to heat than does the reaction tube, so the necessary thermal energy can be put into the system to cause the materials to distill. The hot vapor rises and completely envelops the bulb of the thermometer before passing over it and down toward the receiver. The downward-sloping portion of the distilling head functions as an air condenser. The objective is to observe how the boiling point of the mixture changes during the course of its distillation. (Another simple distillation apparatus is shown in Figure 5.6. Here, the long air condenser will condense even low-boiling liquids, and the receiver is far from the heat.)

■ **FIG. 5.6**
A simple distillation apparatus.

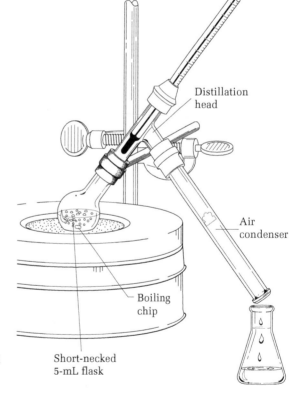

Distillation head

Air condenser

Boiling chip

Short-necked 5-mL flask

■ **FIG. 5.5**
A small-scale simple distillation apparatus. This apparatus can be adapted for fractional distillation by packing the long neck with a copper sponge. The temperature is regulated by either scraping sand away from or piling sand up around the flask.

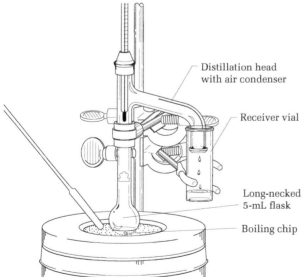

Distillation head with air condenser

Receiver vial

Long-necked 5-mL flask

Boiling chip

The rate of distillation is determined by the heat input to the apparatus. This is most easily and effectively controlled by using a spatula to pile up or scrape away hot sand from around the flask.

(A) Simple Distillation of a Cyclohexane-Toluene Mixture

Online Study Center

Photo: Simple Distillation Apparatus

Viton is resistant to hot aromatic vapors.

The thermometer bulb must be *completely below* the side arm.

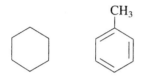

Cyclohexane	Toluene
bp 81°C	bp 111°C
MW 84.16	MW 92.14
n_D^{20} 1.4260	n_D^{20} 1.4960

Throughout this text information regarding the physical properties of substances has been placed with each various structures in the margin. MW is molecular weight, bp is boiling point, den. is density in g/mL, and n_D^{20} is the refractive index.

There are 21 ± 3 drops per milliliter.

To a 5-mL long-necked, round-bottomed flask, add 2.0 mL of dry cyclohexane, 2.0 mL of dry toluene, and a boiling chip (*see* Fig. 5.5). This flask is joined by means of a Viton (black) connector to a distilling head fitted with a thermometer using a rubber connector. The thermometer bulb should be completely below the side arm of the Claisen head so that the mercury reaches the same temperature as the vapor that distills. The end of the distilling head dips well down into a receiving vial, which rests on the bottom of a 30-mL beaker filled with ice. The distillation is started by piling up hot sand to heat the flask. As soon as boiling begins, the vapors can be seen to rise up the neck of the flask. Adjust the rate of heating by piling up or scraping away sand from the flask so that it takes *several minutes* for the vapor to rise to the thermometer. **The rate of distillation should be no faster than 2 drops per minute.**

Record the temperature versus the number of drops during the entire distillation process. If the rate of distillation is as slow as it should be, there will be sufficient time between drops to read and record the temperature. Continue the distillation until only about 0.4 mL remains in the distilling flask. **Never distill to dryness.** On a larger scale explosive peroxides can sometimes accumulate. At the end of the distillation, measure as accurately as possible, perhaps with a syringe, the volume of the distillate and, after it cools, the volume left in the pot; the difference is the holdup of the column if none has been lost by evaporation. Note the barometric pressure, make any thermometer corrections necessary, and make a plot of milliliters (drop number) versus temperature for the distillation.

Cleaning Up. The pot residue should be placed in the organic solvents container. The distillate can also be placed there or recycled.

(B) Simple Distillation of an Ethanol-Water Mixture

In a 5-mL round-bottomed, long-necked flask place 4 mL of a 10% to 20% ethanol-water mixture. Assemble the apparatus as described previously and carry out the distillation until you believe a representative sample of ethanol has collected in the receiver. In the hood place 3 drops of this sample on a Pyrex watch glass and try to ignite it with the blue cone of a microburner flame. Does it burn? Is any unburned residue observed? There was a time when alcohol-water mixtures were mixed with gunpowder and ignited to give proof that the alcohol had not been diluted. One hundred proof alcohol is 50% ethanol by volume.

Cleaning Up. The distillate and pot residue can be disposed in the organic solvents waste container or, if regulations permit, diluted with water and flushed down the drain.

2. Fractional Distillation

IN THIS EXPERIMENT, just as in the last one, you will distill a mixture of two liquids and again record the boiling points as a function of the volume of distillate (drops). The necessity for a very slow rate of distillation cannot be overemphasized.

Apparatus

Online Study Center

Photos: Column Packing with Chore Boy for Fractional Distillation, Fractional Distillation Apparatus

Assemble the apparatus shown in Figure 5.7. The 10-cm column is packed with 1.5 g of copper sponge and connected to the 5-mL short-necked flask using a black (Viton) connector. The column should be vertical and care should be taken to ensure that the bulb of the thermometer does not touch the side of the distilling head. The column, but not the distilling head, will be insulated with glass wool or cotton at the appropriate time to ensure that the process is adiabatic. Alternatively, the column can be insulated with a cut-off 15-mL polyethylene centrifuge tube.

■ **FIG. 5.7**
A small-scale fractional distillation apparatus. The 10-cm column is packed with 1.5 g of copper sponge (Chore Boy).

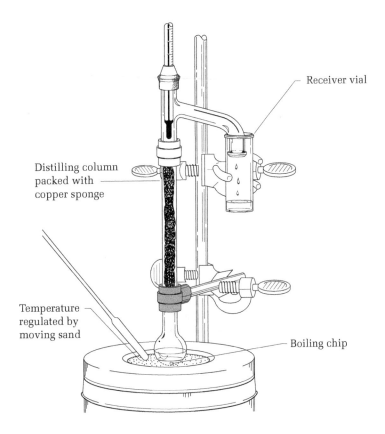

Receiver vial

Distilling column packed with copper sponge

Temperature regulated by moving sand

Boiling chip

(A) Fractional Distillation of a Cyclohexane-Toluene Mixture

To a short-necked flask add 2.0 mL of cyclohexane, 2.0 mL of toluene, and a boiling chip. The distilling column is packed with 1.5 g of copper sponge (Fig. 5.7). The mixture is brought to a boil over a hot sand bath. Observe the ring of condensate that should rise slowly through the column; if you cannot at first see this ring, locate it by touching the column with your fingers. It will be cool above the ring and hot below. Reduce the heat by scraping sand away from the flask and wrap the column, but not the distilling head, with glass wool or cotton if it is not already insulated.

The distilling head and the thermometer function as a small reflux condenser. Again, apply the heat, and as soon as the vapor reaches the thermometer bulb, reduce the heat by scraping away sand. **Distill the mixture at a rate no faster than 2 drops per minute** and record the temperature as a function of the number of drops. If the heat input has been *very* carefully adjusted, the distillation will cease, and the temperature reading will drop after the cyclohexane has distilled. Increase the heat input by piling up the sand around the flask to cause the toluene to distill. Stop the distillation when only about 0.4 mL remains in the flask and measure the volume of distillate and the pot residue as before. Make a plot of the boiling point versus the milliliters of distillate (drops) and compare it to the simple distillation carried out in the same apparatus. Compare your results with those in Figure 5.8.

Cleaning Up. The pot residue should be placed in the organic solvents container. The distillate can also be placed there or recycled.

(B) Fractional Distillation of an Ethanol-Water Mixture

Distill 4 mL of the same ethanol-water mixture used in the simple distillation experiment, following the procedure used for the cyclohexane-toluene mixture

Never distill in an airtight system.

Adjust the heat input to the flask by piling up or scraping away sand around the flask.

Insulate the distilling column but not the Claisen head.

21 ± 3 drops = 1 mL

■ FIG. 5.8
Simple and fractional distillation curves for cyclohexane and toluene.

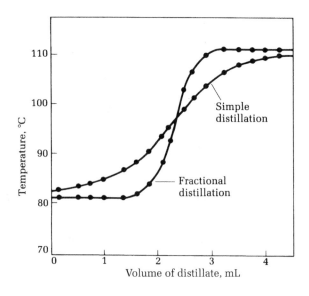

with either the short or the long distilling column. Remove what you regard to be the ethanol fraction and repeat the ignition test. Is any difference noted?

Cleaning Up. The pot residue and distillate can be disposed in the organic solvents waste container or, if regulations permit, diluted with water and flushed down the drain.

3. Instant Microscale Distillation

Online Study Center

Frequently, a very small quantity of freshly distilled material is needed in an experiment. For example, two compounds that need to be distilled freshly are aniline, which turns black because of the formation of oxidation products, and benzaldehyde, a liquid that easily oxides to solid benzoic acid. The impurities that arise in both of these compounds have much higher boiling points than the pure compounds, so a very simple distillation suffices to separate them. This can be accomplished as follows.

Place a few drops of the impure liquid in a reaction tube along with a boiling chip. Clamp the tube in a hot sand bath and adjust the heat so that the liquid refluxes about halfway up the tube. Expel the air from a Pasteur pipette, thrust it down into the hot vapor, and then pull the hot vapor into the cold upper portion of the pipette. The vapor will immediately condense and can then be expelled into another reaction tube that is held adjacent to the hot one (Fig. 5.9). In this way enough pure material can be distilled to determine a boiling point, run a spectrum, make a derivative, or carry out a reaction. Sometimes the first drop or two will be cloudy, which indicates the presence of water. This fraction should be discarded in order to obtain pure dry material.

4. Simple Distillation

Apparatus

In any distillation, the flask should be no more than two-thirds full at the start. Great care should be taken not to distill to dryness because, in some cases, high-boiling explosive peroxides can become concentrated.

Assemble the apparatus for macroscale simple distillation, as shown in Figure 5.10, starting with the support ring followed by the electric flask heater and then the flask. One or two boiling stones are put in the flask to promote even boiling. Each ground joint is greased by putting three or four stripes of grease lengthwise around the male joint and pressing the joint firmly into the other without twisting. The air is thus eliminated, and the joint will appear almost transparent. (Do not use excess grease because it will contaminate the product.) Water enters the condenser at the tublature nearest the receiver. Because of the large heat capacity of water, only a very small stream (3 mm diameter) is needed; too much water pressure will cause the tubing to pop off. A heavy rubber band, or better a Keck clamp, can be used to hold the condenser to the distillation head. Note that the bulb of the thermometer is below the opening into the side arm of the distillation head.

Video: Instant Microscale Distillation

An apparatus for instant
microscale distillation.

An apparatus for macroscale simple
distillation.

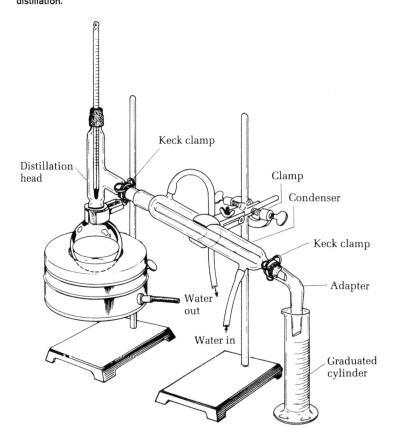

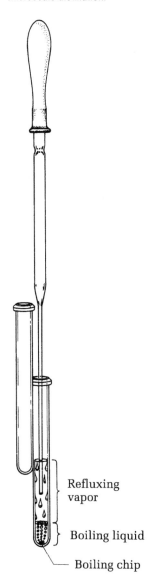

Refluxing
vapor

Boiling liquid

Boiling chip

CAUTION: Cyclohexane and
toluene are flammable; make
sure the distilling apparatus is
tight.

Do not add a boiling chip to a hot
liquid. It may boil over.

(A) Simple Distillation of a Cyclohexane-Toluene Mixture

Place a mixture of 30 mL cyclohexane and 30 mL toluene and a boiling chip in a
dry 100-mL round-bottomed flask and assemble the apparatus for simple distillation. After assuring that all connections are tight, heat the flask strongly until boiling begins. Then adjust the heat until the distillate drops at a regular rate of about
1 drop per second. Record both the temperature and the volume of distillate at

regular intervals. After 50 mL of distillate is collected, discontinue the distillation. Record the barometric pressure, make any thermometer correction necessary, and plot the boiling point versus the volume of distillate. Save the distillate for fractional distillation.

Cleaning Up. The pot residue should be placed in the organic solvents container. The distillate can also be placed there or recycled.

Dispose of cyclohexane and toluene in the container provided. Do not pour them down the drain.

(B) Simple Distillation of an Ethanol-Water Mixture

In a 500-mL round-bottomed flask place 200 mL of a 20% aqueous solution of ethanol. Follow the previous procedure for the distillation of a cyclohexane-toluene mixture. Discontinue the distillation after 50 mL of distillate has been collected. Working in the hood, place 3 drops of distillate on a Pyrex watch glass and try to ignite it with the blue cone of a microburner flame. Does it burn? Is any unburned residue observed?

Cleaning Up. The pot residue and distillate can be disposed in the organic solvents container or, if regulations permit, diluted with water and flushed down the drain.

5. Fractional Distillation

Apparatus

Assemble the apparatus shown in Figures 5.11 and 5.12. The fractionating column is packed with one-fourth to one-third of a metal sponge. The column should be perfectly vertical and be insulated with glass wool covered with aluminum foil (shiny side in). However, insulation is omitted for this experiment so that you can observe what is taking place in the column.

(A) Fractional Distillation of a Cyclohexane-Toluene Mixture

After the flask from the simple macroscale distillation experiment has cooled, pour the 50 mL of distillate back into the distilling flask, add one or two new boiling chips, and assemble the apparatus for fractional distillation. The stillhead delivers into a short condenser fitted with a bent adapter leading into a 10-mL graduated cylinder. Gradually turn up the heat to the electric flask heater until the mixture of cyclohexane and toluene just begins to boil. As soon as boiling starts, turn down the power. Heat slowly at first. A ring of condensate will rise slowly through the column; if you cannot at first see this ring, locate it by cautiously touching the column with your fingers. The rise should be very gradual so that the column can acquire a uniform temperature gradient. Do not apply more heat until you are sure that the ring of condensate has stopped rising; then increase the heat gradually. In a properly conducted operation, the vapor-condensate mixture reaches the top of the column only after several minutes. Once distillation has commenced, it should continue steadily without any drop in temperature at a rate no greater than 1 mL in 1.5–2 min. Observe the flow and keep it steady by slight increases in heat as required. Protect the column from drafts by wrapping it with

An apparatus for macroscale fractional distillation.
The position of the thermometer bulb is critical.

A fractionating column and its
packing. Use one-third of a copper
sponge (Chore Boy).

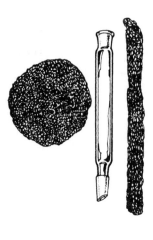

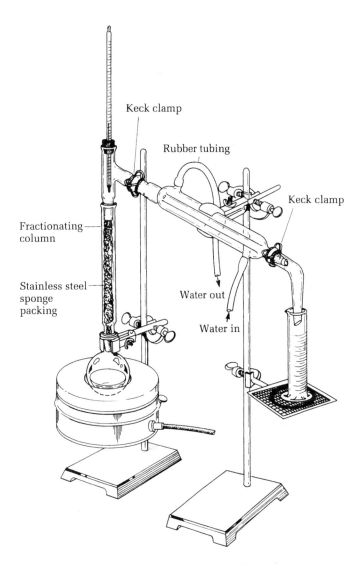

Keck clamp

Rubber tubing

Keck clamp

Fractionating
column

Stainless steel
sponge
packing

Water out

Water in

aluminum foil, glass wool, or even a towel. This insulation will help prevent
flooding of the column, as will slow and steady distillation.

Record the temperature as each milliliter of distillate collects and take more
frequent readings when the temperature starts to rise abruptly. Each time the graduated cylinder fills, quickly empty it into a series of labeled 25-mL Erlenmeyer
flasks. Stop the distillation when a second constant temperature is reached. Plot a
distillation curve and record what you observed inside the column in the course of
the fractionation. Combine the fractions that you think are pure and turn in the

product in a bottle labeled with your name, desk number, the name of the product, the boiling point range, and the weight.

Cleaning Up. The pot residue should be placed in the organic solvents container. The cyclohexane and toluene fractions can also be placed there or recycled.

(B) Fractional Distillation of Ethanol-Water Mixture

Place the 50 mL of distillate from the simple distillation experiment in a 100-mL round-bottomed flask, add one or two boiling chips, and assemble the apparatus for fractional distillation. Follow the previous procedure for the fractional distillation of a cyclohexane-toluene mixture. Repeat the ignition test. Is any difference noted? Alternatively, distill 60 mL of the 10%–20% ethanol-water mixture that results from the fermentation of sucrose (*see* Chapter 64).

Cleaning Up. The pot residue and distillate can be disposed in the organic solvents container or, if regulations permit, diluted with water and flushed down the drain.

6. Fractional Distillation of Unknowns

You will be supplied with an unknown, prepared by your instructor, that is a mixture of two solvents listed in Table 5.2, only two of which form azeotropes. The solvents in the mixture will be mutually soluble and differ in boiling point by more than 20°C. The composition of the mixture (in percentages of the two components) will be either 20:80, 30:70, 40:60, 50:50, 60:40, 70:30, or 80:20. Identify the two compounds and determine the percent composition of each. Perform a fractional distillation on 4 mL of the unknown for microscale; use at least 50 mL of unknown for macroscale. Fractionate the unknown and identify the components from the boiling points. Prepare a distillation curve. You may be directed to analyze your distillate by gas chromatography (*see* Chapter 10) or refractive index (*see* Chapter 14).

TABLE 5.2 • Some Properties of Common Solvents	
Solvent	**Boiling Point (°C)**
Acetone	56.5
Methanol	64.7
Hexane	68.8
1-Butanol	117.2
2-Methyl-2-propanol	82.2
Water	100.0
Toluene*	110.6

*Methanol and toluene form an azeotrope with a boiling point of 63.8°C (69% methanol).

[Handwritten at top of page:]

???

greater area = more (compound)

$\% = 12.026 + 0.001 + 0.103$ $\dfrac{12.02}{12.204} \times 100\%$

Cleaning Up. Organic material goes in the organic solvents container. Water and aqueous solutions can be flushed down the drain.

QUESTIONS

[Handwritten:] Ret. time = time in liquid state, does not pass thru oxygen

1. In either of the simple distillation experiments, can you account for the boiling point of your product in terms of the known boiling points of the pure components of your mixture? If so, how? If not, why not?

2. From the plots of the boiling point versus the volume of distillate in the simple distillation experiments, what can you conclude about the purity of your product?

3. From the plots of the boiling point versus the volume of distillate in either of the fractional distillations of the cyclohexane-toluene mixture, what conclusion can you draw about the homogeneity of the distillate?

4. From the plots of the boiling point versus the volume of distillate in either of the fractional distillations of the ethanol-water mixture, what conclusion can you draw about the homogeneity of the distillate? Does it have a constant boiling point? If constant, is it a pure substance?

5. What is the effect on the boiling point of a solution (e.g., water) produced by a soluble nonvolatile substance (e.g., sodium chloride)? What is the effect of an insoluble substance such as sand or charcoal? What is the temperature of the vapor above these two boiling solutions?

6. In the distillation of a pure substance (e.g., water), why does all of the water not vaporize at once when the boiling point is reached?

7. In fractional distillation, liquid can be seen running from the bottom of the distillation column back into the distilling flask. What effect does this returning condensate have on the fractional distillation?

8. Why is it extremely dangerous to attempt to carry out a distillation in a completely closed apparatus (one with no vent to the atmosphere)?

9. Why is better separation of two liquids achieved by slow rather than fast distillation?

10. Explain why a packed fractionating column is more efficient than an unpacked one.

11. In the distillation of the cyclohexane-toluene mixture, the first few drops of distillate may be cloudy. Explain this occurrence.

12. What effect does the reduction of atmospheric pressure have on the boiling point? Can cyclohexane and toluene be separated if the external pressure is 350 mm Hg instead of 760 mm Hg?

Online Study Center

General Resources
Web Links

13. When water-cooled condensers are used for distillation or for refluxing a liquid, the water enters the condenser at the lowest point and leaves at the highest. Why?

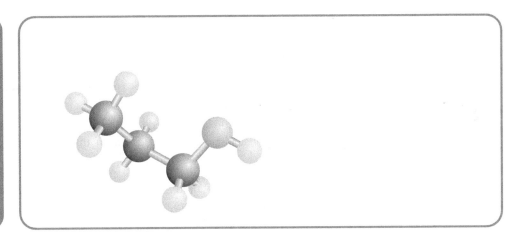

CHAPTER
7

Extraction

Online Study Center

This icon will direct you to
techniques, equipment setups,
and online resources at
http://college.hmco.com/PIC/
williamsonMMOE5e.

PRELAB EXERCISE: Describe how to separate a mixture of
3-toluic acid and 4'-aminoacetophenone using acid-base liquid/
liquid extraction. What species will end up in the aqueous layer if
you mix a solution of benzoic acid and aniline in ether with a solution
of $NaHCO_3$ (aq)? Draw the structure of this species.

Extraction is one of the oldest chemical operations known to humankind. The
preparation of a cup of coffee or tea involves the extraction of flavor and odor
components from dried vegetable matter with hot water. Aqueous extracts of bay
leaves, stick cinnamon, peppercorns, and cloves, along with alcoholic extracts of
vanilla and almond, are used as food flavorings. For the past 150 years or so, organic chemists have extracted, isolated, purified, and characterized the myriad
compounds produced by plants that for centuries have been used as drugs and
perfumes—substances such as quinine from cinchona bark, morphine from the
opium poppy, cocaine from coca leaves, and menthol from peppermint oil. The
extraction of compounds from these natural products is an example of solid/liquid
extraction—the solid being the natural product and the liquid being the solvent
into which the compounds are extracted. In research, a Soxhlet extractor (Fig. 7.1)
is often used for solid/liquid extraction.

Although solid/liquid extraction is the most common technique for brewing
beverages and isolating compounds from natural products, liquid/liquid extraction is a very common method used in the organic laboratory, specifically when
isolating reaction products. Reactions are typically homogeneous liquid mixtures
and can therefore be extracted with either an organic or aqueous solvent. Organic
reactions often yield a number of byproducts—some inorganic and some organic.
Also, because some organic reactions do not go to 100% completion, a small amount
of starting material is present at the end of the reaction. When a reaction is complete, it is necessary to do a *workup*, that is, separate and purify the desired product from the mixture of byproducts and residual starting material. Liquid/liquid
extraction is a common separation step in this workup, which is then followed by

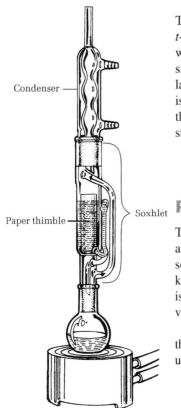

Condenser

Paper thimble — } Soxhlet

purification of the product. There are two types of liquid/liquid extraction: neutral and acid/base. The experiments in this chapter demonstrate solid/liquid extraction and the two types of liquid/liquid extraction.

Organic products are often separated from inorganic substances in a reaction mixture by liquid/liquid extraction with an organic solvent. For example, in the synthesis of 1-bromobutane (*see* Chapter 16), 1-butanol, also a liquid, is heated with an aqueous solution of sodium bromide and sulfuric acid to produce the product and sodium sulfate.

$$2 \; CH_3CH_2CH_2CH_2OH \; + \; 2 \; NaBr \; + \; H_2SO_4 \rightarrow$$
$$2 \; CH_3CH_2CH_2CH_2Br \; + \; 2 \; H_2O \; + \; Na_2SO_4$$

The 1-bromobutane is isolated from the reaction mixture by extraction with *t*-butyl methyl ether, an organic solvent in which 1-bromobutane is soluble and in which water and sodium sulfate are insoluble. The extraction is accomplished by simply adding *t*-butyl methyl ether to the aqueous mixture and shaking it. Two layers will result: an organic layer and an aqueous layer. The *t*-butyl methyl ether is less dense than water and floats on top; it is easily removed/drained away from the water layer and evaporated to leave the bromo product free of inorganic substances, which reside in the aqueous layer.

Partition Coefficient

The extraction of a compound such as 1-butanol, which is slightly soluble in water as well as very soluble in ether, is an equilibrium process governed by the solubilities of the alcohol in the two solvents. The ratio of the solubilities is known as the *distribution coefficient*, also called the *partition coefficient* (k), and is an equilibrium constant with a certain value for a given substance, pair of solvents, and temperature.

The *concentration* of the solute in each solvent can be well correlated with the *solubility* of the solute in the pure solvent, a figure that is readily found in solubility tables in reference books. For substance C

$$k = \frac{\text{concentration of C in } t\text{-butyl methyl ether}}{\text{concentration of C in water}}$$

$$> \frac{\text{solubility of C in } t\text{-butyl methyl ether (g/100 mL)}}{\text{solubility of C in water (g/100 mL)}}$$

Consider compound A that dissolves in *t*-butyl methyl ether to the extent of 12 g/100 mL and dissolves in water to the extent of 6 g/100 mL.

$$k = \frac{12 \text{ g/100 mL } t\text{-butyl methyl ether}}{6 \text{ g/100 mL water}} = 2$$

If a solution of 6 g of A in 100 mL of water is shaken with 100 mL of t-butyl methyl ether, then

$$k = \frac{x \text{ g of A/100 mL } t\text{-butyl methyl ether}}{6 - x \text{ g of A/100 mL water}}$$

from which

$$x = 4.0 \text{ g of A in the ether layer}$$
$$6 - x = 2.0 \text{ g of A left in the water layer}$$

It is, however, more efficient to extract the 100 mL of aqueous solution twice with 50-mL portions of t-butyl methyl ether rather than once with a 100-mL portion.

$$k = \frac{x \text{ g of A/50 mL}}{6 - x \text{ g of A/100 mL}} = 2$$

from which

$$x = 3.0 \text{ g of A in the } t\text{-butyl methyl ether layer}$$
$$6 - x = 3.0 \text{ g of A in the water layer}$$

If this 3.0 g/100 mL of water is extracted again with 50 mL of t-butyl methyl ether, we can calculate that 1.5 g of A will be in the ether layer, leaving 1.5 g in the water layer. So two extractions with 50-mL portions of ether will extract 3.0 g + 1.5 g = 4.5 g of A, whereas one extraction with a 100-mL portion of t-butyl methyl ether removes only 4.0 g of A. Three extractions with 33-mL portions of t-butyl methyl ether would extract 4.7 g. Obviously, there is a point at which the increased amount of A extracted does not repay the effort of multiple extractions, but remember that several small-scale extractions are more effective than one large-scale extraction.

Properties of Extraction Solvents

Liquid/liquid extraction involves two layers: the organic layer and the aqueous layer. The solvent used for extraction should possess many properties, including the following:

- It should readily dissolve the substance to be extracted at room temperature.
- It should have a low boiling point so that it can be removed readily.
- It should not react with the solute or the other solvent.
- It should not be highly flammable or toxic.
- It should be relatively inexpensive.

In addition, it should not be miscible with water (the usual second phase). No solvent meets every criterion, but several come close. Some common liquid/liquid

TABLE 7.1 • Common Solvents Listed by Density	
Solvent	**Density (g/mL)**
Hexane	0.695
Diethyl ether	0.708
t-Butyl methyl ether	0.740
Toluene	0.867
Water	1.000
Dichloromethane	1.325
Chloroform	1.492

extraction solvent pairs are water-ether, water-dichloromethane, and water-hexane. Notice that each combination includes water because most organic compounds are immiscible in water and therefore can be separated from inorganic compounds. Organic solvents such as methanol and ethanol are not good extraction solvents because they are soluble in water.

Identifying the Layers

One common mistake when performing an extraction is to misidentify the layers and discard the wrong one. It is good practice to save all layers until the desired product is in hand. The densities of the solvents will predict the identities of the top and bottom layers. In general, the densities of nonhalogenated organic solvents are less than 1.0 g/mL and those of halogenated solvents are greater than 1.0 g/mL. Table 7.1 lists the densities of common solvents used in extraction.

Although density is the physical property that determines which layer is on top or on bottom, a very concentrated amount of solute dissolved in either layer can reverse the order. The best method to avoid a misidentification is to perform a drop test. Add a few drops of water to the layer in question and watch the drop carefully. If the layer is water, then the drop will mix with the solution. If the solvent is the organic layer, then the water drop will create a second layer.

Identify layers by a drop test.

$$CH_3CH_2-O-CH_2CH_3$$

Diethyl ether
"Ether"
MW 74.12, den. 0.708
bp 34.6°C, n_D^{20} 1.3530

Ethereal Extraction Solvents

In the past, diethyl ether was the most common solvent for extraction in the laboratory. It has high solvent power for hydrocarbons and oxygen-containing compounds. It is highly volatile (bp 34.6°C) and is therefore easily removed from an extract. However, diethyl ether has two big disadvantages: it is highly flammable and poses a great fire threat, and it easily forms peroxides. The reaction of diethyl ether with air is catalyzed by light. The resulting peroxides are higher boiling than the ether and are left as a residue when the ether evaporates. If the residue is heated, it will explode because ether peroxides are treacherously high explosives. In recent years, a new solvent has come on the scene—*tert*-butyl methyl ether.

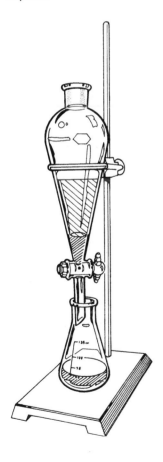

CH₃
|
H₃C—C—O—CH₃
|
CH₃

tert-Butyl methyl ether
MW 88.14, den. 0.741
bp 55.2°C, n_D^{20} 1.369

■ **FIG. 7.2**
A separatory funnel with Teflon stopcock.

tert-Butyl methyl ether, called methyl *tert*-butyl ether (MTBE) in industry, has many advantages over diethyl ether as an extraction solvent. Most important, it does not easily form peroxides, so it can be stored for much longer periods than diethyl ether. And, in the United States, it is less than two-thirds the price of diethyl ether. It is slightly less volatile (bp 55°C), so it does not pose the same fire threat as diethyl ether, although one must be as careful in handling this solvent as in handling any other highly volatile, flammable substance. The explosion limits for *t*-butyl methyl ether mixed with air are much narrower than for diethyl ether, the toxicity is less (it is not a carcinogen), the solvent power is the same, and the ignition temperature is higher (224°C versus 180°C).

The weight percent solubility of diethyl ether dissolved in water is 7.2%, whereas that of *t*-butyl methyl ether is 4.8%. The solubility of water in diethyl ether is 1.2%, while in *t*-butyl methyl ether it is 1.5%. Unlike diethyl ether, *t*-butyl methyl ether forms an azeotrope with water (4% water) that boils at 52.6°C. This means that evaporation of any *t*-butyl methyl ether solution that is saturated with water should leave no water residue, unlike diethyl ether.

The low price and ready availability of *t*-butyl methyl ether came about because it replaced tetraethyl lead as the antiknock additive for high-octane gasoline and as a fuel oxygenate, which helps reduce air pollution, but its water solubility has allowed it to contaminate drinking water supplies in states where leaking underground fuel storage tanks are not well regulated. Consequently, it is being replaced with the much more expensive ethanol. In this text, *t*-butyl methyl ether is *strongly* suggested wherever diethyl ether formerly would have been used in an extraction. It will not, however, work as the only solvent in the Grignard reaction, probably because of steric hindrance. So whenever the word *ether* appears in this text as an extraction solvent, it is suggested that *t*-butyl methyl ether be used and not diethyl ether.

Mixing and Separating the Layers

For microscale separations, mixing and separating the layers with a pipette normally incurs very little product loss. Because the two solvents are typically in a reaction tube for microscale extraction, the two layers can be mixed by drawing up and rapidly expelling them with a pipette. Then the layers are allowed to separate, and the bottom layer is separated by drawing it up into a pipette and transferring it to a different container.

For macroscale separations, a separatory funnel (Fig. 7.2) is used to mix and separate the organic and aqueous layers. In macroscale experiments, a frequently used method of working up a reaction mixture is to dilute the mixture with water and extract it with an organic solvent, such as ether, in a separatory funnel. When the stoppered funnel is shaken to distribute the components between the immiscible solvents *t*-butyl methyl ether and water, pressure always develops through volatilization of ether from the heat of the hands, and liberation of a gas (CO_2) (in acid/base extractions) can increase the pressure. Consequently, the funnel is grasped so that the stopper is held in place by one hand

and the stopcock by the other, as illustrated in Figure 7.3. After a brief shake or two, the funnel is held in the inverted position shown, and the stopcock is opened cautiously (with the funnel stem pointed away from nearby persons) to release pressure. The mixture can then be shaken more vigorously, with pressure released as necessary. When equilibration is judged to be complete, the slight, constant terminal pressure due to ether is released, the stopper is rinsed with a few drops of ether delivered by a Pasteur pipette, and the layers are allowed to separate. The organic reaction product is distributed wholly or largely into the upper ether layer, whereas inorganic salts, acids, and bases pass into the water layer, which can be drawn off. If the reaction was conducted in alcohol or some other water-soluble solvent, the bulk of the solvent is removed in the water layer, and the remainder can be eliminated in two or three washings with 1–2 volumes of water conducted with the techniques used in the first equilibration. The separatory funnel should be supported in a ring stand, as shown in Figure 7.2.

Before adding a liquid to the separatory funnel, check the stopcock. If it is glass, see that it is properly greased, bearing in mind that too much grease will clog the hole in the stopcock and also contaminate the extract. If the stopcock is Teflon, see that it is adjusted to a tight fit in the bore. Store the separatory funnel with the Teflon stopcock loosened to prevent sticking. Because Teflon has a much larger temperature coefficient of expansion than glass, a stuck stopcock can be loosened by cooling the stopcock in ice or dry ice. Do not store liquids in the separatory funnel; they often leak or cause the stopper or stopcock to freeze. To have sufficient room for mixing the layers, fill the separatory funnel no more than three-fourths full. Withdraw the lower layer from the separatory funnel through the stopcock and pour the upper layer out through the neck.

All too often the inexperienced chemist discards the wrong layer when using a separatory funnel. Through incomplete neutralization, a desired component may still remain in the aqueous layer, or the densities of the layers may change. Cautious workers save all layers until the desired product has been isolated. The organic layer is not always the top layer. If in doubt, perform a drop test by adding a few drops of each to water in a test tube.

Practical Considerations When Mixing Layers

Pressure Buildup

The heat of one's hand or heat from acid/base reactions will cause pressure buildup in an extraction mixture that contains a very volatile solvent such as dichloromethane. The extraction container—whether a test tube or a separatory funnel—must be opened carefully to vent this pressure.

Sodium bicarbonate solution is often used to neutralize acids when carrying out acid/base extractions. The result is the formation of carbon dioxide, which can cause foaming and high pressure buildup. Whenever bicarbonate is used, add it very gradually with thorough mixing and frequent venting of the extraction device. If a large amount of acid is to be neutralized with bicarbonate, the process should be carried out in a beaker.

Emulsions

Imagine trying to extract a soap solution (e.g., a nonfoaming dishwashing detergent) into an organic solvent. After a few shakes with an organic solvent, you would have an absolutely intractable emulsion. An emulsion is a suspension of one liquid as droplets in another. Detergents stabilize emulsions, and so any time a detergent-like molecule is in the material being extracted, there is the danger that emulsions will form. Substances of this type are commonly found in nature, so one must be particularly wary of emulsion formation when creating organic extracts of aqueous plant material, such as caffeine from tea. Emulsions, once formed, can be quite stable. You would be quite surprised to open your refrigerator one morning and see a layer of clarified butter floating on the top of a perfectly clear aqueous solution that had once been milk, but milk is the classic example of an emulsion.

Prevention is the best cure for emulsions. This means shaking the solution to be extracted *very gently* until you see that the two layers will separate readily. If a bit of emulsion forms, it may break simply on standing for a sufficient length of time. Making the aqueous layer highly ionic will help. Add as much sodium chloride as will dissolve and shake the mixture gently. Vacuum filtration sometimes works and, when the organic layer is the lower layer, filtration through silicone-impregnated filter paper is helpful. Centrifugation works very well for breaking emulsions. This is easy on a small scale, but often the equipment is not available for large-scale centrifugation of organic liquids.

Shake gently to avoid emulsions.

Drying Agents

The organic solvents used for extraction dissolve not only the compound being extracted but also water. Evaporation of the solvent then leaves the desired compound contaminated with water. At room temperature water dissolves 4.8% of *t*-butyl methyl ether by weight, and the ether dissolves 1.5% of water. But ether is virtually insoluble in water saturated with sodium chloride (36.7 g/100 mL). If ether that contains dissolved water is shaken with a saturated aqueous solution of sodium chloride, water will be transferred from the *t*-butyl methyl ether to the aqueous layer. So, strange as it may seem, ethereal extracts are routinely dried by shaking them with an aqueous saturated sodium chloride solution.

Solvents such as dichloromethane do not dissolve nearly as much water and are therefore dried over a chemical drying agent. Many choices of chemical drying agents are available for this purpose, and the choice of which one to use is governed by four factors: (1) the possibility of reaction with the substance being extracted, (2) the speed with which it removes water from the solvent, (3) the efficiency of the process, and (4) the ease of recovery from the drying agent.

Some very good but specialized and reactive drying agents are potassium hydroxide, anhydrous potassium carbonate, sodium metal, calcium hydride, lithium aluminum hydride, and phosphorus pentoxide. Substances that are essentially neutral and unreactive and are widely used as drying agents include anhydrous calcium sulfate (Drierite), magnesium sulfate, molecular sieves, calcium chloride, and sodium sulfate.

Drierite, a specially prepared form of calcium sulfate, is a fast and effective drying agent. However, it is difficult to ascertain whether enough has been used.

Drierite, $CaSO_4$

Magnesium sulfate, MgSO₄

Molecular sieves, zeolites

Calcium chloride (CaCl₂) pellets
are the drying agent of choice for
small-scale experiments.

Sodium sulfate, Na₂SO₄

An indicating type of Drierite is impregnated with cobalt chloride, which turns from blue to red when it is saturated with water. This works well when gases are being dried, but it should not be used for liquid extractions because the cobalt chloride dissolves in many protic solvents.

Magnesium sulfate is also a fast and fairly effective drying agent, but it is so finely powdered that it always requires careful filtration for removal.

Molecular sieves are sodium alumino-silicates (zeolites) that have well-defined pore sizes. The 4 Å size adsorbs water to the exclusion of almost all organic substances, making them a fast and effective drying agent. Like Drierite, however, it is impossible to ascertain by appearance whether enough has been used. Molecular sieves in the form of 1/16-in. pellets are often used to dry solvents by simply adding them to the container.

Calcium chloride, recently available in the preferred form of pellets (4 to 80 mesh[1]), is a very fast and effective drying agent. It has the advantage that it clumps together when excess water is present, which makes it possible to know how much to add by observing its behavior. Unlike the older granular form, the pellets do not disintegrate into a fine powder. These pellets are admirably suited to microscale experiments where the solvent is removed from the drying agent with a Pasteur pipette. Calcium chloride is much faster and far more effective than anhydrous sodium sulfate; after much experimentation, we have decided that this is the agent of choice, particularly for microscale experiments. These pellets are used for most of the drying operations in this text. Note, however, that calcium chloride reacts with some alcohols, phenols, amides, and some carbonyl-containing compounds. Advantage is sometimes taken of this property to remove not only water from a solvent but also, for example, a contaminating alcohol (*see* Chapter 16—the synthesis of 1-bromobutane from 1-butanol). Because *t*-butyl methyl ether forms an azeotrope with water, its solutions should, theoretically, not need to be dried; evaporation carries away the water. Drying these ether solutions with calcium chloride pellets removes water droplets that get carried into the ether solution.

Sodium sulfate is a very poor drying agent. It has a very high capacity for water but is slow and not very efficient in the removal of water. Like calcium chloride pellets, it clumps together when wet, and solutions are easily removed from it using a Pasteur pipette. Sodium sulfate has been used extensively in the past and should still be used for compounds that react with calcium chloride.

PART 1: The Technique of Neutral Liquid/Liquid Extraction

The workup technique of liquid/liquid extraction has four steps: (1) mixing the layers, (2) separating the layers, (3) drying the organic layer, and (4) removing the solvent. The microscale neutral liquid/liquid extraction technique is described in the following sections.

---

1. These pellets are available from Fisher Scientific, Cat. No. C614–3.

Step 1. Mixing the Layers

Once the organic and aqueous layers are in contact with one another, mixing is required to ensure that the desired compound(s) get extracted into the desired layer. First, place 1–2 mL of an aqueous solution of the compound to be extracted in a reaction tube. Add about 1 mL of extraction solvent, for example, dichloromethane. Note, as you add the dichloromethane, whether it is the top or the bottom layer. (Since dichloromethane is more dense than water, predict what layer it will be.) An effective way to mix the two layers is to flick the tube with a finger. Grasp the tube firmly at the very top between the thumb and forefinger and flick it vigorously at the bottom (Fig. 7.4). You will find that this violent motion mixes the two layers well, but nothing comes out the top. Another good mixing technique is to pull the contents of the reaction tube into a Pasteur pipette and then expel the mixture back into the tube with force. Doing this several times will effect good mixing of the two layers. A stopper can be placed in the top of the tube, and the contents can be mixed by shaking the tube, but the problem with this technique is that the high vapor pressure of the solvent will often force liquid out around the cork or stopper.

Step 2. Separating the Layers

Always draw out the lower layer and place it in another container.

After thoroughly mixing the two layers, allow them to separate. Tap the tube if droplets of one layer are in the other layer or on the side of the tube. After the layers separate completely, draw up the lower dichloromethane layer into a Pasteur pipette. Leave behind any middle emulsion layer. The easiest way to do this is to attach the pipette to a pipette pump (Fig. 7.5). This allows very precise control of the liquid being removed. It takes more skill and practice to remove the lower layer cleanly with a 2-mL rubber bulb attached to a pipette because the high vapor pressure of the solvent tends to make it dribble out. To avoid losing any of the solution, it is best to hold a clean, dry, empty tube in the same hand as the full tube to receive the organic layer (Fig. 7.6).

From the discussion of the partition coefficient, you know that several small extractions are better than one large one, so repeat the extraction process with two further 1-mL portions of dichloromethane. An experienced chemist might summarize all the preceding with the following notebook entry, "Aqueous layer extracted 3 × 1-mL portions CH_2Cl_2," and in a formal report would write, "The aqueous layer was extracted three times with 1-mL portions of dichloromethane."

If you are working on a larger microscale, a microscale separatory funnel (Fig. 7.7) should be used. A separatory funnel, regardless of size, should be filled to only about two-thirds of its capacity so the layers can be mixed by shaking. The microscale separatory funnel has a capacity of 8.5 mL when full, so it is useful for an extraction with a total volume of about 6 mL.

Use a wood boiling stick to poke out the polyethylene frit from the bottom part of the separatory funnel. Store it for later replacement. Close the valve, add up to 5 mL of the solution to be extracted to the separatory funnel, then add the extraction solvent so that the total volume does not exceed 6 mL.

■ FIG. 7.4

Mixing the contents of a reaction tube by flicking it. Grasp the tube firmly at the very top and flick it vigorously at the bottom. The contents will mix without coming out of the tube.

■ FIG. 7.5

The removal of a solvent from a reaction tube with a pipette and pipette pump.

■ FIG. 7.6

Grasp both reaction tubes in one hand when transferring material from one tube to another with a Pasteur pipette.

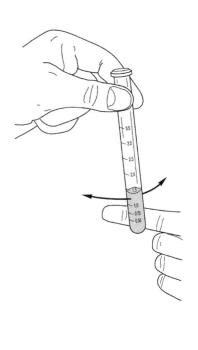

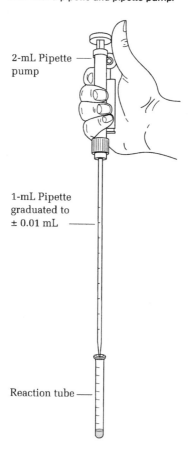

2-mL Pipette pump

1-mL Pipette graduated to ± 0.01 mL

Reaction tube

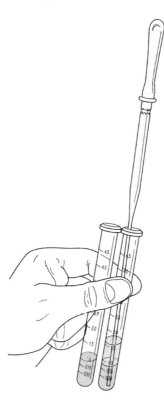

Cap the separatory funnel and mix the contents by inverting the funnel several times. If the two layers separate fairly easily, then the contents can be shaken more thoroughly. If the layers do not separate easily, be careful not to shake the funnel too vigorously because intractable emulsions could form.

Remove the stopper from the funnel, clamp it, and then, grasping the valve with two hands, empty the bottom layer into an Erlenmeyer flask or other container. If the top layer is desired, pour it out through the top of the separatory funnel—don't drain it through the valve, which may have a drop of the lower layer remaining in it.

Step 3. Drying the Organic Layer

Dichloromethane dissolves a very small quantity of water, and microscopic droplets of water are suspended in the organic layer, often making it cloudy. To remove the water, a drying agent, for example, anhydrous calcium chloride pellets, is added to the dichloromethane solution.

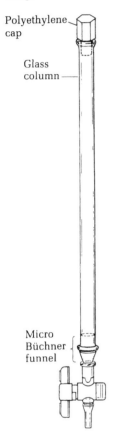

■ **FIG. 7.7**
A microscale separatory funnel.
Remove the polyethylene frit from
the micro Büchner funnel before
using.

Polyethylene
cap

Glass
column

Micro
Büchner
funnel

Record the tare of the final
container.

How Much Drying Agent Should Be Used?

When a small quantity of the drying agent is added, the crystals or pellets become sticky with water, clump together, and fall rapidly as a lump to the bottom of the reaction tube. There will come a point when a new small quantity of drying agent no longer clumps together, but the individual particles settle slowly throughout the solution. As they say in Scandinavia, "Add drying agent until it begins to snow." The drying process takes about 10–15 min, during which time the tube contents should be mixed occasionally by flicking the tube. The solution should no longer be cloudy but clear (although it may be colored).

Once drying is judged complete, the solvent is removed by forcing a Pasteur pipette to the bottom of the reaction tube and pulling the solvent in. Air is expelled from the pipette as it is being pushed through the crystals or pellets so that no drying agent will enter the pipette. It is very important to wash the drying agent left in the reaction tube with several small quantities of pure solvent to transfer all the extract.

Step 4. Removing the Solvent

If the quantity of extract is relatively small, say 3 mL or less, then the easiest way to remove the solvent is to blow a stream of air (or nitrogen) onto the surface of the solution from a Pasteur pipette (Fig. 7.8). Be sure that the stream of air is very gentle before inserting it into the reaction tube. The heat of vaporization of the solvent will cause the tube to become rather cold during the evaporation and, of course, slow down the process. The easiest way to add heat is to hold the tube in your hand.

Another way to remove the solvent is to attach the Pasteur pipette to an aspirator and pull air over the surface of the liquid. This is not quite as fast as blowing air onto the surface of the liquid and runs the danger of sucking up the liquid into the aspirator.

If the volume of liquid is more than about 3 mL, put it into a 25-mL filter flask, put a plastic Hirsch funnel in place, and attach the flask to the aspirator. By placing your thumb in the Hirsch funnel, the vacuum can be controlled, and heat can be applied by holding the flask in the other hand while swirling the contents (Fig. 7.9).

The reaction tube or filter flask in which the solvent is evaporated should be tared (weighed empty), and this weight recorded in your notebook. In this way, the weight of material extracted can be determined by again weighing the container.

EXPERIMENT

 ## Partition Coefficient of Benzoic Acid

> **IN THIS EXPERIMENT** you will shake a solution of benzoic acid in water with the immiscible solvent dichloromethane. The benzoic acid will distribute (partition) itself between the two layers. By removing the organic layer, drying, and evaporating it, the weight of benzoic acid in the dichloromethane can be determined and thus the ratio in the two layers. This ratio is a constant known as the partition coefficient.

■ **FIG. 7.8**
An aspirator tube being used to
remove solvent vapors.

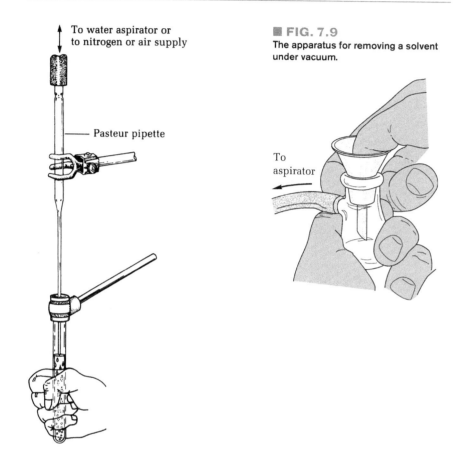

To water aspirator or
to nitrogen or air supply

Pasteur pipette

To
aspirator

In a reaction tube, place about 100 mg of benzoic acid (weighed to the nearest milligram) and add exactly equal volumes of water followed by dichloromethane (about 1.6 mL each). While making this addition, note which layer is organic and which is aqueous. Put a septum on the tube and shake the contents vigorously for at least 2 min. Allow the tube to stand undisturbed until the layers separate and then carefully draw off, using a Pasteur pipette, *all* of the aqueous layer without removing any of the organic layer. It may be helpful to draw out the tip of the pipette to a fine point in a flame and, using this, to tilt the reaction tube on its side to make this separation as clean as possible.

Add anhydrous calcium chloride pellets to the dichloromethane in very small quantities until it no longer clumps together. Mix the contents of the tube by flicking it and allow it to stand for about 5 min to complete the drying process. Using a dry Pasteur pipette, transfer the dichloromethane to a tared dry reaction tube or a 10-mL Erlenmeyer flask containing a boiling chip. Complete the transfer by washing the drying agent with two more portions of solvent that are added to the original solution and then evaporate the solvent. This can be done by boiling off the solvent while removing solvent vapors with an aspirator tube or by blowing a stream of air or nitrogen into the container while warming it in one's hand (*see* Fig. 7.8). This operation should be performed in a hood.

From the weight of the benzoic acid in the dichloromethane layer, the weight in the water layer can be obtained by difference. The ratio of the weight in dichloromethane to the weight in water is the distribution coefficient because the volumes of the two solvents were equal. Report the value of the distribution coefficient in your notebook.

Cleaning Up. The aqueous layer can be flushed down the drain. Dichloromethane goes into the halogenated organic solvents container. After allowing the solvent to evaporate from the sodium sulfate in the hood, place the sodium sulfate in the non-hazardous solid waste container. If local regulations do not allow for the evaporation of solvents in a hood, dispose of the wet sodium sulfate in a special container.

Part 2: Acid/Base Liquid/Liquid Extraction

Acid/base liquid/liquid extraction involves carrying out simple acid/base reactions to separate strong organic acids, weak organic acids, neutral organic compounds, and basic organic substances. The chemistry involved is given in the following equations, using benzoic acid, phenol, naphthalene, and aniline as examples of the four types of compounds.

Here is the strategy (refer to the flow sheet in Fig. 7.10): The four organic compounds are dissolved in *t*-butyl methyl ether. The ether solution is shaken with a saturated aqueous solution of sodium bicarbonate, a weak base. This will react only with the strong acid, benzoic acid (**1**), to form the ionic salt, sodium benzoate (**5**), which dissolves in the aqueous layer and is removed. The ether solution now contains just phenol (**2**), naphthalene (**4**), and aniline (**3**). A 3 *M* aqueous solution of sodium hydroxide is added, and the mixture is shaken. The hydroxide, a strong base, will react only with the phenol (**2**), a weak acid, to form sodium phenoxide (**6**), an ionic compound that dissolves in the aqueous layer and is removed. The ether now contains only naphthalene (**4**) and aniline (**3**). Shaking it with dilute hydrochloric acid removes the aniline, a base, as the ionic anilinium chloride (**7**). The aqueous layer is removed. Evaporation of the *t*-butyl methyl ether now leaves naphthalene (**4**), the neutral compound. The other three compounds are recovered by adding acid to the sodium benzoate (**5**) and sodium phenoxide (**6**) and base to the anilinium chloride (**7**) to regenerate the covalent compounds benzoic acid (**1**), phenol (**2**), and aniline (**3**).

The ability to separate strong acids from weak acids depends on the acidity constants of the acids and the basicity constants of the bases. In the first equation consider the ionization of benzoic acid, which has an equilibrium constant (K_a) of 6.8×10^{25}. The conversion of benzoic acid to the benzoate anion in equation 4 is governed by the equilibrium constant, K (equation 5), obtained by combining equations 3 and 4.

> The pK_a of carbonic acid, H_2CO_3, is 6.35.

$$C_6H_5COOH + H_2O \rightleftharpoons C_6H_5COO^- + H_3O^+ \qquad (1)$$

$$K_a = \frac{[C_6H_5COO^-][H_3O^+]}{[C_6H_5COOH]} = 6.8 \times 10^{-5}, pK_a = 4.17 \qquad (2)$$

■ **FIG. 7.10**

A flow sheet for the separation of a strong acid, a weak acid, a neutral compound, and a base: benzoic acid, phenol, naphthalene, and aniline (this page). Acid/base reactions of the acidic and basic compounds (opposite page).

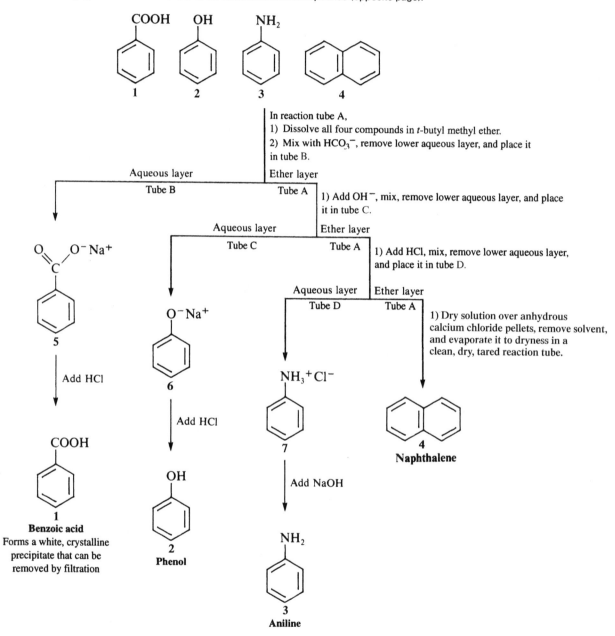

Benzoic acid
Forms a white, crystalline
precipitate that can be
removed by filtration

Phenol and aniline each form oily layers on
top of the aqueous layer. Extract each with
t-butyl methyl ether: Add ether to the tube, mix, separate
layers, dry the ether layer over anhydrous
calcium chloride pellets, remove solution from drying
agent, and evaporate the solvent.

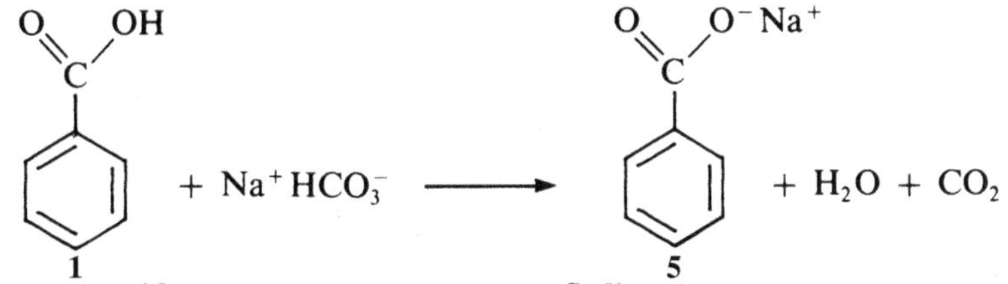

1
Benzoic acid
pK_a = 4.17
Covalent, sol. in org. solvents

5
Sodium benzoate
Ionic, sol. in water

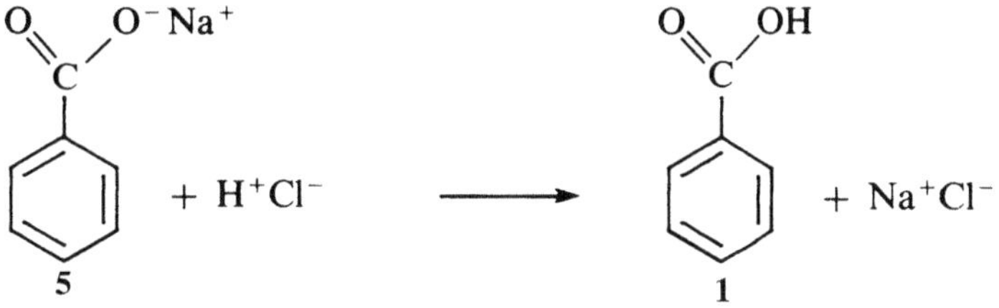

5 → 1

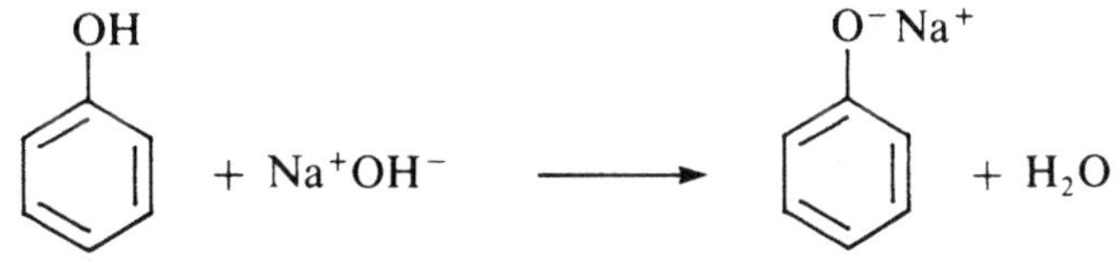

2
Phenol
pK_a = 10
Covalent, sol. in org. solvents

6
Sodium phenoxide
Ionic, sol. in water

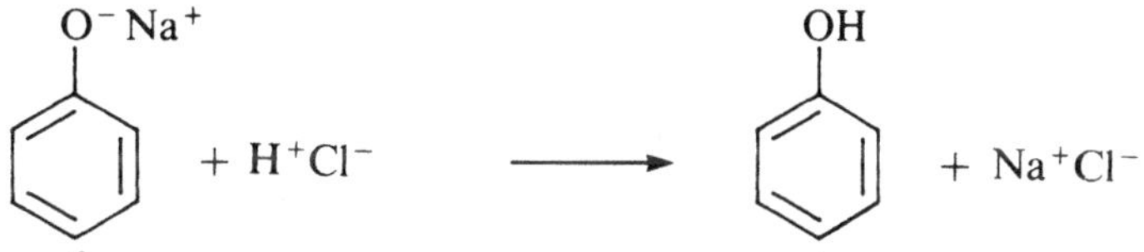

6 → 2

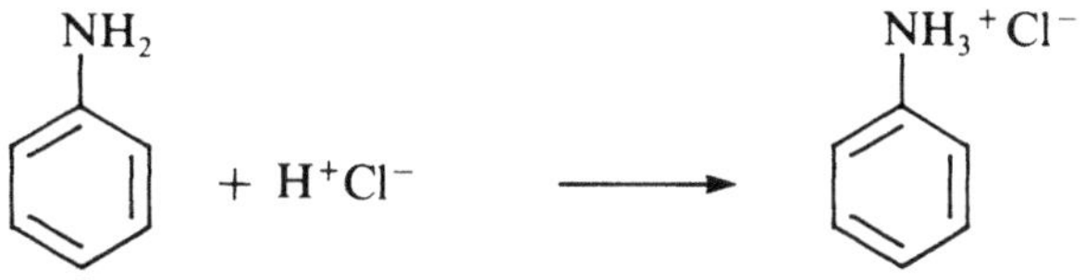

3
Aniline
pK_b = 9.30
Covalent, sol. in org. solvents

7
Anilinium chloride
Ionic, sol. in water

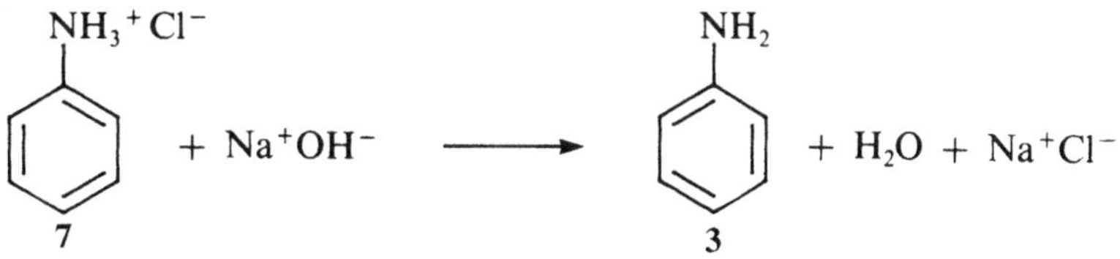

7 → 3

$$K_w = [H_3O^+][OH^-] = 10^{-14} \tag{3}$$

$$C_6H_5COOH + OH^- \rightleftharpoons C_6H_5COO^- + H_2O \tag{4}$$

$$K = \frac{[C_6H_5COO^-]}{[C_6H_5COOH][OH^-]} = \frac{K_a}{K_w} = \frac{6.8 \times 10^{-5}}{10^{-14}} = 3.2 \times 10^8 \tag{5}$$

If 99% of the benzoic acid is converted to $C_6H_5COO^-$,

$$\frac{[C_6H_5COO^-]}{[C_6H_5COOH]} = \frac{99}{1} \tag{6}$$

then from equation 5 the hydroxide ion concentration would need to be 6.8×10^{-7} M. Because saturated $NaHCO_3$ has $[OH^- = 3 \times 10^{-4}$ M, the hydroxide ion concentration is high enough to convert benzoic acid completely to sodium benzoate.

For phenol, with a K_a of 10^{-10}, the minimum hydroxide ion concentration that will produce the phenoxide anion in 99% conversion is 10^{-2} M. The concentration of hydroxide in 10% sodium hydroxide solution is 10^{-1} M, and so phenol in strong base is entirely converted to the water-soluble salt.

General Considerations

If acetic acid was used as the reaction solvent, it would also be distributed largely into the aqueous phase; if the reaction product is a neutral substance, however, the residual acetic acid in the ether can be removed by one washing with excess 5% sodium bicarbonate solution. If the reaction product is a higher molecular weight acid, for example, benzoic acid (C_6H_5COOH), it will stay in the ether layer, while acetic acid is being removed by repeated washing with water; the benzoic acid can then be separated from neutral byproducts by extraction with sodium bicarbonate or sodium hydroxide solution and acidification of the extract. Acids of high molecular weight are extracted only slowly by sodium bicarbonate, so sodium carbonate is used in its place; however, carbonate is more prone than bicarbonate to produce emulsions. Sometimes an emulsion in the lower layer can be settled by twirling the separatory funnel by its stem. An emulsion in the upper layer can be broken by grasping the funnel by the neck and swirling it. Because the tendency to emulsify increases with the removal of electrolytes and solvents, a little sodium chloride solution is added with each portion of wash water. If the layers are largely clear but an emulsion persists at the interface, the clear part of the water layer can be drawn off, and the emulsion run into a second funnel and shaken with fresh ether.

Liquid/liquid extraction and acid/base extraction are employed in the majority of organic reactions because it is unusual to have the product crystallize from the reaction mixture or to be able to distill the reaction product directly from the reaction mixture. In the research literature, one will often see the statement "the reaction mixture was worked up in the usual way," which implies an extraction process of the type described here. Good laboratory practice dictates, however, that the details of the process be written out.

EXPERIMENTS

1. Separation of a Carboxylic Acid, a Phenol, and a Neutral Substance

O=C—OH (attached to benzene ring)

Benzoic acid
mp 123°C, pK_a 4.17

OCH$_3$ (on benzene ring) ... OCH$_3$

**1,4-Dimethoxybenzene
(Hydroquinone dimethyl ether)**
mp 57°C

Online Study Center

Photo: Extraction with Ether;
Video: Extraction with Ether

OH (on benzene ring)

H$_3$C—C—CH$_3$
          |
         CH$_3$

4-*tert*-Butylphenol
mp 101°C, pK_a 10.17

A mixture of equal parts of a carboxylic acid, a phenol, and a neutral substance is to be separated by extraction from an ether solvent. Note carefully the procedure for this extraction. In the next experiment you are to work out your own extraction procedure. Your unknown will consist of either benzoic acid or 2-chlorobenzoic acid (the carboxylic acid), 4-*t*-butyl phenol or 4-bromophenol, and biphenyl or 1,4-dimethoxybenzene (the neutral substance). The object of this experiment is to identify the three substances in the mixture and to determine the percent recovery of each from the mixture.

Procedure

> **IN THIS EXPERIMENT** three organic solids are separated by reaction with base followed by extraction. Bicarbonate converts carboxylic acids (but not phenols) to ions. Hydroxide ion converts phenols (as well as carboxylic acids) to ions. Ionic substances are soluble in water. The addition of hydrochloric acid to the aqueous ionic solutions regenerates nonionic substances. At each step you should ask yourself, "Have I converted a nonionic substance to an ionic one?" (or vice versa). The ionic substances will be in the aqueous layer; the nonionic ones will be in the organic layer.

Dissolve about 0.18 g of the mixture (record the exact weight) in 2 mL of *t*-butyl methyl ether or diethyl ether in a reaction tube (tube 1). Then add 1 mL of a saturated aqueous solution of sodium bicarbonate to the tube. Use the graduations on the side of the tube to measure the amounts because they do not need to be exact. Mix the contents of the tube thoroughly by pulling the two layers into a Pasteur pipette and expelling them forcefully into the reaction tube. Do this for about 3 min. Allow the layers to separate completely and then draw off the lower layer into another reaction tube (tube 2). Add another 0.15 mL of sodium bicarbonate solution to the tube, mix the contents as before, and add the lower layer to tube 2. Exactly which chemical species is in tube 2? Add 0.2 mL of ether to tube 2, mix it thoroughly, remove the ether layer, and discard it. This is called *backwashing* and serves to remove any organic material that might contaminate the contents of tube 2.

Add 1.0 mL of 3 *M* aqueous sodium hydroxide to tube 1, shake the mixture thoroughly, allow the layers to separate, draw off the lower layer using a clean Pasteur pipette, and place it in tube 3. Extract tube 1 with two 0.15-mL portions of water, and add these to tube 3. Backwash the contents of tube 3 with 0.15 mL of ether and discard the ether wash just as was done for tube 2. Exactly which chemical species is in tube 3?

To tube 1 add saturated sodium chloride solution, mix, remove the aqueous layer, and then add to the ether anhydrous calcium chloride pellets until the drying

Online Study Center

Videos: Filtration of Crystals Using the Pasteur Pipette, Microscale Filtration on the Hirsch Funnel

OH

Br

4-Bromophenol
mp 66°C, pK_a 10.2

Biphenyl
mp 71°C

COOH

Cl

2-Chlorobenzoic acid
mp 141°C, pK_a 2.92

The best way to remove the solvent: under a gentle stream of air or nitrogen.

agent no longer clumps together. Wash it off with ether after the drying process is finished. Allow 5–10 min for the drying of the ether solution.

Using the concentration information given in the inside back cover of this book, calculate exactly how much concentrated hydrochloric acid is needed to neutralize the contents of tube 2. Then, by dropwise addition of concentrated hydrochloric acid, carry out this neutralization while testing the solution with litmus paper. An excess of hydrochloric acid does no harm. This reaction must be carried out with *extreme care* because much carbon dioxide is released in the neutralization. Add a boiling stick to the tube and very cautiously heat the tube to bring most of the solid carboxylic acid into solution. Allow the tube to cool slowly to room temperature and then cool it in ice. Remove the solvent with a Pasteur pipette and recrystallize the residue from boiling water. Again, allow the tube to cool slowly to room temperature and then cool it in ice. At the appropriate time, stir the crystals and collect them on a Hirsch funnel using the procedures detailed in Chapter 4. The crystals can be transferred and washed on the funnel using a small quantity of ice water. The solubility of benzoic acid in water is 1.9 g/L at 0°C and 68 g/L at 95°C. The solubility of chlorobenzoic acid is similar. Turn the crystals out onto a tared piece of paper, allow them to dry thoroughly, and determine the percent recovery of the acid. Assess the purity of the product by checking its melting point.

In exactly the same way, neutralize the contents of tube 3 with concentrated hydrochloric acid. This time, of course, there will be no carbon dioxide evolution. Again, heat the tube to bring most of the material into solution, allow it to cool slowly, remove the solvent, and recrystallize the phenol from boiling water. At the appropriate time, after the product has cooled slowly to room temperature and then in ice, it is also collected on a Hirsch funnel, washed with a very small quantity of ice water, and allowed to dry. The percent recovery and melting point are determined.

The neutral compound is recovered using the Pasteur pipette to remove the ether from the drying agent and to transfer it to a tared reaction tube. The drying agent is washed two or three times with additional ether to ensure complete transfer of the product.

Evaporate the solvent by placing the tube in a warm water bath and directing a stream of nitrogen or air onto the surface of the ether in the hood (*see* Fig. 7.8 on page 146). An aspirator tube can also be used for this purpose. Determine the weight of the crude product and then recrystallize it from methanol-water if it is the low-melting compound. Reread Chapter 4 for detailed instructions on carrying out the process of recrystallization from a mixed solvent. The product is dissolved in about 0.5–1 mL of methanol, and water is added until the solution gets cloudy, which indicates that the solution is saturated. This process is best carried out while heating the tube in a hot water bath at 50°C. Allow the tube to cool slowly to room temperature and then cool it thoroughly in ice. If you have the high-melting compound, recrystallize it from ethanol (8 mL/g).

The products are best isolated by collection on a Hirsch funnel using an ice-cold alcohol–water mixture to transfer and wash the compounds. Determine the percent recovery and the melting point. Turn in the products in neatly labeled $1\frac{1}{2}$-in. × $1\frac{1}{2}$-in. (4 cm × 4 cm) ziplock plastic bags attached to the laboratory report. If the yield on recrystallization is low, concentrate the filtrate (the mother liquor) and obtain a second crop of crystals.

2. Separation of Neutral and Basic Substances

> **IN THIS EXPERIMENT** remember that hydrochloric acid will convert a nonionic amine to an ionic substance and that base will regenerate the nonionic substance from the ionic form. The nonionic and neutral substances are ether soluble, and the ionic substances will be found in the aqueous layer.

Ether vapors are heavier than air and can travel along bench tops, run down drain troughs, and collect in sinks. Be extremely careful to avoid flames when working with volatile ethers.

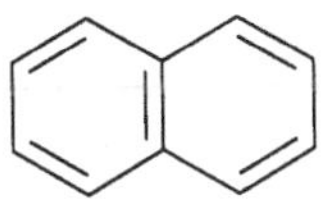

A mixture of equal parts of a neutral substance (naphthalene or 1,4-dichlorobenzene) and a basic substance (4-chloroaniline or ethyl 4-aminobenzoate) is to be separated by extraction from an ether solution. Naphthalene and 1,4-dichlorobenzene are completely insoluble in water. The bases will dissolve in hydrochloric acid, while the neutral compounds will remain in ether solution. The bases are insoluble in cold water but will dissolve to some extent in hot water and are soluble in ethanol. Naphthalene and 1,4-dichlorobenzene can be purified as described in Chapter 4. They also sublime very easily. Keep the samples covered.

Naphthalene	**1,4-Dichlorobenzene**	**4-Chloroaniline**	**Ethyl 4-aminobenzoate**
mp 82°C	mp 56°C	mp 68–71°C, pK_b 4.15	mp 90°C, pK_b 4.92

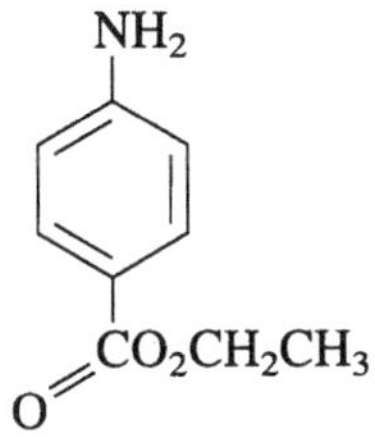

Plan a step-by-step procedure for separating 200 mg of the mixture into its components and have the plan checked by your instructor before proceeding. A flow sheet is a convenient way to present the plan. Select the correct solvent or mixture of solvents for the recrystallization of the bases on the basis of solubility tests. Determine the weights and melting points of the isolated and purified products and calculate the percent recovery of each. Turn in the products in neatly labeled vials or 1 1/2-in. × 1 1/2-in. ziplock plastic bags attached to the report.

Cleaning Up. Combine all aqueous filtrates and solutions, neutralize them, and flush the resulting solution down the drain. Used ether should be placed in the organic solvents container, and the drying agent, once the solvent has evaporated from it, can be placed in the nonhazardous solid waste container. If local regulations do not allow for the evaporation of solvents in a hood, dispose of the wet sodium sulfate in a special container. Any 4-chloroaniline or 1,4-dichlorobenzene should be placed in the halogenated waste container.

3. Separation of Acidic and Neutral Substances

$pH = -\log [H^+]$
pK_a = acidity constant
pK_b = basicity constant

A mixture of equal proportions of benzoic acid, 4-*t*-butylphenol, and 1,4-dimethoxybenzene is to be separated by extraction from *t*-butyl methyl ether. Note the detailed directions for extraction carefully. Prepare a flow sheet (*see* Fig. 7.10 on page 148) for this sequence of operations. In the next experiment you will work out your own extraction procedure.

Procedure

Extinguish all flames when working with *t*-butyl methyl ether! The best method for removing the ether is by simple distillation. Dispose of waste ether in the container provided.

Dissolve 3 g of the mixture in 30 mL of *t*-butyl methyl ether and transfer the mixture to a 125-mL separatory funnel (*see* Fig. 7.2 on page 139) using a little *t*-butyl methyl ether to complete the transfer. Add 10 mL of water and note which layer is organic and which is aqueous. Add 10 mL of a 3 *M* aqueous solution of sodium bicarbonate to the funnel. Swirl or stir the mixture to allow carbon dioxide to escape. Stopper the funnel and cautiously mix the contents. Vent the liberated carbon dioxide and then shake the mixture thoroughly with frequent venting of the funnel. Repeat the process with another 10 mL of bicarbonate solution. Allow the layers to separate completely and then draw off the lower layer into a 50-mL Erlenmeyer flask (labeled flask 1). What does this layer contain?

Add 10 mL of 1.5 *M* aqueous sodium hydroxide to the separatory funnel, shake the mixture thoroughly, allow the layers to separate, and draw off the lower layer into a 50-mL Erlenmeyer flask (labeled flask 2). Repeat the process with another 10 mL of base. Then add an additional 5 mL of water to the separatory funnel, shake the mixture as before, and add this to flask 2. What does flask 2 contain?

Add 15 mL of a saturated aqueous solution of sodium chloride to the separatory funnel, shake the mixture thoroughly, allow the layers to separate, and draw off the lower layer, which can be discarded. What is the purpose of adding saturated sodium chloride solution? Carefully pour the ether layer into a 50-mL Erlenmeyer flask (labeled flask 3) from the top of the separatory funnel, taking great care not to allow any water droplets to be transferred. Add about 4 g of anhydrous calcium chloride pellets to the ether extract and set it aside.

Acidify the contents of flask 2 by dropwise addition of concentrated hydrochloric acid while testing with litmus paper. Cool the flask in an ice bath.

Cautiously add concentrated hydrochloric acid dropwise to flask 1 until the contents are acidic to litmus and then cool the flask in ice.

Decant (pour off) the ether from flask 3 into a tared flask, making sure to leave all of the drying agent behind. Wash the drying agent with additional ether to ensure complete transfer of the product. If decantation is difficult, then remove the drying agent by gravity filtration (*see* Fig. 4.6 on page 69). Put a boiling stick in the flask and evaporate the ether in the hood. An aspirator tube can be used for

Benzoic acid
mp 123°C, pK_a 4.17

4-*tert*-Butylphenol
mp 101°C, pK_a 10.17

Online Study Center

Video: Macroscale Crystallization

■ **FIG. 7.11**
An aspirator tube in use.

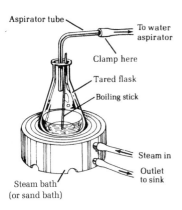

■ **FIG. 7.11**
An aspirator tube in use.

this purpose (Fig. 7.11). Determine the weight of the crude *p*-dimethoxybenzene and then recrystallize it from methanol. See Chapter 4 for detailed instructions on how to carry out recrystallization.

Isolate the *t*-butylphenol from flask 2, employing vacuum filtration on a Hirsch funnel, and wash it on the filter with a small quantity of ice water. Determine the weight of the crude product and then recrystallize it from ethanol. Similarly isolate, weigh, and recrystallize from boiling water the benzoic acid in flask 1. The solubility of benzoic acid in water is 1.9 g/L at 0°C and 68 g/L at 95°C.

Dry the purified products, determine their melting points and weights, and calculate the percent recovery of each substance, bearing in mind that the original mixture contained 1 g of each compound. Hand in the three products in neatly labeled vials.

Cleaning Up. Combine all aqueous layers, washes, and filtrates. Dilute with water, neutralize using either sodium carbonate or dilute hydrochloric acid. This material can then be flushed down the drain with excess water. Methanol filtrate and any ether go in the organic solvents container. Allow ether to evaporate from the calcium chloride in the hood. Then place the calcium chloride in the nonhazardous solid waste container. If local regulations do not allow for the evaporation of solvents in a hood, dispose of the wet sodium sulfate in a special container.

Online Study Center

Video: Microscale Filtration
on the Hirsch Funnel

OCH₃

**1,4-Dimethoxybenzene
(Hydroquinone dimethyl ether)**
mp 57°C

Naphthalene
mp 82°C

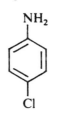

NH₂

Cl

4-Chloroaniline
mp 68–71°C, pK_b 10.0

4. Separation of Neutral and Basic Substances

A mixture of equal parts of a neutral substance (naphthalene) and a basic substance (4-chloroaniline) is to be separated by extraction from *t*-butyl methyl ether solution. The base will dissolve in hydrochloric acid, whereas the neutral naphthalene will remain in the *t*-butyl methyl ether solution. 4-Chloroaniline is insoluble in cold water but will dissolve to some extent in hot water and is soluble in ethanol. Naphthalene can be purified as described in Chapter 4.

Plan a procedure for separating 2.0 g of the mixture into its components and have the plan checked by your instructor before proceeding. A flow sheet is a convenient way to present the plan. Using solubility tests, select the correct solvent or mixture of solvents to recrystallize 4-chloroaniline. Determine the weights and melting points of the isolated and purified products and calculate the percent recovery of each. Turn in the products in neatly labeled vials.

Cleaning Up. Combine all aqueous filtrates and solutions, neutralize them, and flush the resulting solution down the drain with a large excess of water. Used *t*-butyl methyl ether should be placed in the organic solvents container, and the drying agent, once the solvent has evaporated from it, can be placed in the nonhazardous solid waste container. If local regulations do not allow for the evaporation of solvents in a hood, dispose of the wet sodium sulfate in a special container. Any 4-chloroaniline should be placed in the chlorinated organic compounds container.

Handle aromatic amines with care. Most are toxic, and some are carcinogenic. Avoid breathing the dust and vapor from the solid and keep the compounds off the skin, which is best done by wearing nitrile gloves.

5. Extraction and Purification of Components in an Analgesic Tablet

IN THIS EXPERIMENT a powdered analgesic tablet, Excedrin, is boiled with dichloromethane and filtered. The solid on the filter is boiled with ethanol, which dissolves everything but the binder. The ethanol is evaporated and from the hot solution acetaminophen recrystallizes. The dichloromethane solution is shaken with base that converts aspirin to the water-soluble carboxylate anion. The dichloromethane is then evaporated to give caffeine that is purified by sublimation. The aqueous carboxylate anion solution is made acidic, which frees aspirin; warming the mixture and allowing it to cool allows aspirin to recrystallize. It is isolated by filtration.

Excedrin contains aspirin, caffeine, and acetaminophen as determined by thin-layer chromatography (TLC; *see* Chapter 8) or high performance liquid chromatography. A tablet is held together with a binder to prevent the components from crumbling when stored or while being swallowed. A close reading of the contents on the package will disclose the nature of the binder. Starch is commonly used, as is microcrystalline cellulose or silica gel. All of these have one property in common: They are insoluble in water and common organic solvents.

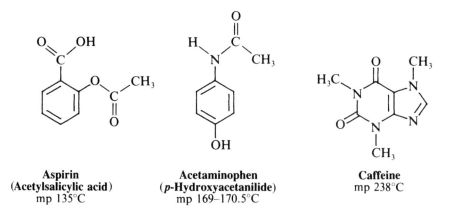

Aspirin
(Acetylsalicylic acid)
mp 135°C

Acetaminophen
(*p*-Hydroxyacetanilide)
mp 169–170.5°C

Caffeine
mp 238°C

TABLE 7.2 • Solubilities					
	Water	**Ethanol**	**Chloroform**	**Diethyl ether**	**Ligroin**
Aspirin	0.33 g/100 mL at 25°C; 1 g/100 mL at 37°C	1 g/5 mL	1 g/17 mL	1 g/13 mL	
Acetaminophen	v. sl. sol. cold; sol. hot	sol.	ins.	sl. sol.	ins.
Caffeine	1 g/46 mL at 25°C; 1 g/5.5 mL at 80°C; 1 g/1.5 mL at 100°C	1 g/66 mL at 25°C; 1 g/22 mL at 60°C	1 g/5.5 mL	1 g/530 mL	sl. sol.

Inspection of the structures of caffeine, acetylsalicylic acid, and acetaminophen reveals that one is a base, one is a strong organic acid, and one is a weak organic acid. It might be tempting to separate this mixture using exactly the same procedure employed in separating benzoic acid, 4-*t*-butylphenol, and 1,4-dimethoxybenzene (experiment 3)—that is, dissolve the mixture in dichloromethane; separate the strongly acidic component by reaction with bicarbonate ion, a weak base; then remove the weakly acidic component by reaction with hydroxide, a strong base. This process would leave the neutral compound in the dichloromethane solution.

In the present experiment the solubility data (see Table 7.2) reveal that the weak acid, acetaminophen, is not soluble in ether, chloroform, or dichloromethane, so it cannot be extracted by a strong base. We can take advantage of this lack of solubility by dissolving the other two components, caffeine and aspirin, in dichloromethane and removing the acetaminophen by filtration. The binder is also insoluble in dichloromethane, but treatment of the solid mixture with ethanol will dissolve the acetaminophen and not the binder. These can then be separated by filtration, with the acetaminophen isolated by evaporation of the ethanol.

This experiment is a test of technique. It is not easy to separate and recrystallize a few milligrams of a compound that occurs in a mixture.

Microscale Procedure

Handle dichloromethane in the hood. It is a suspected carcinogen.

Online Study Center

Photos: Micro Büchner Funnel, Vacuum Filtration into Reaction Tube through Hirsch Funnel, Use of the Wilfilter

The binder can be starch, microcrystalline cellulose, or silica gel.

In a mortar, grind an Extra Strength Excedrin tablet to a very fine powder. The label states that this analgesic contains 250 mg of aspirin, 250 mg of acetaminophen, and 65 mg of caffeine per tablet. Place 300 mg of this powder in a reaction tube and add 2 mL of dichloromethane. Warm the mixture briefly and note that a large part of the material does not dissolve. Filter the mixture on a microscale Büchner funnel (the base of a chromatography column; Fig. 7.12) into another reaction tube. This is done by transferring the slurry to the funnel with a Pasteur pipette and completing the transfer with a small portion of dichloromethane. This filtrate is solution 1. A Hirsch funnel (Fig. 7.13) or a Wilfilter (Fig. 7.14) can also be used for this procedure. Pressure filtration is another alternative.

Transfer the powder on the filter to a reaction tube, add 1 mL of ethanol, and heat the mixture to boiling on the sand bath (with a boiling stick). Not all the material will go into solution. That which does not is the binder. Filter the mixture on

FIG. 7.12
A microscale Büchner funnel assembly.

FIG. 7.12
A microscale Büchner funnel assembly.

Thermometer adapter

Micro Büchner funnel with fritted disk

To aspirator

Vacuum-tight connector

Reaction tube with filtrate

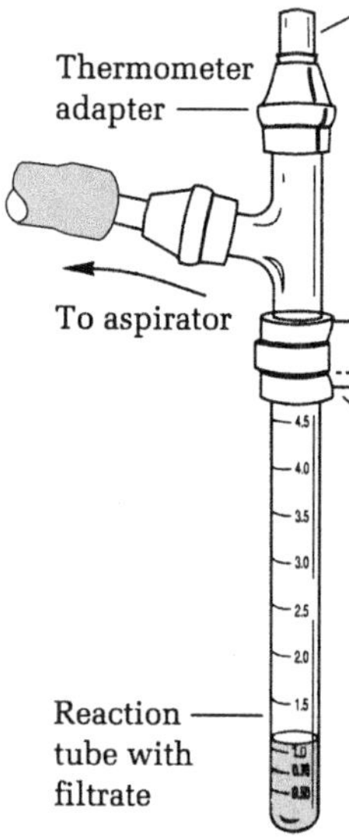

FIG. 7.13
A Hirsch funnel with an integral adapter, a polyethylene frit, and a 25-mL filter flask.

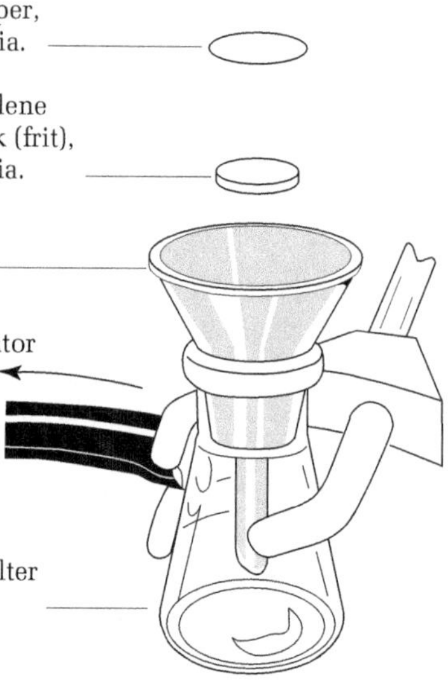

Filter paper, 12 mm dia.

Polyethylene filter disk (frit), 10 mm dia.

Hirsch funnel

To aspirator

25-mL Filter flask

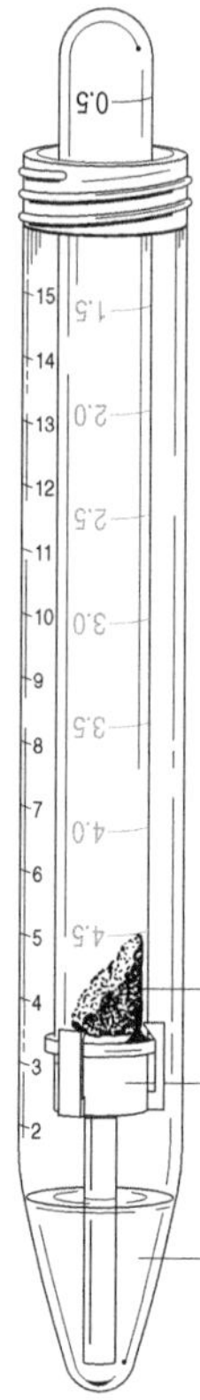

FIG. 7.14
The Wilfilter filtration apparatus. *See* Chapter 4 for usage information.

Crystals

Wilfilter

Filtrate

Online Study Center

Photos: Vacuum Filtration into Reaction Tube through Hirsch Funnel, Micro Büchner Funnel

Acetaminophen

Online Study Center

Video: Filtration of Crystals Using the Pasteur Pipette

Check product purity by TLC (Chapter 8) using 25:1 ethyl acetate–acetic acid to elute the silica gel plates.

the same microscale Büchner funnel into a tared reaction tube and complete the transfer and washing using a few drops of hot ethanol.

Evaporate about two-thirds of the filtrate by boiling off the ethanol or, better, by warming the solution and blowing a stream of air into the reaction tube. Heat the residue to boiling (add a boiling stick to prevent bumping) and, if necessary, add more ethanol to bring the solid into solution. Allow the saturated solution to cool slowly to room temperature to deposit crystals of acetaminophen, which is reported to melt at 169°C–170.5°C. After the mixture has cooled to room temperature, cool it in ice for several minutes, remove the solvent with a Pasteur pipette, wash the crystals once with 2 drops of ice-cold ethanol, remove the ethanol, and dry the crystals under aspirator vacuum while heating the tube on a steam or sand bath.

Alternatively, the original ethanol solution can be evaporated to dryness, and the residue recrystallized from boiling water. The crystals can be collected on a Hirsch funnel (Fig. 7.13) or by use of a Wilfilter (Fig. 7.14). Once the crystals are dry, determine their weight and melting point. TLC analysis (see Chapter 8) and a determination of the melting points of these crystals and the two other components of this mixture will indicate their purity.

Online Study Center

Video: Recrystallization; Photos: Vacuum Filtration into Reaction Tube through Hirsch Funnel, Use of the Wilfilter

Video: Extraction with Dichloromethane

Photos: Sublimation Apparatus, Filtration Using a Pasteur Pipette; Videos: Recrystallization, Filtration Using a Pasteur Pipette

Caffeine

See Figure 4.13 on page 73 for drying crystals under vacuum.

Online Study Center

Photos: Filtration Using a Pasteur Pipette, Use of the Wilfilter; Videos: Recrystallization, Filtration of Crystals Using the Pasteur Pipette

Aspirin

Alternative procedure: Use a Hirsch funnel or a Wilfilter to isolate the aspirin.

The dichloromethane filtered from the binder and acetaminophen mixture (solution 1) should contain caffeine and aspirin. These can be separated by extraction either with acid (which will remove the caffeine as a water-soluble salt) or with base (which will remove the aspirin as a water-soluble salt). We shall use the latter procedure.

To the dichloromethane solution in a reaction tube, add 1 mL of 3 *M* sodium hydroxide solution and shake the mixture thoroughly. Remove the aqueous layer, add 0.2 mL more water, shake the mixture thoroughly, and again remove the aqueous layer, which is combined with the first aqueous extract.

To the dichloromethane, add calcium chloride pellets until the drying agent no longer clumps together. Shake the mixture over a 5-min to 10-min period to complete the drying process; then remove the solvent, wash the drying agent with more solvent, and evaporate the combined extracts to dryness under a stream of air to leave crude caffeine.

The caffeine can be purified by sublimation (Fig. 7.15) or by recrystallization. Recrystallize the caffeine by dissolving it in the minimum quantity of 30% ethanol in tetrahydrofuran. It also can be recrystallized by dissolving the product in a minimum quantity of hot toluene or acetone and adding to this solution ligroin (hexanes) until the solution is cloudy while at the boiling point. In any case, allow the solution to cool slowly to room temperature; then cool the mixture in ice and remove the solvent from the crystals with a Pasteur pipette. Remove the remainder of the solvent under aspirator vacuum and determine the weight of the caffeine and its melting point.

The aqueous hydroxide extract contains aspirin as the sodium salt of the carboxylic acid. To the aqueous solution, add 3 *M* hydrochloric acid dropwise until the solution tests strongly acid with indicator paper; then add 2 more drops of acid. This will give a suspension of white acetylsalicylic acid in the aqueous solution. It could be filtered off and recrystallized from boiling water, but this would cause transfer losses. An easier procedure is to heat the aqueous solution that contains the precipitated aspirin.

Add a boiling stick and heat the mixture to boiling (Fig. 7.16), at which time the aspirin should dissolve completely. If it does not, add more water. Long boiling will hydrolyze the aspirin to salicylic acid (mp 157°C–159°C). Once completely dissolved, the aspirin should be allowed to recrystallize slowly as the solution cools to room temperature in an insulated container. Once the tube has reached room temperature, it should be cooled in ice for several minutes, and then the solvent is removed with a Pasteur pipette. Wash the crystals with a few drops of ice-cold water and isolate them with a Wilfilter or scrape them out onto a piece of filter paper. Squeezing the crystals between sheets of filter paper will hasten drying. Once these crystals are completely dry, determine the weight of the acetylsalicylic acid and its melting point.

Cleaning Up. Place any dichloromethane-containing solutions in the halogenated organic waste container and the other organic liquids in the organic solvents container. The aqueous layers should be diluted and neutralized with sodium carbonate before being flushed down the drain. After it is free of solvent, the calcium

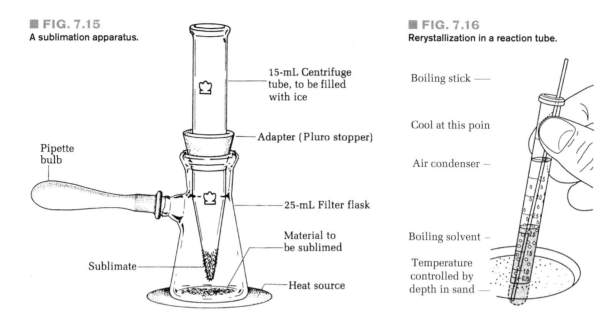

■ FIG. 7.15
A sublimation apparatus.

15-mL Centrifuge tube, to be filled with ice

Adapter (Pluro stopper)

Pipette bulb

25-mL Filter flask

Material to be sublimed

Sublimate

Heat source

■ FIG. 7.16
Rerystallization in a reaction tube.

Boiling stick

Cool at this poin

Air condenser

Boiling solvent

Temperature controlled by depth in sand

chloride can be placed in the nonhazardous solid waste container. If local regulations do not allow for the evaporation of solvents in a hood, dispose of the wet sodium sulfate in a special container.

Macroscale Procedure

Handle dichloromethane in the hood. It is a suspected carcinogen.

The binder can be starch, microcrystalline cellulose, or silica gel.

Acetaminophen

In a mortar, grind two Extra Strength Excedrin tablets to a very fine powder. The label states that this analgesic contains 250 mg of aspirin, 250 mg of acetaminophen, and 65 mg of caffeine per tablet. Place this powder in a test tube and add 7.5 mL of dichloromethane. Warm the mixture briefly and note that a large part of the material does not dissolve. Filter the mixture into another test tube. This can be done by transferring the slurry to a funnel equipped with a piece of filter paper. Use a Pasteur pipette and complete the transfer with a small portion of dichloromethane. This filtrate is solution 1.

Transfer the powder on the filter to a test tube, add 4 mL of ethanol, and heat the mixture to boiling (with a boiling stick). Not all of the material will go into solution. That which does not is the binder. Filter the mixture into a tared test tube and complete the transfer and washing by using a few drops of hot ethanol. This is solution 2.

Evaporate about two-thirds of solution 2 by boiling off the ethanol (with a boiling stick) or, better, by warming the solution and blowing a stream of air into the test tube. Heat the residue to boiling (add a boiling stick to prevent bumping) and add more ethanol, if necessary, to bring the solid into solution. Allow the saturated solution to cool slowly to room temperature to deposit crystals of acetaminophen, which is reported to melt at 169°C–170.5°C. After the mixture has

Check product purity by TLC using 25:1 ethyl acetate–acetic acid to elute the silica gel plates.

An alternative to shaking is pipette mixing or using a vortex stirrer, if available.

Caffeine

**Sodium
acetylsalicylate**
(soluble in water)

Aspirin

Alternative procedure: Use a Hirsch funnel (*see* Fig. 4.14 on page 74) to isolate the aspirin.

cooled to room temperature, cool it in ice for several minutes, remove the solvent with a Pasteur pipette, wash the crystals once with 2 drops of ice-cold ethanol, remove the ethanol, and dry the crystals under aspirator vacuum while heating the tube on a steam or sand bath.

Alternatively, the original ethanol solution is evaporated to dryness, and the residue is recrystallized from boiling water. The crystals are best collected and dried on a Hirsch funnel (*see* Fig. 7.13 on page 158). Once the crystals are dry, determine their weight and melting point. TLC analysis (*see* Chapter 8) and a determination of the melting points of these crystals and the two other components of this mixture will indicate their purity.

The dichloromethane filtered from the binder and acetaminophen mixture (solution 1) should contain caffeine and aspirin. These can be separated by extraction either with acid (which will remove the caffeine as a water-soluble salt) or with base (which will remove the aspirin as a water-soluble salt). We shall use the latter procedure.

To the dichloromethane solution in a test tube, add 4 mL of 3 *M* sodium hydroxide solution and shake the mixture thoroughly. Remove the aqueous layer, add 1 mL more water, shake the mixture thoroughly, and again remove the aqueous layer, which is combined with the first aqueous extract.

To the dichloromethane add anhydrous calcium chloride pellets until the drying agent no longer clumps together. Shake the mixture over a 5–10-min period to complete the drying process, then remove the solvent, wash the drying agent with more solvent, and evaporate the combined extracts to dryness under a stream of air to leave crude caffeine.

The caffeine can be purified by sublimation or by recrystallization. Recrystallize the caffeine by dissolving it in the minimum quantity of 30% ethanol in tetrahydrofuran. It can also be recrystallized by dissolving the product in a minimum quantity of hot toluene or acetone and adding hexanes to this solution until the solution is cloudy while at the boiling point. In any case, allow the solution to cool slowly to room temperature; then cool the mixture in ice and remove the solvent from the crystals with a Pasteur pipette. Remove the remainder of the solvent under aspirator vacuum and determine the weight of the caffeine and its melting point.

The aqueous hydroxide extract contains aspirin as the sodium salt of the carboxylic acid. To the aqueous solution add 3 *M* hydrochloric acid dropwise until the solution tests strongly acid with indicator paper; then add 2 more drops of acid. This will give a suspension of white acetylsalicylic acid in the aqueous solution. It could be filtered off and recrystallized from boiling water, but this would cause transfer losses. An easier procedure is to simply heat the aqueous solution that contains the precipitated aspirin and allow it to recrystallize on slow cooling.

Add a boiling stick and heat the mixture to boiling, at which time the aspirin should dissolve completely. If it does not, add more water. Long boiling will hydrolyze the aspirin to salicylic acid (mp 157°C–159°C). Once completely dissolved, the aspirin should be allowed to recrystallize slowly as the solution cools to room temperature in an insulated container. Once the tube has reached room temperature, it should be cooled in ice for several minutes, and then the solvent is

removed with a Pasteur pipette. The crystals are to be washed with a few drops of
ice-cold water and then scraped out onto a piece of filter paper. Squeezing the
crystals between sheets of the filter paper will hasten drying. Once these crystals
are completely dry, determine the weight of the acetylsalicylic acid and its melting point.

Cleaning Up. Place any dichloromethane-containing solutions in the halogenated
organic waste container and the other organic liquids in the organic solvents
container. The aqueous layers should be diluted and neutralized with sodium carbonate before being flushed down the drain. After it is free of solvent, the calcium
chloride can be placed in the nonhazardous solid waste container. If local regulations do not allow for the evaporation of solvents in a hood, dispose of the wet
sodium sulfate in a special container.

Extractions from Common Items

6. Extraction of Caffeine from Tea

Tea and coffee have been popular beverages for centuries, primarily because
they contain caffeine, a stimulant. Caffeine stimulates respiration, the heart, and the
central nervous system, and it is a diuretic (i.e., it promotes urination). It can cause
nervousness and insomnia and, like many drugs, can be addictive, making it difficult to reduce the daily dose. A regular coffee drinker who consumes just 4 cups per
day can experience headache, insomnia, and even nausea upon withdrawal from the
drug. On the other hand, it helps people to pay attention and can sharpen moderately
complex mental skills as well as prolong the ability to exercise.

Caffeine may be the most widely abused drug in the United States. During the
course of a day, an average person may unwittingly consume up to 1 g of caffeine.
The caffeine content of some common foods and drugs is given in Table 7.3.

Caffeine belongs to a large class of compounds known as alkaloids. These are
of plant origin, contain basic nitrogen, often have a bitter taste and a complex
structure, and usually have physiological activity. Their names usually end in -ine;
many are quite familiar by name if not chemical structure—for example, nicotine,
cocaine, morphine, and strychnine.

Tea leaves contain tannins, which are acidic, as well as a number of colored
compounds and a small amount of undecomposed chlorophyll (soluble in
dichloromethane). To ensure that the acidic substances remain water soluble and
that the caffeine will be present as the free base, sodium carbonate is added to the
extraction medium.

The solubility of caffeine in water is 2.2 mg/mL at 25°C, 180 mg/mL at 80°C,
and 670 mg/mL at 100°C. It is quite soluble in dichloromethane, the solvent used
in this experiment to extract the caffeine from water.

Caffeine can be easily extracted from tea bags. The procedure one would use
to make a cup of tea—simply "steeping" the tea with very hot water for about
7 min—extracts most of the caffeine. There is no advantage to boiling the tea
leaves with water for 20 min. Because caffeine is a white, slightly bitter, odorless,

TABLE 7.3 • Caffeine Content of Common Foods and Drugs	
Espresso	120 mg per 2 oz
Coffee, regular, brewed	103 mg per cup
Instant coffee	57 mg per cup
Coffee, decaffeinated	2–4 mg per cup
Tea	30–75 mg per cup
Cocoa	5–40 mg per cup
Milk chocolate	6 mg per oz
Baking chocolate	35 mg per oz
Coca-Cola, Classic	46 mg per 12 oz
Jolt Cola	72 mg per 12 oz
Anacin, Bromo-Seltzer, Midol	32 mg per pill
Excedrin, Extra Strength	65 mg per pill
Dexatrim, Dietac, Vivarin	200 mg per pill
Dristan	16 mg per pill
No-Doz	100 mg per pill

crystalline solid, it is obvious that water extracts more than just caffeine. When the brown aqueous solution is subsequently extracted with dichloromethane, caffeine primarily dissolves in the organic solvent. Evaporation of the solvent leaves crude caffeine, which on sublimation yields a relatively pure product. When the concentrated tea solution is extracted with dichloromethane, emulsions can form very easily. There are substances in tea that cause small droplets of the organic layer to remain suspended in the aqueous layer. This emulsion formation results from vigorous shaking. To avoid this problem, it might seem that one could boil the tea leaves with dichloromethane first and then extract the caffeine from the dichloromethane solution with water. In fact, this does not work. Boiling 25 g of tea leaves with 50 mL of dichloromethane gives only 0.05 g of residue after evaporation of the solvent. Subsequent extractions yield even less material. Hot water causes the tea leaves to swell and is obviously a far more efficient extraction solvent. An attempt to sublime caffeine directly from tea leaves is also unsuccessful.

Microscale Procedure

Online Study Center
Video: Caffeine from Tea

In a 30-mL beaker place 15 mL of water, 2 g of sodium carbonate, and a wooden boiling stick. Bring the water to a boil on the sand bath, remove the boiling stick, and brew a very concentrated tea solution by immersing a tea bag (2.4 g tea) in the very hot water for 5 min. After the tea bag cools enough to handle, and being careful not to break the bag, squeeze as much water from the bag as possible. Again bring the water to a boil and add a new tea bag to the hot solution. After 5 min, remove the tea bag and squeeze out as much water as possible. This can be done easily on a Hirsch funnel. Rinse the bag with a few mL of very hot water but be

CAUTION: Do not breathe the vapors of dichloromethane and, if possible, work with this solvent in the hood.

Balance the centrifuge tubes.

sure the total volume of aqueous extract does not exceed 12 mL. Pour the extract into a 15-mL centrifuge tube and cool the solution in ice to below 40°C (the boiling point of dichloromethane).

Using three 2-mL portions of dichloromethane, extract the caffeine from the tea. Cork the tube and use a gentle rocking motion to carry out the extraction. Vigorous shaking will produce an intractable emulsion, whereas extremely gentle mixing will fail to extract the caffeine. If you have ready access to a centrifuge, the shaking can be very vigorous because any emulsions formed can be broken fairly well by centrifugation for about 90 s. After each extraction, remove the lower organic layer into a reaction tube, leaving any emulsion layer behind. Dry the combined extracts over anhydrous calcium chloride pellets for 5–10 min in an Erlenmeyer flask. Add the drying agent in portions with shaking until it no longer clumps together. Transfer the dry solution to a tared 25-mL filter flask, wash the drying agent twice with 2-mL portions of dichloromethane, and evaporate it to dryness (*see* Fig. 7.9 on page 146). The residue will be crude caffeine (determine its weight), which is to be purified by sublimation.

Fit the filter flask with a Pluro stopper or no. 2 neoprene adapter through which is thrust a 15-mL centrifuge tube. Put a pipette bulb on the side arm. Clamp the flask with a large three-prong clamp, fill the centrifuge tube with ice and water, and heat the flask on a hot sand bath (*see* Fig. 7.15 on page 160). Caffeine is reported to sublime at about 170°C. Tilt the filter flask and rotate it in a hot sand bath to drive more caffeine onto the centrifuge tube. Use a heat gun to heat the upper walls of the filter flask. When sublimation ceases, remove the ice water from the centrifuge tube and allow the flask to cool somewhat before removing the centrifuge tube. Scrape the caffeine onto a tared weighing paper, weigh and, using a plastic funnel, transfer it to a small vial or a plastic bag. At the discretion of your instructor, determine the melting point with a sealed capillary. The melting point of caffeine is 238°C. Using the centrifugation technique to separate the extracts; about 30 mg of crude caffeine can be obtained. This will give you 10–15 mg of sublimed material, depending on the caffeine content of the particular tea being used. The isolated caffeine can be used to prepare caffeine salicylate (experiment 9).

Cleaning Up. Discard the tea bags in the nonhazardous solid waste container. Allow the solvent to evaporate from the drying agent and discard in the same container. If local regulations do not allow for the evaporation of solvents in a hood, dispose of the wet sodium sulfate in a special container. Place any unused and unrecovered dichloromethane in the chlorinated organic compounds container. The apparatus can be cleaned with soap and hot water. Caffeine can be flushed down the drain because it is biodegradable.

Macroscale Procedure

To an Erlenmeyer flask containing 25 g of tea leaves (or 10 tea bags) and 20 g of sodium carbonate, add 225 mL of vigorously boiling water. Allow the mixture to stand for 7 min and then decant into another Erlenmeyer flask. To the hot tea leaves,

CAUTION: Carry out work with dichloromethane in the hood.

Rock the separatory funnel very gently to avoid emulsions.

Dispose of used dichloromethane in the container provided.

Online Study Center

Video: Recrystallization

add another 50 mL of hot water and then immediately decant and combine with the first extract. Very little, if any, additional caffeine is extracted by boiling the tea leaves for 20 min. Decantation works nearly as well as vacuum filtration and is much faster.

Cool the aqueous solution to near room temperature and extract it twice with 30-mL portions of dichloromethane. Do not shake the separatory funnel so vigorously as to cause emulsion formation, bearing in mind that if it is not shaken vigorously enough the caffeine will not be extracted into the organic layer. Use a gentle rocking motion of the separatory funnel. Drain off the dichloromethane layer on the first extraction; include the emulsion layer on the second extraction. Dry the combined dichloromethane solutions and any emulsion layer with anhydrous calcium chloride pellets. Add sufficient drying agent until it no longer clumps together on the bottom of the flask. Carefully decant or filter the dichloromethane solution into a tared Erlenmeyer or distilling flask. Silicone-impregnated filter paper passes dichloromethane and retains water. Wash the drying agent with a further portion of solvent and evaporate or distill the solvent. A wood applicator stick is better than a boiling chip to promote smooth boiling because it is easily removed once the solvent is gone. The residue of greenish-white crystalline caffeine should weigh about 0.25 g.

Recrystallization of Caffeine

To recrystallize the caffeine, dissolve it in 5 mL of hot acetone, transfer it with a Pasteur pipette to a small Erlenmeyer flask and, while it is hot, add ligroin to the solution until a faint cloudiness appears. Set the flask aside and allow it to cool slowly to room temperature. This mixed-solvent method of recrystallization depends on the fact that caffeine is far more soluble in acetone than ligroin, so a combination of the two solvents can be found where the solution is saturated in caffeine and will appear cloudy. Cool the solution containing the crystals and remove them by vacuum filtration, employing a Hirsch funnel or a very small Büchner funnel. Use a few drops of ligroin to transfer and wash the crystals. If you wish to obtain a second crop of crystals, collect the filtrate in a test tube, concentrate it to the cloud point using an aspirator tube (*see* Fig. 7.11 on page 155), and repeat the recrystallization process.

Cleaning Up. The filtrate can be diluted with water and washed down the drain. Any dichloromethane collected goes into the halogenated organic waste container. After the solvent is allowed to evaporate from the drying agent in the hood, the drying agent can be placed in the nonhazardous solid waste container; otherwise it goes in the hazardous waste container. If local regulations do not allow for the evaporation of solvents in a hood, dispose of the wet sodium sulfate in a special container. The tea leaves go in the nonhazardous solid waste container.

7. Extraction of Caffeine from Cola Syrup

Coca-Cola was originally flavored with extracts from the leaves of the coca plant and the kola nut. Coca is grown in northern South America; the Indians of Peru and Bolivia have for centuries chewed the leaves to relieve the pangs of hunger and sensitivity to high mountain cold. The cocaine from the leaves causes local anesthesia of the stomach. It has limited use as a local anesthetic for surgery on the eye, nose, and throat. Unfortunately, it is now a widely abused and illicit drug. Kola nuts contain about 3% caffeine as well as a number of other alkaloids. The kola tree is in the same family as the cacao tree from which cocoa and chocolate are obtained. Modern cola drinks do not contain cocaine; however, Coca-Cola contains 46 mg of caffeine per 12-oz serving. The acidic taste of many soft drinks comes from citric, tartaric, phosphoric, and benzoic acids.

Caffeine
mp 238°C

Cocaine

Automatic soft drink dispensing machines mix a syrup with carbonated water. In the following experiment caffeine is extracted from concentrated cola syrup.

Microscale Procedure

Add 1 mL of concentrated ammonium hydroxide to a mixture of 5 mL of commercial cola syrup and 5 mL of water in a 15-mL centrifuge tube. Add 1 mL of dichloromethane and tip the tube gently back and forth for 5 min. Do not shake the mixture as in a normal extraction because an emulsion will form, and the layers will not separate. After the layers have separated as much as possible, remove the clear lower layer, leaving the emulsion behind. Using 1.5 mL of dichloromethane, repeat the extraction in the same way two more times. At the final separation, include the emulsion layer with the dichloromethane. If a centrifuge is available, the mixture can be shaken vigorously, and the emulsion broken by centrifugation for 90 s. Combine the extracts in a reaction tube and dry the solution with anhydrous calcium chloride pellets. Add the drying agent with shaking until it no longer clumps together. After 5–10 min, remove the solution with a Pasteur pipette and place it in a tared filter flask. Wash off the drying agent with more dichloromethane and evaporate the mixture to dryness. Determine the crude weight of caffeine; then sublime it as described in the preceding experiment.

⚠️ CAUTION: Do not breathe the vapor of dichloromethane. Work with this solvent in the hood.

Macroscale Procedure

Add 10 mL of concentrated ammonium hydroxide to a mixture of 50 mL of commercial cola syrup and 50 mL of water. Place the mixture in a separatory funnel, add 50 mL of dichloromethane, and swirl the mixture and tip the funnel back and forth for at least 5 min. Do not shake the solutions together as in a normal extraction because an emulsion will form, and the layers will not separate. An emulsion is made up of droplets of one phase suspended in the other. (Milk is an emulsion.) Separate the layers. Repeat the extraction with a second 50-mL portion of dichloromethane. From your knowledge of the density of dichloromethane and water, you should be able to predict which is the top layer and which is the bottom layer. If in doubt, add a few drops of each layer to water. The aqueous layer will be soluble; the organic layer will not. Combine the dichloromethane extracts and any emulsion that has formed in a 125-mL Erlenmeyer flask; then add anhydrous calcium chloride pellets to remove water from the solution. Add the drying agent until it no longer clumps together at the bottom of the flask but swirls freely in solution. Swirl the flask with the drying agent from time to time over a 10-min period. Carefully decant (pour off) the dichloromethane or remove it by filtration through a fluted filter paper, add about 5 mL more solvent to the drying agent to wash it, and decant this also. Combine the dried dichloromethane solutions in a tared flask and remove the dichloromethane by distillation or evaporation on a steam or sand bath. Remember to add a wood applicator stick to the solution to promote even boiling. Determine the weight of the crude product.

⚠ Chlorinated solvents are toxic, insoluble in water, and expensive and should never be poured down the drain.

Recrystallization of Caffeine

To recrystallize the caffeine, dissolve it in 5 mL of hot acetone, transfer it with a Pasteur pipette to a small Erlenmeyer flask, and, while it is hot, add ligroin to the solution until a faint cloudiness appears. Set the flask aside and allow it to cool slowly to room temperature. This mixed-solvent method of recrystallization depends on the fact that caffeine is far more soluble in acetone than ligroin, so a combination of the two solvents can be found where the solution is saturated in caffeine (the cloud point). Cool the solution containing the crystals and remove them by vacuum filtration, employing a Hirsch funnel or a very small Büchner funnel. Use a few drops of ligroin to transfer and wash the crystals. If you wish to obtain a second crop of crystals, collect the filtrate in a test tube, concentrate it to the cloud point using an aspirator tube (*see* Fig. 7.11 on page 155), and repeat the recrystallization process.

Cleaning Up. Combine all aqueous filtrates and solutions, neutralize them, and flush the resulting solution down the drain. Used dichloromethane should be placed in the halogenated waste container, and the drying agent, once the solvent has evaporated from it, can be placed in the nonhazardous solid waste container. If local regulations do not allow for the evaporation of solvents in a hood, dispose of the wet sodium sulfate in a special container. The ligroin-acetone filtrates should be placed in the organic solvents container.

Sublimation of Caffeine. Sublimation is a fast and easy way to purify caffeine. Using the apparatus depicted in Figure 7.15 (on page 160), sublime the crude caffeine at atmospheric pressure following the procedure in part 3 of Chapter 6.

Online Study Center

Photos of Extraction Procedure

8. *Isolation of Caffeine from Instant Coffee*

Instant coffee, according to manufacturers, contains between 55 mg and 62 mg of caffeine per 6-oz cup, and a cup is presumably made from a teaspoon of the powder, which weighs 1.3 g; so 2 g of the powder should contain 85–95 mg of caffeine. Unlike tea, however, coffee contains other compounds that are soluble in dichloromethane, so obtaining pure caffeine from coffee is not easy. The objective of this experiment is to extract instant coffee with dichloromethane (the easy part) and then to try to devise a procedure for obtaining pure caffeine from the extract.

From TLC analysis (*see* Chapter 8), you may deduce that certain impurities have a high R_f value in hydrocarbons (in which caffeine is insoluble). Consult reference books (see especially the *Merck Index*[2]) to determine the solubility (and lack of solubility) of caffeine in various solvents. You might try trituration (grinding the crude solid with a solvent) to dissolve impurities preferentially. Column chromatography is another possible means of purifying the product. Or you might convert all of it to the salicylate and then regenerate the caffeine from the salicylate. Experiment! Or you can simply use the following procedure.

Procedure

> **IN THIS EXPERIMENT** a very concentrated aqueous solution of coffee is prepared and shaken vigorously with an organic solvent to make an intractable emulsion that can be broken (separated into two layers) by centrifugation. Caffeine is isolated by recrystallization.

In a 10-mL Erlenmeyer flask, place 2 g of sodium carbonate and 2 g of instant coffee powder. Add 9 mL of boiling water, stir the mixture well, bring it to a boil again with stirring, cool it to room temperature, and then pour it into a 15-mL plastic centrifuge tube fitted with a screw cap. Add 2 mL of dichloromethane, cap the tube, shake it vigorously for 60 s; then centrifuge it at high speed for 90 s. Remove the clear yellow dichloromethane layer and place it in a 10-mL Erlenmeyer flask. Repeat this process twice more. To the combined extracts add anhydrous calcium chloride pellets until they no longer clump together, allow the solution to dry for a few minutes; then transfer it to a tared 25-mL filter flask and wash the drying agent with more solvent. Remove the solvent as was done in the tea extraction experiment and determine the weight of the crude caffeine. You should obtain about 60 mg of crude product. Sublimation of this orange powder gives an impure orange sublimate that smells strongly of coffee, so sublimation is not a good way to purify this material.

Dissolve a very small quantity of the product in a drop of dichloromethane and perform a TLC analysis of the crude material. Dissolve the remainder of the material in 1 mL of boiling 95% ethanol; then dilute the mixture with 1 mL of

Caffeine has no odor.

2. O'Neill, M. J.; Smith, A.; Heckelman, P. E.; Budavari, S., eds. *The Merck Index*, 13th ed.; Merck and Co., Inc.: Rahway, NJ, 2001.

Video: Recrystallization

t-butyl methyl ether, heat to boiling, and allow to cool slowly to room temperature. Long, needlelike crystals should form in the orange solution. Alternatively, recrystallize the product from a 1:1 mixture of ligroin (hexanes) and 2-propanol, using about 2 mL. Cool the mixture in ice for at least 10 min and then collect the product on a Hirsch funnel. Complete the transfer with the filtrate and then wash the crystals twice with cold 50/50 ethanol/*t*-butyl methyl ether. The yield of white fluffy needles of caffeine should be more than 30 mg.

Cleaning Up. Allow the solvent to evaporate from the drying agent and discard it in the nonhazardous waste container. If local regulations do not allow for the evaporation of solvents in a hood, dispose of the wet sodium sulfate in a special container. Place any unused and unrecovered dichloromethane in the chlorinated organic solvents container.

Preparation of a derivative of caffeine.

9. Caffeine Salicylate

One way to confirm the identity of an organic compound is to prepare a derivative of it. Caffeine melts and sublimes at 238°C. It is an organic base and can therefore accept a proton from an acid to form a salt. The salt formed when caffeine combines with hydrochloric acid, like many amine salts, does not have a sharp melting point; it merely decomposes when heated. But the salt formed from salicylic acid, even though ionic, has a sharp melting point and can thus be used to help characterize caffeine. Figure 7.17 is the ^{1}H NMR spectrum of caffeine.

Caffeine   **Salicylic acid**   **Caffeine salicylate**

CAUTION: Petroleum ether is very flammable. Extinguish all flames.

Recrystallization from mixed solvents

Online Study Center
Video: Filtration of Crystals Using the Pasteur Pipette; Photo: Drying Crystals Under Vacuum

Procedure

The quantities given can be multiplied by 5 or 10, if necessary. To 10 mg of sublimed caffeine in a tared reaction tube, add 7.5 mg of salicylic acid and 0.5 mL of dichloromethane. Heat the mixture to boiling and add petroleum ether (a poor solvent for the product) dropwise until the mixture just turns cloudy, indicating that the solution is saturated. If too much petroleum ether is added, then clarify it by adding a very small quantity of dichloromethane. Insulate the tube to allow it to cool slowly to room temperature; then cool it in ice. The needlelike crystals are isolated by removing the solvent with a Pasteur pipette while the reaction tube is in the ice bath. Evaporate the last traces of solvent under vacuum (Fig. 7.18) and

FIG. 7.17
The ^{1}H NMR spectrum of caffeine
(250 MHz).

FIG. 7.18
The drying of crystals under
vacuum in beaker of warm water.

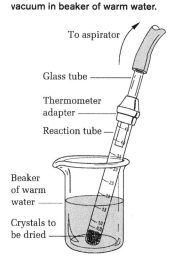

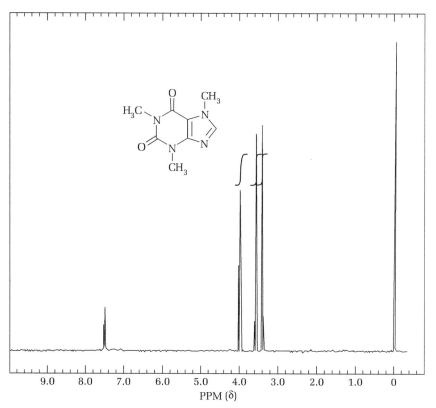

determine the weight of the derivative and its melting point. Caffeine salicylate is
reported to melt at 137°C.

Cleaning Up. Place the filtrate in the halogenated organic solvents container.

QUESTIONS

1. Suppose a reaction mixture, when diluted with water, afforded 300 mL of an
   aqueous solution of 30 g of the reaction product malononitrile [$CH_2(CN)_2$],
   which is to be isolated by extraction with ether. The solubility of
   malononitrile in ether at room temperature is 20.0 g/100 mL and in water is
   13.3 g/100 mL. What weight of malononitrile would be recovered by
   extraction with (a) three 100-mL portions of ether and (b) one 300-mL
   portion of ether? *Suggestion*: For each extraction let x equal the weight
   extracted into the ether layer. In part (a) the concentration in the ether layer
   is $x/100$ and in the water layer is $(30 - x)/300$; the ratio of these quantities
   is equal to $k = 20/13.3$.

2. Why is it necessary to remove the stopper from a separatory funnel when liquid is being drained from it through the stopcock?

3. The pK_a of *p*-nitrophenol is 7.15. Would you expect this to dissolve in sodium bicarbonate solution? The pK_a of 2,5-dinitrophenol is 5.15. Will it dissolve in bicarbonate solution?

4. The distribution coefficient, k = conc. in ligroin ÷ conc. in water, between ligroin and water for solute A is 7.5. What weight of A would be removed from a solution of 10 g of A in 100 mL of water by a single extraction with 100 mL of ligroin? What weight of A would be removed by four successive extractions with 25-mL portions of ligroin? How much ligroin would be required to remove 98.5% of A in a single extraction?

5. In experiment 1, how many moles of benzoic acid are present? How many moles of sodium bicarbonate are contained in 1 mL of a 10% aqueous solution? (A 10% solution has 1 g of solute in 9 mL of solvent.) Is the amount of sodium bicarbonate sufficient to react with all of the benzoic acid?

6. To isolate benzoic acid from a bicarbonate solution, it is acidified with concentrated hydrochloric acid, as in experiment 1. What volume of acid is needed to neutralize the bicarbonate? The concentration of hydrochloric acid is expressed in various ways on the inside back cover of this laboratory manual.

7. How many moles of 4-*t*-butylphenol are in the mixture to be separated in experiment 1? How many moles of sodium hydroxide are contained in 1 mL of 5% sodium hydroxide solution? (Assume the density of the solution is 1.0.) What volume of concentrated hydrochloric acid is needed to neutralize this amount of sodium hydroxide solution?

8. Draw a flow sheet to show how you would separate the components of a mixture containing an acid substance, toluic acid, a basic substance, *p*-bromo-aniline, and anthracene, a neutral substance.

9. Write equations showing how caffeine could be extracted from an organic solvent and subsequently isolated.

10. Write equations showing how acetaminophen might be extracted from an organic solvent such as an ether, if it were soluble.

11. Write detailed equations showing the mechanism by which aspirin is hydrolyzed in boiling, slightly acidic water.

Online Study Center

General Resources
Web Links

CHAPTER

8

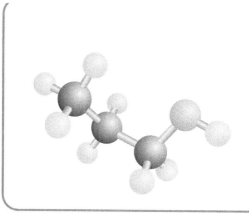

Thin-Layer Chromatography: Analyzing Analgesics and Isolating Lycopene from Tomato Paste

Online Study Center

This icon will direct you to techniques, equipment setups, and online resources at http://college.hmco.com/PIC/ williamsonMMOE5e.

PRELAB EXERCISE: Based on the number and polarity of the functional groups in aspirin, acetaminophen, ibuprofen, and caffeine, whose structures are shown on page 184, predict which of these four compounds has the highest R_f value and which has the lowest.

Chromatography is the separation of two or more compounds or ions caused by their molecular interactions with two phases—one moving and one stationary. These two phases can be a solid and a liquid, a liquid and a liquid, a gas and a solid, or a gas and a liquid. You very likely have seen chromatography carried out on paper towels or coffee filters to separate inks and food dyes. The cellulose paper is the stationary or solid phase, and a propanol-water mixture is the mobile or liquid phase. The samples are spotted near one edge of the paper, and this edge is dipped into the liquid phase. The solvent is drawn through the paper by capillary action, and the molecules are separated based on how they interact with the paper. Although there are several different forms of chromatography, the principles are essentially the same.

Thin-layer chromatography (TLC) is a sensitive, fast, simple, and inexpensive analytical technique that you will use repeatedly while carrying out organic experiments. It is a micro technique; as little as 10^{-9} g of material can be detected, although the usual sample size is from 1×10^{-6} g to 1×10^{-8} g. The stationary phase is normally a polar solid adsorbent, and the mobile phase can be a single solvent or a combination of solvents.

TLC requires micrograms of material.

Uses of Thin-Layer Chromatography

1. **To determine the number of components in a mixture.** TLC affords a quick and easy method for analyzing such things as a crude reaction mixture, an extract from a plant substance, or the ingredients in a pill. Knowing the

number and relative amounts of the components aids in planning further analytical and separation steps.

2. To determine the identity of two substances. If two substances spotted on the same TLC plate give spots in identical locations, they *may* be identical. If the spot positions are not the same, the substances cannot be the same. It is possible for two or more closely related but not identical compounds to have the same positions on a TLC plate. Changing the stationary or mobile phase will usually effect their separation.

3. To monitor the progress of a reaction. By sampling a reaction at regular intervals, it is possible to watch the reactants disappear and the products appear using TLC. Thus, the optimum time to halt the reaction can be determined, and the effect of changing such variables as temperature, concentrations, and solvents can be followed without having to isolate the product.

4. To determine the effectiveness of a purification. The effectiveness of distillation, crystallization, extraction, and other separation and purification methods can be monitored using TLC, with the caveat that a single spot does not guarantee a single substance.

5. To determine the appropriate conditions for a column chromatographic separation. In general, TLC is not satisfactory for purifying and isolating macroscopic quantities of material; however, the adsorbents most commonly used for TLC—silica gel and alumina—are also used for column chromatography, which is discussed in Chapter 9. Column chromatography is used to separate and purify up to 1 g of a solid mixture. The correct adsorbent and solvent to use for column chromatography can be rapidly determined by TLC.

6. To monitor column chromatography. As column chromatography is carried out, the solvent is collected in a number of small flasks. Unless the desired compound is colored, the various fractions must be analyzed in some way to determine which ones have the desired components of the mixture. TLC is a fast and effective method for doing this.

The Principles of Chromatography

To thoroughly understand the process of TLC (and other types of chromatography), we must examine the process at the molecular level. All forms of chromatography involve a dynamic and rapid equilibrium of molecules between the liquid and the stationary phases. For the chromatographic separation of molecules A and B shown in Figure 8.1, there are two states:

1. Free—dissolved in the liquid or gaseous mobile phase
2. Adsorbed—sticking to the surface of the solid stationary phase

Molecules A and B are continuously moving back and forth between the dissolved (free) and adsorbed states, with billions of molecules adsorbing and billions of other molecules desorbing from the solid stationary phase each second. The equilibrium between the free and adsorbed states depends on the relative strength of the

■ FIG. 8.1
The mixture of molecules A and B is in a dynamic equilibrium between the free and adsorbed states.

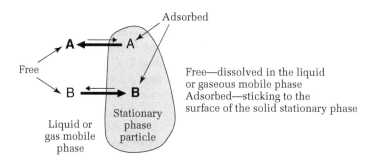

■ FIG. 8.2
The mixture of molecules A and B is in a dynamic equilibrium between the stationary adsorbent and a *flowing* mobile phase.

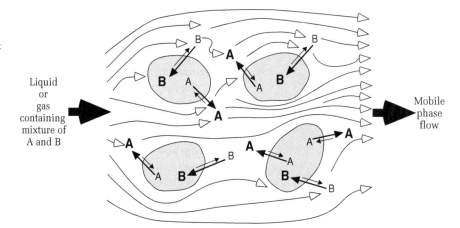

attraction of A and B to the liquid phase molecules *versus* the strength of attraction of A and B to the stationary phase structure. As discussed in the introduction to Chapter 3, the strength of these attractive forces depends on the following factors:

- Size, polarity, and hydrogen bonding ability of molecules A and B
- Polarity and hydrogen bonding ability of the stationary phase
- Polarity and hydrogen bonding ability of the mobile phase solvent

Molecules distribute themselves, or *partition*, between the mobile and stationary phases depending on these attractive forces. As implied by the equilibrium arrows in Figure 8.1, the A molecules are less polar and are thus weakly attracted to a polar stationary phase, spending most of their time in the mobile phase. In contrast, equilibrium for the more polar B molecules lies in the direction of being adsorbed onto the polar stationary phase. The equilibrium constant k (also called the *partition coefficient*) is a measure of the distribution of molecules between the mobile phase and the stationary phase and is similar to the distribution coefficient for liquid/liquid extraction. This constant changes with structure.

Simply adding a mixture to a combination of a liquid phase and a stationary phase will not separate it into its pure components. For separation to happen, the liquid phase must be mobile and be flowing past the stationary phase, as depicted in Figure 8.2. Because the A molecules spend more time in the mobile phase, they will be carried through the stationary phase and be eluted faster and move farther

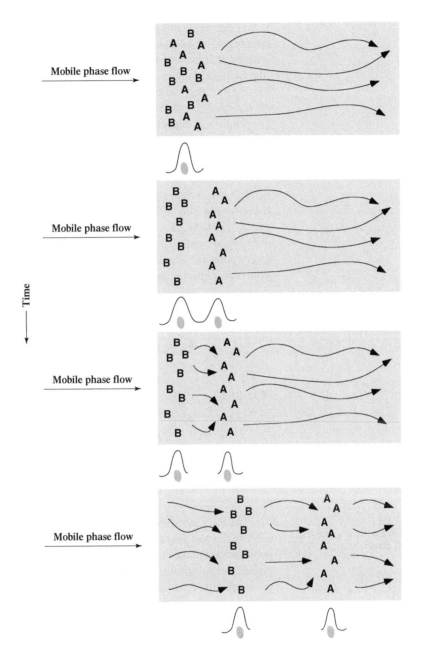

■ FIG. 8.3
A chromatographic separation. Over time, the mobile phase carries the less weakly adsorbed A molecules ahead of the more strongly adsorbed B molecules.

in a given amount of time. Because the B molecules are adsorbed on the stationary phase more than A molecules, the B molecules spend less time in the mobile phase and therefore migrate through the stationary phase more slowly and are eluted later. The B molecules do not migrate as far in the same amount of time. The consequence of this difference is that A is gradually separated from B by moving ahead in the flowing mobile phase as time passes, as shown in Figure 8.3.

A simple analogy may help to illustrate these concepts. Imagine a group of hungry and not-so-hungry people riding a moving sidewalk (the mobile phase in this analogy) that moves beside a long buffet table covered with all sorts of delicious food (the stationary phase). Hungry people, attracted to the food, will step off and on the moving belt many times in order to fill their plates. The not-so-hungry people will step off and on to get food far less often. Consequently, the more strongly attracted hungry people will lag behind, while the not-so-hungry ones will move ahead. The two types of people are thus separated based on the strength of their attraction for the food.

Stationary Phase Adsorbents

In TLC, the stationary phase is a polar adsorbent, usually finely ground alumina $[(Al_2O_3)_x]$ or silica $[(SiO_2)_x]$ particles, coated as a thin layer on a glass slide or plastic sheet. Silica, commonly called *silica gel* in the laboratory, is simply very pure white sand. The extended covalent network of these adsorbents creates a very polar surface. Partial structures of silica and alumina are shown below. The silicon or aluminum atoms are the smaller, darker spheres:

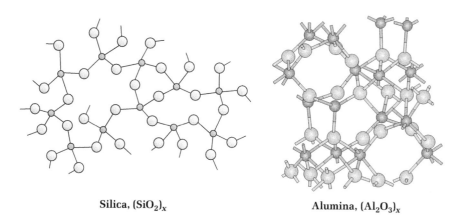

Silica, $(SiO_2)_x$ Alumina, $(Al_2O_3)_x$

FIG. 8.4
A partial silica structure showing polar Si–O bonds.

$$\overset{\delta-}{O} \quad \overset{\delta-}{O}$$
$$\overset{\delta+}{Si}$$

The electropositive character of the aluminum or silicon and the electronegativity of oxygen create a very polar stationary phase (Fig. 8.4). Therefore, the more polar the molecules to be separated, the stronger the attraction to the stationary phase. Nonpolar molecules will tend to stay in the mobile phase. In general, the more polar the functional group, the stronger the adsorption on the stationary phase and the more slowly the molecules will move. In an extreme situation, the molecules will not move at all. This problem can be overcome by increasing the polarity of the mobile phase so that the equilibrium between the free and adsorbed state is shifted toward the free state.

Although silica is the most common stationary phase used for TLC, many other types are used, ranging from paper to charcoal, nonpolar to polar, and reverse phase to normal phase. Several different types of stationary phases are listed according to polarity in Table 8.1.

Silica gel and alumina are commonly used in column chromatography for the purification of macroscopic quantities of material (*see* Chapter 9). Of the two,

TABLE 8.1 • Common Stationary Phases Listed by Increasing Polarity
Polydimethyl siloxane*
Methyl- or Phenylsiloxane*
Cyanopropylsiloxane*
Carbowax [poly(ethyleneglycol)]*
Reverse phase (hydrocarbon-coated silica, e.g., C$_{18}$)
Paper
Cellulose
Starch
Calcium sulfate
Silica (silica gel)
Florisil (magnesium silicate)
Magnesium oxide
Alumina (aluminum oxide; acidic, basic, or neutral)
Activated carbon (charcoal or Norit pellets)

*Stationary phase for gas chromatography

(left margin, vertical: Increasing polarity ↓)

alumina, when anhydrous, is the more active; that is, it will adsorb substances more strongly. It is thus the adsorbent of choice for the separation of relatively nonpolar substrates, such as hydrocarbons, alkyl halides, ethers, aldehydes, and ketones. To separate more polar substrates, such as alcohols, carboxylic acids, and amines, the less active adsorbent, silica gel, is often used.

Molecular Polarity and Elution Sequence

Elution sequence is the order in which the components of a mixture move during chromatography.

Assuming we are using a polar adsorbent, how can we determine how rapidly the compounds in our particular mixture move, that is, their elution sequence? Because the more polar compounds will adsorb more strongly to the polar stationary phase, they will move the slowest and the shortest distance on a TLC plate. Non-polar compounds will move rapidly; they will elute first or move the greatest distance on the TLC plate. Table 8.2 lists several common compound classes according to how they move or elute on silica or alumina.

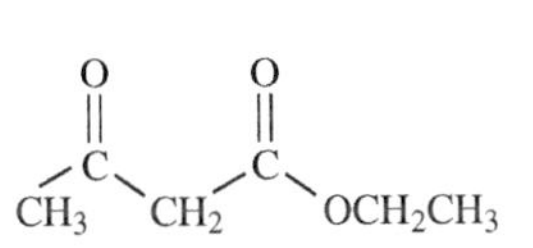

You should be able to look at a molecular structure, identify its functional group(s), and easily determine whether it is more or less polar than another structure with different functional groups. Note that the polarity of a molecule increases as the number of functional groups in that molecule increases. Thus, ethyl acetoacetate, with both ketone and ester groups, is more polar than ethyl pentanoate, which has only an ester group. However, it should be noted that chromatography is not an exact science. The rules discussed here can be used to help predict the order of elution; however, only performing an experiment will give definitive answers.

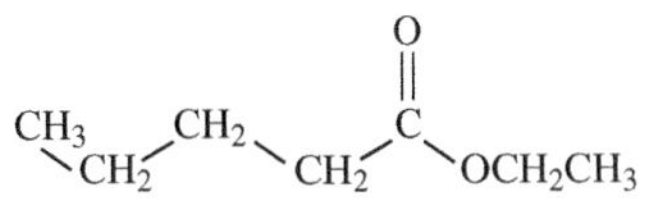

Ethyl acetoacetate

Ethyl pentanoate

TABLE 8.2 • Elution Order for Some Common Functional Groups with a Silica or Alumina Stationary Phase
Highest/fastest (elute with nonpolar mobile phase)
Alkane hydrocarbons
Alkyl halides (halocarbons)
Alkenes (olefins)
Dienes
Aromatic hydrocarbons
Aromatic halides
Ethers
Esters
Ketones
Aldehydes
Amines
Alcohols
Phenols
Carboxylic acids
Sulfonic acids
Lowest/slowest (need polar mobile phase to elute)

Increasing polarity of functional group ↓

Mobile Phase Solvent Polarity

The key to a successful chromatographic separation is the mobile phase. You cannot change the polarities of the compounds in your mixture, and you normally use silica gel or alumina as the stationary phase. In extreme situations very polar substances chromatographed on alumina will not migrate very far from the starting point (i.e., give low R_f values), and nonpolar compounds chromatographed on silica gel will travel with the solvent front (i.e., give high R_f values). These extremes of behavior are markedly affected, however, by the solvents used to carry out the chromatography. A polar solvent will carry along with it polar substrates, and nonpolar solvents will do the same with nonpolar compounds—another example of the generalization "like dissolves like." By using different solvents, either alone or as mixtures, we can adjust the polarity of the mobile phase and affect the equilibria between the free and adsorbed states. Changing the polarity of the mobile phase can optimize the chromatographic separation of mixtures of compounds with a wide variety of polarities.

Table 8.3 lists, according to increasing polarity, some solvents that are commonly used for both TLC and column chromatography. Because the polarities of benzene, carbon tetrachloride, or chloroform can be matched by other, less toxic solvents, these three solvents are seldom used. In general, the solvents for TLC and column chromatography are characterized by having low boiling points that allow them to be easily evaporated and low viscosities that allow them to migrate

Avoid using benzene, carbon tetrachloride, and chloroform. Benzene is known to be a carcinogen when exposure is prolonged; the others are suspected carcinogens.

TABLE 8.3 • Common Mobile Phases Listed by Increasing Polarity
Helium
Nitrogen
Pentanes (petroleum ether)
Hexanes (ligroin)
Cyclohexane
Carbon tetrachloride*
Toluene
Chloroform*
Dichloromethane (methylene chloride)
t-Butyl methyl ether
Diethyl ether
Ethyl acetate
Acetone
2-Propanol
Pyridine
Ethanol
Methanol
Water
Acetic acid

Suspected carcinogens

(Left margin, with downward arrow: Increasing polarity)

rapidly. A solvent more polar than methanol is seldom needed. Often, two solvents are used in a mixture of varying proportions; the polarity of the mixture is a weighted average of the two. Hexane and ether mixtures are often employed.

Finding a good solvent system is usually the most critical aspect of TLC. If the mobile phase has not been previously determined, start with a nonpolar solvent such as hexane or ligroin and observe the separation. If the mixture's components do not move very far, try adding a polar solvent such as ether or ethyl acetate to the hexane. Compare the separation to the previous plate. In most cases, a combination of two solvents is the best choice. If the spots stay at the bottom of the plate, add more of the polar solvent. If they run with the solvent front (move to the top), increase the proportion of the nonpolar solvent. Unfortunately, some trial and error is usually involved in determining which solvent system is the best. There is a large amount of literature on the solvents and adsorbents used in the separation of a wide variety of substances.

Spotting the TLC Plate

It is recommended that you use commercially available TLC plates, poly(ethylene terephthalate) (Mylar) sheets coated with silica gel using polyacrylic acid as a binder; these fluoresce under ultraviolet (UV) light.[1] The TLC plates must be

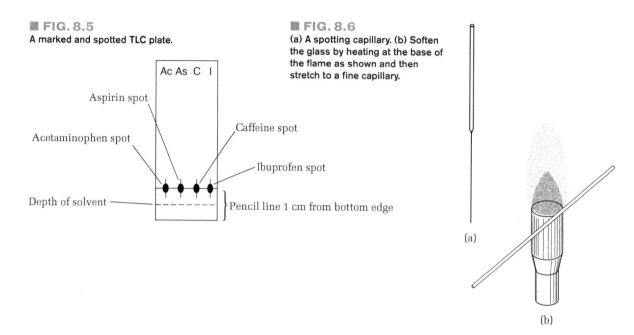

■ **FIG. 8.5**
A marked and spotted TLC plate.

Ac As C I

Aspirin spot

Acetaminophen spot

Caffeine spot

Ibuprofen spot

Depth of solvent

Pencil line 1 cm from bottom edge

■ **FIG. 8.6**

(a) A spotting capillary. (b) Soften the glass by heating at the base of the flame as shown and then stretch to a fine capillary.

(a)

(b)

handled gently, or the 100-mm-thick coating of silica gel can be easily scratched off. With a pencil, lightly draw a faint line 1 cm from the end and then three or four short hash marks to guide spotting. Lightly write identifying letters at the top of the plate to keep track of the placement of the compound spots (Fig. 8.5). Note that a pencil is always used to mark TLC plates because the graphite (carbon) is inert. If ink is used to mark the plate, it will chromatograph just as any other organic compound and give flawed results.

You need to dissolve only a few milligrams of material because one can detect a few micrograms of compound on a TLC plate. Choose a volatile solvent. Even if the material is only partially soluble, you will normally be able to observe the compound because only low concentrations are needed. It is extremely important that the spots be as small as possible and that they be applied using a 1% (not more than 2%) solution of the compounds being separated.

Once the sample is prepared, a spotting capillary must be used to add the sample to the plate. Spotting capillaries can be made by drawing out open-end melting point tubes or Pasteur pipette stems in a burner flame (Fig. 8.6).[2] The bore of these capillaries should be so small that once a liquid is drawn into them, it will not flow out to form a drop. Practice spotting just pure solvent onto an unmarked TLC plate. Dip the capillary into the solvent and let a 2–3 cm column of solvent flow into it by

Too much sample is a frequent problem. Use a 1% solution of the mixture. Apply very small spots.

CAUTION: Bunsen burners should be used only in lab areas that are far from flammable organic solvents.

---

1. Whatman flexible plates for TLC, cat. no. 4410 222 (Fisher cat. no. 05-713-162); cut with scissors to 1″ × 3″ or 2.5 × 7.5 cm Baker-flex, Silica Gel IB-F (J.T. Baker, Phillipsburg, N.J.). Unlike student-prepared plates, these coated sheets give very consistent results. A supply of these plates makes it a simple matter to examine most of the reactions in this book for completeness of reaction, purity of product, and side reactions.

2. Three-inch pieces of old and unusable gas chromatography capillary columns are also effective spotting capillaries.

Using a wide-mouth bottle
to develop a TLC plate.

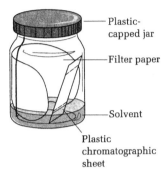

— Plastic-capped jar

— Filter paper

— Solvent

Plastic
chromatographic
sheet

■ FIG. 8.8
Using a foil-covered beaker
to develop a TLC plate.

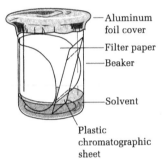

— Aluminum
foil cover

— Filter paper

— Beaker

— Solvent

Plastic
chromatographic
sheet

■ FIG. 8.9
A UV lamp used to visualize spots.

capillary action, hold this vertically over the *coated* side of the plate, and lower the pipette until the tip just touches the adsorbent. Only then will liquid flow onto the plate; quickly withdraw the capillary when the spot is about 1 mm in diameter. The center of the letter *o* on this page is more than 1 mm in diameter. The distance between the sides of the letter *n* is 1 mm. The solvent should evaporate quickly, leaving your mixture behind on the plate. You may have to spot the plate a couple of times to ensure that sufficient material is present; do not spot too much sample because this will lead to a poor separation. Smearing, smudging, and overlapping of spots will make the identification of separated components difficult. Practice spotting a number of times until you develop good spotting technique. You are now ready to spot the mixture solutions as described in Experiments 1 and 2.

Development

Once the dilute solution of the mixture has been spotted on the plate, the next step is the actual chromatographic separation, called *plate development*. The marked and spotted TLC plate is inserted into a 4-oz wide-mouth bottle (Fig. 8.7) or beaker (Fig. 8.8) containing 4 mL of an organic solvent or solvent mixture. The bottle is lined with filter paper that is wet with solvent to saturate the atmosphere within the container. Use tweezers to place the plate in the development chamber; oils from your fingers can sometimes smear or ruin a TLC plate. Also make sure that the origin spots are not below the solvent level in the chamber. If the spots are submerged in the solvent, they are washed off the plate and lost. The top of the bottle is put in place and the time noted. (If a beaker is used, the beaker is to be covered with aluminium foil.) The solvent travels up the thin layer by capillary action, and if the substance is a pure colored compound, one soon sees a spot traveling either along with the solvent front or, more commonly, at some distance behind the solvent front. Once the solvent has run within a centimeter of the top of the plate, remove the plate with tweezers. Immediately, before the solvent evaporates, use a pencil to draw a line across the plate where the solvent front can be seen. The proper location of this solvent front line is important for R_f calculations.

Visualization

If you are fortunate enough to be separating organic molecules that are colored, such as dyes, inks, or indicators, then visualizing the separated spots is easy. However, because most organic compounds are colorless, this is rarely the case.

For most compounds a UV light works well for observing the separated spots. TLC plates normally contain a fluorescent indicator that makes them glow green under UV light of wavelength 254 nm. Compounds that adsorb UV light at this wavelength will quench the green fluorescence, yielding dark purple or bluish spots on the plate. Simply hold the plate by its edges under a UV lamp as shown in Figure 8.9, and the compound spots become visible to the naked eye. Lightly circle the spots with a pencil so that you will have a permanent record of their location for later calculations.

Another useful visualizing technique is to use an iodine (I_2) chamber. Certain compounds, such as alkanes, alcohols, and ethers, do not absorb UV light sufficiently

⚠️
Never look into a UV lamp.

to quench the fluorescence of the TLC plate and therefore will not show up under a UV lamp. However, they will adsorb iodine vapors and can be detected (after any residual solvent has evaporated) by placing the plate for a few minutes in a capped 4-oz bottle containing some crystals of iodine. Iodine vapor is adsorbed by the organic compound to form brown spots. These brown spots should be outlined with a pencil immediately after removing the plate from the iodine bottle because they will soon disappear as the iodine sublimes away; a brief return to the iodine chamber will regenerate the spots. Using both the UV lamp and iodine vapor visualization methods will ensure the location of all spots on the TLC plate.

Many specialized spray reagents have also been developed to give specific colors for certain types of compounds.

R_f Values

R_f is the ratio of the distance the spot travels from the origin to the distance the solvent travels.

In addition to qualitative results, TLC can also provide a chromatographic parameter known as an R_f value. The R_f value is the retardation factor or the ratio-to-front value expressed as a decimal fraction. The R_f value is the ratio of the distance the spot travels from the point of origin to the distance the solvent travels. The R_f value can be calculated as follows:

$$R_f = \frac{\text{distance spot travels}}{\text{distance solvent travels}}$$

This number should be calculated for each spot observed on a TLC plate. Figure 8.10 shows a diagram of a typical TLC plate and how the distances are measured to calculate the R_f value. The best separations are usually achieved when the R_f values fall between 0.3 and 0.7.

■ **FIG. 8.10**
A developed TLC plate with spots visualized and R_f values determined.

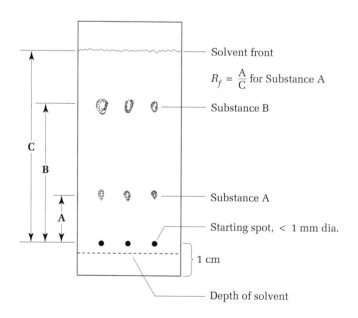

Solvent front

$R_f = \dfrac{A}{C}$ for Substance A

Substance B

C

B

Substance A

A

Starting spot, < 1 mm dia.

1 cm

Depth of solvent

If two spots travel the same distance or have the same R_f value, then it might be concluded that the two components are the same molecule. Just as many organic molecules have the same melting point and color, many can have the same R_f value, so identical R_f values do not necessarily mean identical compounds. For comparisons of R_f values to be valid, TLC plates must be run under the exact same conditions for stationary phase, mobile phase, and temperature. Even then, additional information such as a mixed melting point or an IR spectrum should be obtained before concluding that two substances are identical.

Comparison of Different Types of Chromatography

Table 8.4 summarizes the terminology used in chromatography and how these apply to different types of chromatography. All of the chromatographic types involve the same principles but vary in the nature of the stationary phase and the mobile phase and the measure of separation.

EXPERIMENTS

 ## 1. Analgesics

Analgesics are substances that relieve pain. The most common of these is aspirin, a component of more than 100 nonprescription drugs. In Chapter 41, the history of this most popular drug is discussed. In this experiment, analgesic tablets will be analyzed by TLC to determine which analgesics they contain and whether they contain caffeine, which is often added to counteract the sedative effects of the analgesic.

In addition to aspirin and caffeine, the most common components of the currently used analgesics are acetaminophen and ibuprofen. In addition to one or

TABLE 8.4 • Chromatography Terms and Their Definitions with Examples

Chromatography Term	Definition	Examples
Mixture	A collection of different compounds	Aspirin, ibuprofen, caffeine, and fluorene/fluorenone
Stationary phase	A fixed material that can adsorb compounds	Alumina, silica gel, and silicone gum
Mobile phase	A moving liquid or gas that dissolves compounds and carries them along	Hexane, CH_2Cl_2, and ethyl acetate (TLC and column chromatography); helium gas (gas chromatography)
Adsorption	The strength of attraction between the compounds and the stationary phase	London forces, hydrogen bonds, and dipole-dipole attractive forces
Separation	A measure of the elution or migration rate of compounds	R_f (TLC); elution volume (column chromatography); retention time (gas chromatography)

more of these substances, each tablet contains a binder—often starch, microcrystalline cellulose, or silica gel. And to counteract the acidic properties of aspirin, an inorganic buffering agent is added to some analgesics. An inspection of analgesic labels will reveal that most cold remedies and decongestants contain both aspirin and caffeine in addition to the primary ingredient.

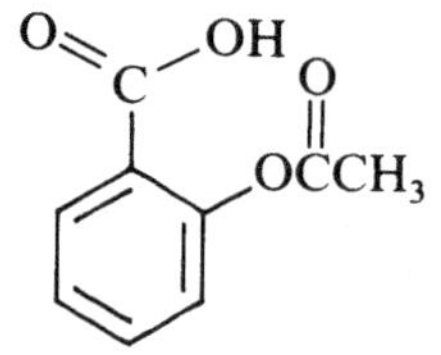

Aspirin
Acetylsalicylic acid

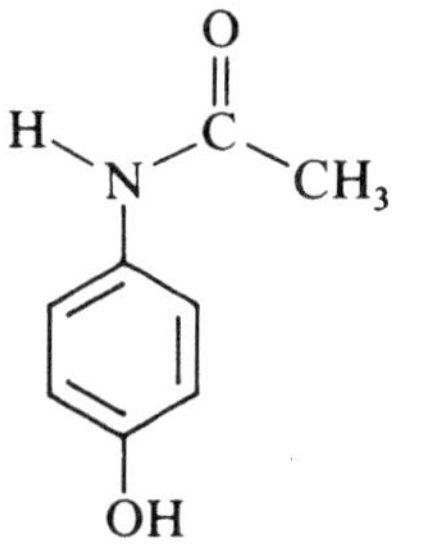

Acetaminophen
4-Acetamidophenol

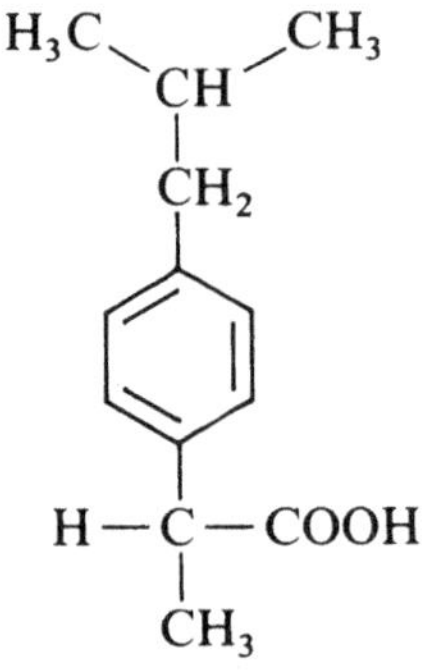

Ibuprofen
2-(4-Isobutylphenyl)propionic acid

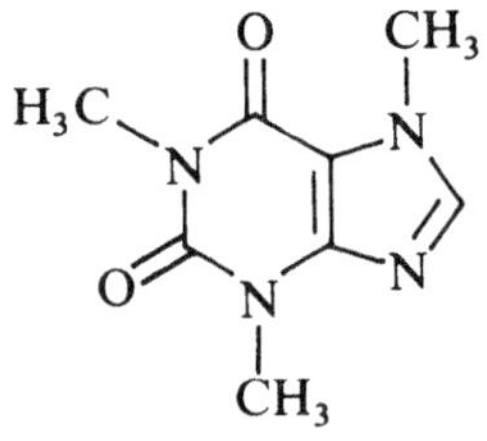

Caffeine

To identify an unknown by TLC, the usual strategy is to run chromatograms of known substances (the standards) and the unknown at the same time. If the unknown has one or more spots that correspond to spots with the same R_f values as the standards, then those substances are probably present.

Proprietary drugs that contain one or more of the common analgesics and sometimes caffeine are sold under some of the following brand names: Bayer Aspirin, Anacin, Datril, Advil, Excedrin, Extra Strength Excedrin, Tylenol, and Vanquish. Note that ibuprofen has a chiral carbon atom. The S-(+)-enantiomer is more effective than the other.

Procedure

Before proceeding, practice the TLC spotting technique described earlier. Following that procedure, draw a light pencil line about 1 cm from the end of a chromatographic plate. On this line spot aspirin, acetaminophen, ibuprofen, and caffeine, which are available as reference standards. Use a separate capillary for each standard (or rinse the capillary carefully before reusing). Make each spot as small as possible, preferably less than 0.5 mm in diameter. Examine the plate under UV light to see that enough of each compound has been applied; if not, add more. On a separate plate, run the unknown and one or more of the standards.

The unknown sample is prepared by crushing a part of a tablet, adding this powder to a test tube or small vial along with an appropriate amount of ethanol, and then mixing the suspension. Not all of the crushed tablet will dissolve, but enough will go into solution to spot the plate. The binder—starch or silica—will not dissolve. Weigh out only part of the tablet to try to prepare a 1% solution of the unknown. Typically, ibuprofen tablets contain 200 mg of the active ingredient, aspirin tablets contain 325 mg, and acetaminophen tablets contain 500 mg.

Because of the insoluble binder, not all of the unknown will dissolve.

To the developing jar or beaker (*see* Fig. 8.7 or Fig. 8.8 on page 181), add 4 mL of the mobile phase, a mixture of 95% ethyl acetate and 5% acetic acid. Insert the spotted TLC plates with a tweezers. After the solvent has risen nearly to the top of the plate, remove the plate from the developing chamber, mark the solvent front with a pencil, and allow the solvent to dry. Examine the plate under UV light to see the components as dark spots against a bright green-blue background. Outline the spots with a pencil. The spots can also be visualized by putting the plate in an iodine chamber made by placing a few crystals of iodine in the bottom of a capped 4-oz jar. Calculate the R_f values for the spots and identify the components in the unknown.

Cleaning Up. Solvents should be placed in the organic solvents container; dry, used chromatographic plates can be discarded in the nonhazardous solid waste container.

2. Plant Pigments

The botanist Michael Tswett discovered the technique of chromatography and applied it, as the name implies, to colored plant pigments. The leaves of plants contain, in addition to chlorophyll-a and chlorophyll-b, other pigments that are revealed in the fall when the leaves die and the chlorophyll rapidly decomposes. Among the most abundant of the other pigments are the carotenoids, which include the carotenes and their oxygenated homologs, the xanthophylls. The bright orange β-carotene is the most important of these because it is transformed in the liver to vitamin A, which is required for night vision.

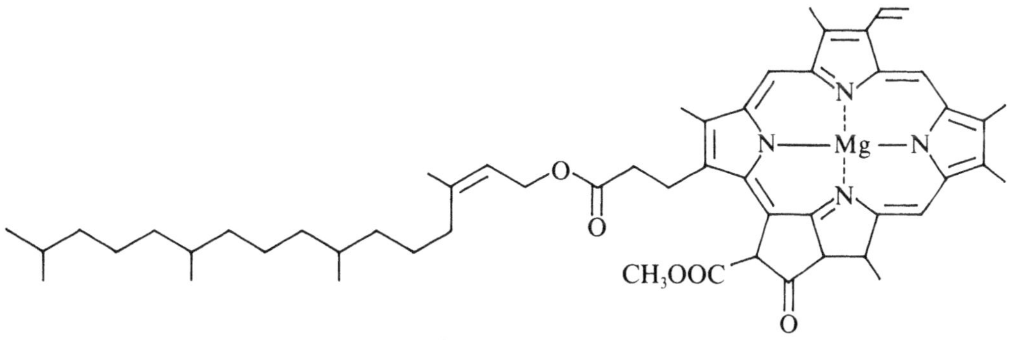

Chlorophyll-a

β-Carotene (C₄₀H₅₆)
mp 183°C, λ_max^hexane 451 nm

β-Carotene (C$_{40}$H$_{56}$)
mp 183°C, λ_{max}^{hexane} 451 nm

FIG. 8.11
An energy-minimized, space-filling model of lycopene. The molecule is flat, but steric hindrance of the methyl groups causes the molecule to bend into an S shape.

HO

3′-Dehydrolutein (a xanthophyll)

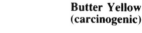

**Butter Yellow
(carcinogenic)**

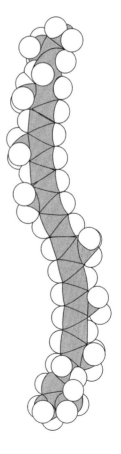

Lycopene (C$_{40}$H$_{56}$)
MW 536.85
mp 173°C, λ_{max}^{hexane} 475 nm

Cows eat fresh, green grass that contains carotene, but they do not metabolize the carotene entirely, so it ends up in their milk. Butter made from this milk is therefore yellow. In the winter the silage cows eat does not contain carotene because that compound readily undergoes air oxidation, and the butter made at that time is white. For some time an azo dye called Butter Yellow was added to winter butter to give it the accustomed color, but the dye was found to be a carcinogen. Now winter butter is colored with synthetic carotene, as is all margarine.

Isoprene

Lycopene from tomato paste and β-carotene from strained carrots

As an interesting variation, try extracting lycopene from commercial ketchup.

CAUTION: Do not breathe the vapors of dichloromethane. Carry out the extraction in the hood.

Lycopene (Fig. 8.11), the red pigment of the tomato, is a C_{40}-carotenoid made up of eight isoprene units. β-Carotene, the yellow pigment of the carrot, is an isomer of lycopene in which the double bonds at C_1—C_2 and C'_1—C'_2 are replaced by bonds extending from C_1 to C_6 and from C'_1 to C'_6 to form rings. The chromophore in each case is a system of 11 all-*trans* conjugated double bonds; the closing of the two rings causes β-carotene to absorb at shorter wavelengths than lycopene does, shifting its color from red to yellow.

In 1911, Richard Willstätter and Heinrich R. Escher isolated 20 mg of lycopene per kilogram of fresh tomatoes, which contain about 96% water.[3] They then found a more convenient source in commercial tomato paste: the seeds and skin were eliminated, and the water content was reduced by evaporation in vacuum to a content of 26% solids. From this they isolated 150 mg of lycopene per kilogram of paste. The expected yield in the following experiment is 0.075 mg, which is not enough to weigh on a balance.

A jar of strained carrots sold as baby food serves as a convenient source of β-carotene. The German investigators isolated 1 g of β-carotene per kilogram of dried, shredded carrots of unstated water content.

The following procedure calls for the dehydration of tomato or carrot paste with ethanol and extraction with dichloromethane, an efficient solvent for lipids.

Experimental Considerations

Carotenoids are very sensitive to light-catalyzed air oxidation. Perform this experiment as rapidly as possible; keep the solutions as cool and dark as possible. This extraction produces a mixture of products that can be analyzed by both TLC and column chromatography (*see* Chapter 9). If enough material for TLC only is desired, use one-tenth the quantities of starting material and solvents employed in the following procedure. This extraction can also be carried out with hexane if the ventilation is not adequate enough to use dichloromethane. However, hexane is more prone to form emulsions than the chlorinated solvent.

 ### Procedure

> IN THIS EXPERIMENT some tomato or carrot paste is treated with acetone, which will remove water and lipids but not the highly colored carotenoid hydrocarbons. The carotenoids are extracted by dichloromethane and analyzed by TLC.

A 5-g sample of fresh tomato or carrot paste (baby food) is transferred to the bottom of a 25 × 150 mm test tube, followed by 10 mL of acetone. The mixture is stirred and shaken before being filtered on a Hirsch funnel. Scrape as much of the material from the tube as possible and press it dry on the funnel. Let the tube drain thoroughly. Place the filtrate in a 125-mL Erlenmeyer flask.

3. Willstätter, R.; Escher, H. R. *Z. Physiol. Chem.* **1911,** *64,* 47–61.

Return the solid residue to the test tube, shake it with a 10-mL portion of dichloromethane, and again filter the material on a Hirsch funnel. Add the filtrate to the 125-mL flask. Repeat this process two more times and then pour the combined filtrates into a separatory funnel. Add water and sodium chloride solution (which aids in the breaking of emulsions) and shake the funnel gently. This aqueous extraction will remove the acetone and any water-soluble components from the mixture, leaving the hydrocarbon carotenoids in the dichloromethane. Dry the colored organic layer over anhydrous calcium chloride and filter the solution into a dry flask. Remove about 0.5 mL of this solution and store it under nitrogen in the dark until it can be analyzed by TLC. Evaporate the remainder of the dichloromethane solution to dryness under a stream of nitrogen or under vacuum on a rotary evaporator. This material can be used for column chromatography (*see* Chapter 9). If it is to be stored, fill the flask with nitrogen and store it in a dark place.

Air can be used for the evaporation, but nitrogen is better because these hydrocarbons air oxidize with great rapidity.

Thin-Layer Chromatography

Spot the mixture on a TLC plate about 1 cm from the bottom and 8 mm from the edge. Make one spot concentrated by repeatedly touching the plate, but ensure that the spot is as small as possible, certainly less than 1 mm in diameter. The other spot can be of lower concentration. Develop the plate with an 80:20 hexane-acetone mixture. With other plates you could try cyclohexane and toluene as eluents and also hexane-ethanol mixtures of various compositions.

Many spots may be seen. There are two common carotene and chlorophyll isomers and four xanthophyll isomers.

The container in which the chromatography is carried out should be lined with filter paper that is wet with the solvent so that the atmosphere in the container will be saturated with the solvent vapor. After elution is completed, remove the TLC plate and mark the solvent front with a pencil and outline the colored spots. Examine the plate under UV light. Also place the plate in an iodine chamber to visualize the spots.

Cleaning Up. The aqueous saline filtrate containing acetone can be flushed down the drain. Recovered and unused dichloromethane should be placed in the halogenated organic waste container; the solvents used for TLC should be placed in the organic solvents container. If local regulations allow, evaporate any residual solvent from the drying agents in the hood and place the dried solid in the nonhazardous waste container. Otherwise, place the wet drying agent in a waste container designated for this purpose. Used plant material and dry TLC plates can be discarded in the nonhazardous waste container.

 ### Procedure

In a small mortar grind 2 g of green or brightly colored fall leaves (do not use ivy or waxy leaves) with 10 mL of ethanol, pour off the ethanol (which serves to break up and dehydrate the plant cells), and grind the leaves successively with three 1-mL portions of dichloromethane that are decanted or withdrawn with a Pasteur pipette and placed in a test tube. The pigments of interest are extracted by the dichloromethane. Alternatively, place 0.5 g of carrot paste (baby food) or tomato paste in a test tube, stir and shake the paste with 3 mL of ethanol until the paste has a somewhat dry or fluffy appearance, remove the ethanol, and extract the dehydrated paste with three 1-mL portions of dichloromethane. Stir and shake the plant material with the solvent to extract as much of the pigments as possible.

Fill the tube containing the dichloromethane extract from leaves or vegetable paste with a saturated sodium chloride solution and shake the mixture. Remove the aqueous layer; add anhydrous calcium chloride pellets to the dichloromethane solution until the drying agent no longer clumps together. Shake the mixture with the drying agent for about 5 min and then withdraw the solvent with a Pasteur pipette and place it in a test tube. Add to the solvent a few pieces of Drierite to complete the drying process. Gently stir the mixture for about 5 min, transfer the solvent to a test tube, wash off the drying agent with more solvent, and then evaporate the combined dichloromethane solutions under a stream of nitrogen while warming the tube in your hand or in a beaker of warm water. Carry out this evaporation in the hood.

These hydrocarbons air oxidize with great rapidity.

Immediately cork the tube filled with nitrogen and then add 1 or 2 drops of dichloromethane to dissolve the pigments for TLC analysis. Carry out the analysis without delay by spotting the mixture on a TLC plate about 1 cm from the bottom and 8 mm from the edge. Make one spot concentrated by repeatedly touching the plate, but ensure that the spot is as small as possible—less than 1.0 mm in diameter. The other spot can be of lower concentration. Develop the plate with a 70:30 hexane-acetone mixture. With other plates try cyclohexane and toluene as eluents and also hexane-ethanol mixtures of various compositions. The container in which the chromatography is carried out should be lined with filter paper that is wet with the solvent so that the atmosphere in the container will be saturated with solvent vapor. After elution is completed, mark the solvent front with a pencil and outline the colored spots. Examine the plate under the UV light. Are any new spots seen? Report colors and R_f values for all of your spots and identify each as lycopene, carotene, chlorophyll, or xanthophyll.

Cleaning Up. The ethanol used for the dehydration of the plant material can be flushed down the drain along with the saturated sodium chloride solution. Recovered and unused dichloromethane should be placed in the halogenated organic waste container. The solvents used for TLC should be placed in the organic solvents container. If local regulations allow, evaporate any residual solvent from the drying agents in the hood and place the dried solid in the nonhazardous waste container. Otherwise, place the wet drying agent in a waste container designated for this purpose. Used plant material and dry TLC plates can be discarded in the nonhazardous waste container.

3. For Further Investigation

Many of the pigments in plants are made up of compounds called *anthocyanins*. Grind about 4 g of colored plant tissue (flower petals, blueberries, strawberries, cranberries, apple skins, red cabbage, red or purple grapes, etc.) and a small amount of alumina or fine sand with about 4 mL of a mixture of 99% methanol and 1% hydrochloric acid. Spot the extract on cellulose TLC plates and elute with a solvent mixture of 20% concentrated hydrochloric acid, 40% water, and 4% formic acid. Note the number and color of the spots. Look up the structures of the possible anthocyanins, of which many are glycosides of the aglycones delphinidin, peonidin, malvidin, and cyanidin, among others.

4. Colorless Compounds

You will now apply the thin-layer technique to a group of colorless compounds. The spots can be visualized under UV light if the plates have been coated with a fluorescent indicator; chromatograms can also be developed in a 4-oz bottle containing crystals of iodine. During development, spots will appear rapidly but remember that they also disappear rapidly. Therefore, outline each spot with a pencil immediately on withdrawal of the plate from the iodine chamber. Some suggested solvents are pure cyclohexane, pure toluene, toluene (3 mL) plus dichloromethane (1 mL), or toluene (4.5 mL) plus methanol (1/2 mL).

The compounds for trial are to be selected from the following list (all 1% solutions in toluene; those compounds with an asterisk are fluorescent under UV light):

1. Anthracene*
2. Cholesterol
3. 2,7-Dimethyl-3,5-octadiyne-2,7-diol
4. Diphenylacetylene
5. *trans,trans*-1,4-Diphenyl-1,3-butadiene*
6. *p*-Di-*t*-butylbenzene
7. 1,4-Di-*t*-butyl-2,5-dimethoxybenzene
8. *trans*-Stilbene
9. 1,2,3,4-Tetraphenylnaphthalene*
10. Tetraphenylthiophene
11. *p*-Terphenyl*
12. Triphenylmethanol
13. Triptycene

Except for tetraphenylthiophene, the structures for all of these molecules will be found in this book.

Make your own selections.

It is up to you to make selections and to plan your own experiments. Do as many as time permits. One plan would be to select a pair of compounds that are estimated to be separable and that have R_f values determinable with the same solvent. One can assume that a hydroxyl compound will travel less rapidly with a hydrocarbon solvent than a hydroxyl-free compound; you will therefore know what to expect if the solvent contains a hydroxylic component. An aliphatic solvent should carry along an aromatic compound with aliphatic substituents better than one without such groups. However, instead of relying on assumptions, you can do brief preliminary experiments on used plates on which previous spots are visible or outlined. If you spot a pair of compounds on such a plate and let the solvent rise about 3 cm from the starting line before development, you might be able to tell if a certain solvent is appropriate for a given sample. Alternatively, make some spots on a plate (new or used) and then touch each spot with a different solvent held in a capillary. In Figure 8.12a, the mixture did not move away from the point of origin; in Figure 8.12b, two concentric rings are seen between the origin and the solvent front. This is how a good solvent behaves. In Figure 8.12c, the mixture of compounds traveled with the solvent front.

Preliminary trials on used plates

■ **FIG. 8.12**
A fast method for determining
the correct solvent for TLC. See
the text for the procedure.

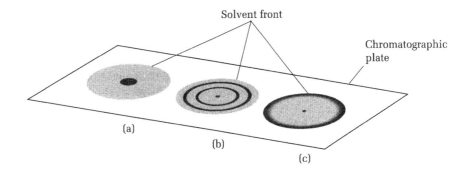

Once a solvent is chosen, run a complete chromatogram on the two compounds
on a fresh plate. If separation of the two compounds seems feasible, put two spots
of one compound on a plate, let the solvent evaporate, and put spots of the second
compound over the first ones. Run a chromatogram and see if you can detect two
spots in either lane (with colorless compounds, it is advisable not to attempt a three-
lane chromatogram until you have acquired considerable practice and skill).

Cleaning Up. Solvents should be placed in the organic solvents container, and dry,
used chromatographic plates can be discarded in the nonhazardous solid waste
container.

Discussion

If you have investigated hydroxylated compounds, you doubtless have found that
it is reasonably easy to separate a hydroxylated from a nonhydroxylated compound or a diol from a mono-ol. How, by a simple reaction followed by a thin-
layer chromatogram, could you separate cholesterol from triphenylmethanol?
Heating a sample of each with acetic anhydride and a trace of pyridine catalyst for
5 min on a steam bath, followed by chromatography, should do it. A first trial of a
new reaction leaves questions about what has happened and how much, if any,
starting material is present. A comparative chromatogram of the reaction mixture
with starting material may tell the story. How crude is a crude reaction product?
How many components are present? The thin-layer technique may give the answers to these questions and suggest how best to process the product. A preparative column chromatogram may afford a large number of fractions of eluent (say,
1 to 30). Some fractions probably contain nothing and should be discarded, while
others should be combined for evaporation and workup. How can you identify the
good and the useless fractions? Take a few used plates and put numbered circles
on clean places of each; spot samples of each of the fractions; and, without any
chromatography, develop the plates with iodine. Negative fractions for discard
will be obvious, and the pattern alone of positive fractions may allow you to infer
which fractions can be combined. Thin-layer chromatograms of the first and last
fractions of each suspected group would then show whether or not your inferences
are correct.

Fluorescence

Never look into a UV lamp.

Four of the compounds listed in Experiment 4 are fluorescent under UV light. These compounds give colorless spots that can be picked up on a chromatogram by fluorescence (after removal from the UV-absorbing glass bottle). If a UV light source is available, spot the four compounds on a used plate and observe the fluorescence.

Take this opportunity to examine a white shirt or handkerchief under UV light to see if it contains a brightener, that is, a fluorescent white dye or optical bleach. These substances are added to counteract the yellow color that repeated washing gives to cloth. Brighteners of the type of Calcofluor White MR, a sulfonated *trans*-stilbene derivative, are commonly used in detergent formulations for cotton; the substituted coumarin derivative is typical of brighteners used for nylon, acetate, and wool. Detergents normally contain 0.1%–0.2% of optical bleach. The amount of dye on a freshly laundered shirt is approximately 0.01% of the weight of the fabric.

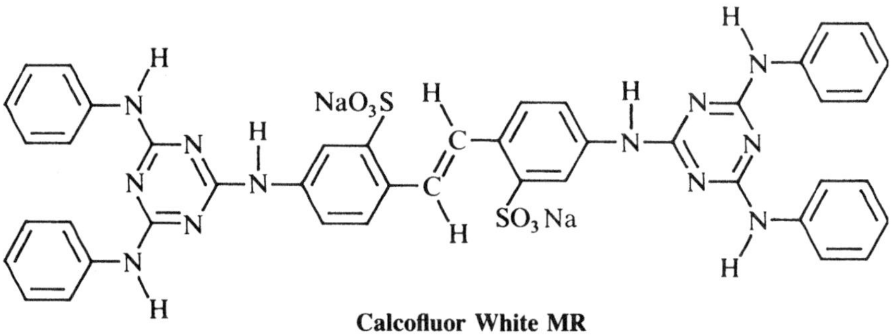

Calcofluor White MR

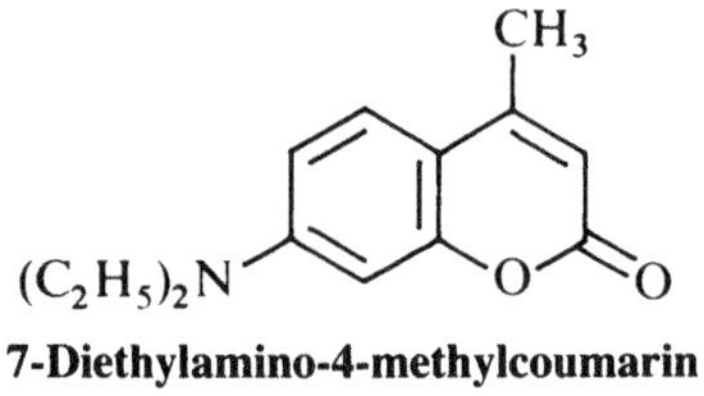

7-Diethylamino-4-methylcoumarin

Online Study Center
General Resources
Additional Experiments

QUESTIONS

1. Why might it be very difficult to visualize the separation of *cis*- and *trans*-2-butene by TLC?
2. What error is introduced into the determination of an R_f value if the top is left off the developing chamber?
3. What problem will ensue if the level of the developing liquid is higher than the applied spot in a TLC analysis?

4. In what order (from top to bottom) would you expect to find naphthalene, butyric acid, and phenyl acetate on a silica gel TLC plate developed with dichloromethane?

5. In carrying out an analysis of a mixture, what do you expect to see when the TLC plate has been allowed to remain in the developing chamber too long, so that the solvent front has reached the top of the plate?

6. Arrange the following in order of increasing R_f with TLC: acetic acid, acetaldehyde, 2-octanone, decane, and 1-butanol.

7. What will be the result of applying too much compound to a TLC plate?

8. Why is it necessary to run TLC in a closed container and to have the interior vapor saturated with the solvent?

9. What will be the appearance of a TLC plate if a solvent of too low polarity is used for the development? a solvent of too high polarity?

10. A TLC plate showed two spots with R_f values of 0.25 and 0.26. The plate was removed from the developing chamber, the residual solvent was allowed to evaporate from the plate, and then the plate was returned to the developing chamber. What would you expect to see after the second development was complete?

11. One of the analgesics has a chiral center. Which compound is it? One of the two enantiomers is far more effective at reducing pain than the other.

12. Using a ruler to measure distances, calculate the R_f value for substance B in Figure 8.10.

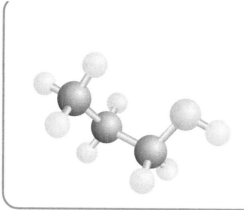

CHAPTER 9

Column Chromatography: Fluorenone, Cholesteryl Acetate, Acetylferrocene, and Plant Pigments

PRELAB EXERCISE: Compare column chromatography and thin-layer chromatography (TLC) with regard to the (1) quantity of material that can be separated, (2) time needed for the analysis, (3) solvent systems used, and (4) ability to separate compounds.

Column chromatography is one of the most useful methods for the separation and purification of both solids and liquids when carrying out microscale experiments. It becomes expensive and time consuming, however, when more than about 10 g of material must be purified. Column chromatography involves the same chromatographic principles as detailed for TLC in Chapter 8, so be sure that you understand those before doing the experiments in this chapter.

As discussed in Chapter 1, organic chemists obtain new compounds by synthesizing or isolating natural products that have been biosynthesized by microbes, plants, or animals. In most cases, initial reaction products or cell extracts are complex mixtures containing many substances. As you have seen, recrystallization, distillation, liquid/liquid extraction, and sublimation can be used to separate and purify a desired compound from these mixtures. However, these techniques are frequently not adequate for removing impurities that are closely related in structure. In these cases, column chromatography is often used. The broad applicability of this technique becomes obvious if you visit any organic chemistry research lab, where chromatography columns are commonplace.

Three of the five experiments in this chapter involve synthesis and may be your first experience in running an organic reaction. Experiments 1 and 2 involve the synthesis of a ketone. In Experiment 3 an ester of cholesterol is prepared. Experiment 4 demonstrates the separation of colored compounds. Experiment 5 involves the isolation and separation of natural products (plant pigments), which is analogous to Experiment 2 in Chapter 8 but on a larger scale.

The most common adsorbents for column chromatography—silica gel and alumina—are the same stationary phases as used in TLC. The sample is dissolved in a small quantity of solvent (the eluent) and applied to the top of the column. The eluent, instead of rising by capillary action up a thin layer, flows down through the column filled with the adsorbent. Just as in TLC, there is an equilibrium established between the solute adsorbed on the silica gel or alumina and the eluting solvent flowing down through the column, with the less strongly absorbed solutes moving ahead and eluting earlier.

Three mutual interactions must be considered in column chromatography: the activity of the stationary adsorbent phase, the polarity of the eluting mobile solvent phase, and the polarity of the compounds in the mixture being chromatographed.

Additional Principles of Column Chromatography

Adsorbents

A large number of adsorbents have been used for column chromatography, including cellulose, sugar, starch, and inorganic carbonates; but most separations employ alumina $[(Al_2O_3)_x]$ or silica gel $[(SiO_2)_x]$. Alumina comes in three forms: acidic, neutral, and basic. The neutral form of Brockmann activity grade II or III, 150 mesh, is most commonly employed. The surface area of this alumina is about 150 m^2/g. Alumina as purchased will usually be activity grade I, meaning that it will strongly adsorb solutes. It must be deactivated by adding water, shaking, and allowing the mixture to reach equilibrium over an hour or so. The amount of water needed to achieve certain activities is given in Table 9.1. The activity of the alumina on TLC plates is usually about III. Silica gel for column chromatography, 70–230 mesh, has a surface area of about 500 m^2/g and comes in only one activity.

TABLE 9.2 • Elutropic Series for Solvents
n-Pentane (least polar)
Petroleum ether
Cyclohexane
Hexanes
Carbon disulfide
t-Butyl methyl ether
Dichloromethane
Tetrahydrofuran
Dioxane
Ethyl acetate
2-Propanol
Ethanol
Methanol
Acetic acid (most polar)

TABLE 9.1 • Alumina Activity					
Brockmann activity grade	I	II	III	IV	V
Percent by weight of water	0	3	6	10	15

Solvents

Solvent systems for use as mobile phases in column chromatography can be determined from TLC, the scientific literature, or experimentally. Normally, a separation will begin with a nonpolar or low-polarity solvent, allowing the compounds to adsorb to the stationary phase; then the polarity of the solvent is *slowly* switched to desorb the compounds and allow them to move with the mobile phase. The polarity of the solvent should be changed gradually. A sudden change in solvent polarity will cause heat evolution as the alumina or silica gel adsorbs the new solvent. This will vaporize the solvent, causing channels to form in the column that severely reduce its separating power.

Several solvents are listed in Table 9.2, arranged in order of increasing polarity (elutropic series), with n-pentane being the least polar. The order shown

Petroleum ether: mostly
isometric pentanes

in the table reflects the ability of these solvents to dislodge a polar substance adsorbed onto either silica gel or alumina, with *n*-pentane having the lowest solvent power.

As a practical matter, the following sequence of solvents is recommended in an investigation of unknown mixtures: elute first with petroleum ether (pentanes); then hexanes; followed by hexanes containing 1%, 2%, 5%, 10%, 25%, and 50% ether; pure ether; ether and dichloromethane mixtures; followed by dichloromethane and methanol mixtures. Either diethyl ether or *t*-butyl methyl ether can be used, but *t*-butyl methyl ether is recommended. Solvents such as methanol and water are normally not used because they can destroy the integrity of the stationary phase by dissolving some of the silica gel. Some typical solvent combinations are hexanes-dichloromethane, hexanes–ethyl acetate, and hexanes-toluene. An experimentally determined ratio of these solvents can sufficiently separate most compounds.

Compound Mobility

The ease with which different classes of compounds elute from a column is indicated in Table 9.3. Molecules with nonpolar functional groups are least adsorbed and elute first, while more polar or hydrogen-bonding molecules are more strongly adsorbed and elute later. The order is similar to that of the eluting solvents—another application of "like dissolves like."

TABLE 9.3 • Elution Order for Solutes
Alkanes (first)
Alkenes
Dienes
Aromatic hydrocarbons
Ethers
Esters
Ketones
Aldehydes
Amines
Alcohols
Phenols
Acids (last)

Sample and Column Size

Chromatography columns can be as thin as a pencil for milligram quantities to as big as a barrel for the industrial-scale separation of kilogram quantities. A microscale column for the chromatography of about 50 mg of material is shown in Figure 9.1; columns with larger diameters, as shown in Figures 9.2 and 9.3, are used for macroscale procedures. The amount of alumina or silica gel used should generally weigh at least 30 times as much as the sample, and the column, when packed, should have a height at least 10 times the diameter. The density of silica gel is 0.4 g/mL, and the density of alumina is 0.9 g/mL, so the optimum size for any column can be calculated.

Packing the Column

Microscale Procedure

Before you pack the column, tare several Erlenmeyer flasks, small beakers, or 20-mL vials to use as receivers. Weigh each one carefully and mark it with a number on the etched circle.

Uniform packing of the chromatography column is critical to the success of this technique. Two acceptable methods for packing a column are dry packing and slurry packing, which normally achieve the best results. Assemble the column as depicted in Figure 9.1. To measure the amount of adsorbent, fill the column one-half to two-thirds full; then pour the powder out into a small beaker or flask. Clamp the column in a vertical position and close the valve. Always grasp the

Online Study Center
Photo: Column Chromatography;
Video: Column Chromatography

FIG. 9.1
A microscale chromatographic column.

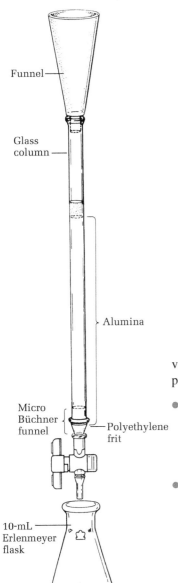

Funnel

Glass column

Alumina

Micro Büchner funnel

Polyethylene frit

10-mL Erlenmeyer flask

FIG. 9.2
A macroscale chromatographic column.

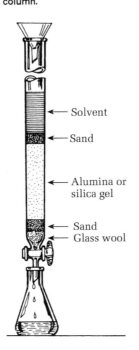

Solvent

Sand

Alumina or silica gel

Sand
Glass wool

FIG. 9.3
A chromatographic tube on ring stand.

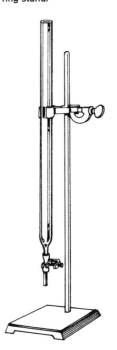

valve with one hand while turning it with the other. Fill the column with a non-polar solvent such as hexanes almost to the top.

● **Dry Packing Method.**   This is the simplest method for preparing a microscale column. Slowly add the powedered alumina or silica gel through the funnel while gently tapping the side of the column with a pencil. The solid should "float" to the bottom of the column. Try to pack the column as evenly as possible; cracks, air bubbles, and channels will lead to a poor separation.

● **Slurry Packing Method.**   To slurry pack a column, add about 8 mL of hexanes to the adsorbent in a flask or beaker, stir the mixture to eliminate air bubbles, and then (this is the hard part) swirl the mixture to get the adsorbent suspended in the solvent and immediately pour the entire slurry into the funnel. Open the valve, drain some solvent into the flask that had the adsorbent in it and finish transferring the slurry to the column. Place an empty flask under the column and allow the solvent to drain to about 5 mm above the top surface of the adsorbent. Tap the column with a pencil until the packing settles to a minimum height. Try to pack the column as evenly as possible; cracks, air bubbles, and channels will lead to a poor separation.

The slurry method normally gives the best column packing, but it is also the more difficult technique to master. Whether the dry packing or slurry packing method is chosen, the most important aspect of packing the column is creating an evenly distributed and packed stationary phase. The slurry method is often used for macroscale separations.

Once the column is loaded with solvent and adsorbent, place a flask under it, open the stopcock (use two hands for the microscale column), and allow the solvent level to drop to the *top* of the packing. Avoid allowing the solvent level to go below the stationary phase (known as letting the column "run dry") because this allows air bubbles and channel formation to occur, which leads to a poor separation.

Macroscale Procedure

Extinguish all flames; work in laboratory hood.

Before you pack the column, prepare several Erlenmeyer flasks to use as receivers by taring (weighing) each one carefully and marking each with a number on the etched circle.

Dry Packing Method

The column can be prepared using a 50-mL burette such as the one shown in Figure 9.2 or using the less expensive and equally satisfactory chromatographic tube shown in Figure 9.3, in which the flow of solvent is controlled by a screw pinchclamp. Weigh the required amount of silica gel (12.5 g in the first experiment), close the pinchclamp on the tube, and fill about half full with a 90:10 mixture of hexanes and ether. With a wooden dowel or glass rod, push a small plug of glass wool through the liquid to the bottom of the tube, dust in through a funnel enough sand to form a 1-cm layer over the glass wool, and level the surface by tapping the tube. Unclamp the tube. With your right hand grasp both the top of the tube and the funnel so that the whole assembly can be shaken to dislodge silica gel that may stick to the walls; with your left hand pour in the silica gel slowly (Fig. 9.4) while tapping the column with a rubber stopper fitted on the end of a pencil. If necessary, use a Pasteur pipette full of a 90:10 mixture of hexanes and ether to wash down any silica gel that adheres to the walls of the column above the liquid. When the silica gel has settled, add a little sand to provide a protective layer at the top. Open the pinchclamp, let the solvent level fall until it is just slightly above the upper layer of sand, and then stop the flow.

■ FIG. 9.4
A useful technique for filling a chromatographic tube with silica gel.

Slurry Packing Method

Alternatively, the silica gel can be added to the column (half filled with hexanes) by slurrying the silica gel with a 90:10 mixture of hexanes and ether in a beaker. The powder is stirred to suspend it in the solvent and immediately poured through a wide-mouth funnel into the chromatographic tube. Rap the column with a rubber stopper to cause the silica gel to settle and to remove bubbles. Add a protective layer of sand to the top. The column is now ready for use.

Cleaning Up. After use, the tube is conveniently emptied by pointing the open end into a beaker, opening the pinchclamp, and applying gentle air pressure to the tip. If the plug of glass wool remains in the tube after the alumina leaves, wet it with acetone and reapply air pressure. Allow the adsorbent to dry in the hood and then dispose of it in the nonhazardous waste container.

Adding the Sample

Dissolve the sample completely in a very minimum volume of dichloromethane (just a few drops) in a small flask or vial. Add to this solution 300 mg of the adsorbent, stir, and evaporate the solvent completely by heating the slurry *very gently* with *constant* stirring to avoid bumping. Remember that dichloromethane boils at 41°C. Pour this dry powder into the funnel of the chromatography column, wash it down onto the column with a few drops of hexane, and then tap the column to remove air bubbles from the layer of adsorbent-solute mixture just added. Open the valve and carefully add new solvent in such a manner that the top surface of the column is not disturbed. A thin layer of fine sand can be added to the column after the sample to avoid disturbance of the column surface when the solvent is being added. Run the solvent down near to the surface several times to apply the sample as a narrow band at the top of the column.

Eluting the Column

Fill the column with solvent, open the stopcock, and continue to add more solvent while collecting 1–3 mL fractions in small tared containers. Collecting small fractions is important to the success of your column separation. Fractions that are too small can always be pooled together; however, if the collected fractions are too large, you may get more than one compound in any particular fraction. If this occurs, the only way to attain separation is to redo the chromatography. Column chromatography is a lengthy process, so collecting large fractions is discouraged.

Isolating the Separated Compounds

If the mixture to be separated contains colored compounds, then monitoring the column is very simple. The colored bands will move down the column along with the solvent, and as they approach the end of the column, you can collect the separated colors in individual containers. However, most organic molecules are colorless. In this case, the separation must be monitored by TLC. Spot each fraction on a TLC plate (Fig. 9.5). Four or five fractions can be spotted on a single plate. Before you develop the plate, do a quick examination under UV light to see if there is any compound where you spotted. If not, you can spot the next fraction in that location. Note which fraction is in which lane. Develop the plate and use the observed spot(s) to determine which compound is in each of the collected fractions. Spotting some of the starting material or the product (if available) on the TLC plate as a standard will help in the identification.

The colors of the fractions or the results from analyzing the fractions by TLC will indicate which fractions contain the compound(s) you are interested in isolating. Combine fractions containing the same compound and evaporate the

■ FIG. 9.5
Spot each fraction on a TLC plate. Examine under UV light to see which fractions contain the compound.

solvent. Recrystallization may be used to further purify a solid product. However, on a milligram scale, there is usually not enough material to do this.

Other Types of Chromatography

High-performance liquid chromatography

Reverse-phase chromatography

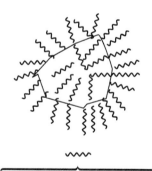

C_{18} **reverse-phase packing**

Chiral chromatography

Relying only on gravity, liquid flow through a column can be quite slow, especially if the column is tightly packed. One method to speed up the process is *flash chromatography*. This method uses a pressure of about 10 psi of air or nitrogen on top of the column to force the mobile phase through the column. Normally, doing this would give a poorer separation. However, it has been found that with a finer mesh of alumina or silica gel, flash chromatography can increase the speed without lowering the quality of the separation. Go to this book's website for an illustrated set of instructions for packing and using a flash chromatography column.

High-performance liquid chromatography (HPLC) is a high-tech version of column chromatography, which is capable of separating complex mixtures with dozens of components. A high-pressure pump forces solvent at pressures up to 10,000 psi through a stainless steel tube packed tightly with extremely small adsorbent particles. The eluent flows from the column to a detector, such as a tiny UV absorbance cell or a mass spectrometer that is able to detect extremely small amounts of separated components, as little as a picogram (10^{-12} g).

HPLC is used extensively in biochemistry to separate cellular components such as proteins, lipids, and nucleic acids. Mixtures of these types of compounds can be dissolved only in a predominantly aqueous mobile phase such as methanol-water or acetonitrile-water, and normal silica gel or alumina stationary phases do not work well with high concentrations of water. Rather than polar stationary phases, highly nonpolar ones called *reverse-phase packings* are used. These are manufactured by bonding lots of hydrocarbon molecules to the surfaces of silica gel particles, which convert the particles into highly nonpolar grease balls. With this packing, the order of elution is the reverse of that observed for a normal silica gel phase. On a reverse-phase column, the more nonpolar compounds will adhere to the nonpolar stationary phase more strongly, and the polar compounds will elute first.

Is it possible to separate two enantiomers (optical isomers), both of which have the same intermolecular attractive forces? Chiral stationary phases can be used to separate enantiomers. Giving the stationary phase an asymmetry or handedness allows one enantiomer to be specifically retained on the column. Such columns are quite expensive and are limited to a particular type of separation, but they have led to great achievements in separation science. This separation technique is of great importance in the pharmaceutical industry because the U.S. Food and Drug Administration (FDA) specifies the amounts of impurities, including enantiomers, that can be found in drugs. For example, Thalidomide, a drug prescribed as a sedative and an antidepressant in the 1960s, was found to be a potent teratogen that caused birth defects when pregnant women took the drug. It was quickly pulled from the market. Thalidomide has two enantiomers, and further research demonstrated that only one of the enantiomers caused the birth defects.

EXPERIMENTS

1. Air Oxidation of Fluorene to Fluorenone

The 9-position of fluorene is unusually reactive for a hydrocarbon. The protons on this carbon atom are acidic by virtue of being doubly benzylic, and, consequently, this carbon can be oxidized by several reagents, including oxygen from the air. Here the oxidation is carried out using a phase-transfer catalyst called methyltricapryl ammonium chloride, commonly known as Stark's catalyst or Aliquat 336. In the presence of this catalyst, the hydroxide is carried into the organic layer, where it can remove one of the acidic fluorene protons, creating a carbanion that can react with oxygen in the air. An intermediate hydroperoxide is formed and loses water to give the ketone.

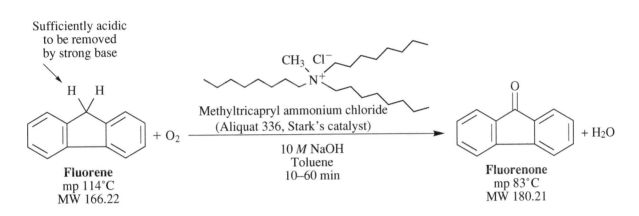

Microscale Procedure

> **IN THIS EXPERIMENT** fluorene is oxidized to fluorenone by oxygen from the air in a base-catalyzed reaction involving a phase-transfer catalyst. The reaction is monitored by TLC and stopped when about a 50:50 mixture of product and starting material has formed. The toluene layer is separated and washed with 5% hydrochloric acid and saturated sodium chloride. After drying over calcium chloride, the toluene is removed and the fluorene-fluorenone mixture is separated by microscale column chromatography on alumina, using hexanes as the mobile phase. Fractions are analyzed by TLC; common fractions are pooled and weighed after solvent removal to give pure fluorene and fluorenone.

CAUTION: Sodium hydroxide is a strong base and is very corrosive. Wash your hands immediately if contact occurs.

This experiment requires a 50-mL to 125-mL separatory funnel. To a 25-mL Erlenmeyer flask clamped to a ring stand above a magnetic stirrer, add 5 mL of 10 M NaOH and 70 mg of fluorene while stirring with a $\frac{1}{2}$-inch stir bar. Add 5 mL of toluene and stir until all of the solid has dissolved. (Observe the color and identify which layer is organic and which is aqueous.) Add approximately 3 drops of Stark's catalyst (Aliquat 336) to the solution. Stir vigorously but without splashing the solution. The reaction can take anywhere from 10 min to 30 min. Follow the rate of the reaction by TLC. Develop the TLC plate by using 20% dichloromethane in hexanes and use a UV lamp to visualize the products. Also spot the plate with a 1% fluorene standard. When approximately half of the fluorene has been converted to fluorenone (as evidenced by the fact that the product spot is about the same size and intensity), pour the reaction mixture into a separatory funnel, rinsing the beaker with an additional 3 mL of toluene, which is also added to the separatory funnel.

Separate the organic layer from the aqueous layer. Wash the organic layer in the separatory funnel with 1.5 *M* hydrochloric acid (three separate times with 5 mL each time) and then saturated sodium chloride (three separate times with 5 mL each time). After each washing, drain the aqueous layer from the separatory funnel into a waste beaker. Dry the remaining toluene layer over anhydrous calcium chloride pellets in a 125-mL Erlenmeyer flask. Add the pellets until they no longer clump together (3–5 scoops). Allow the product to dry for 5–10 min before decanting the toluene from the solid calcium chloride and transferring the toluene to a 100-mL tared beaker. Wash the solid calcium chloride with 3 mL of toluene, adding this to the main portion of toluene to complete the transfer of product. The crude mixture of fluorene and fluorenone will be separated by column chromatography in the next lab period. The beaker containing the toluene extract can be allowed to stand in your hood (labeled properly) until the next lab period. The toluene will evaporate in the interim. Alternatively, if you have time, reduce the volume of solvent (~1 mL) by heating gently on a sand bath. Insert a boiling stick if you do this.

Cleaning Up. Carefully dilute the strongly basic aqueous reaction layer 10- to 20-fold with water in a large beaker, add the hydrochloric acid washes, and then neutralize by slowly adding additional hydrochloric acid. Flush this and the sodium chloride washes down the drain with lots of water. If local regulations allow, evaporate any residual solvent from the drying agents in the hood and place the dried solid in the nonhazardous waste container. Otherwise, place the wet drying agent in a waste container designated for this purpose.

Column Chromatography of the Fluorene-Fluorenone Mixture

Set up a sand bath at a heating setting of 50 before you do anything else.

Figure 9.1 on page 197 shows the typical setup you will use for the chromatographic separation of fluorene and fluorenone, and a general outline of the procedure is given on page 196. It is essential to have at least 10 clean 10-mL Erlenmeyer flasks, reaction tubes, small beakers, test tubes, or vials available to collect the chromatography fractions as they elute. At least two of these should be weighed, with their tare weight recorded. Once you have this done and the column is assembled, you can pack the column with alumina according to the following instructions.

Packing the Column

Before you assemble the column, check the small plug that fits into the bottom of the column to make sure that it has a small fritted disk inside. Next, make sure that the plug fits snugly into the glass column and is not easy to pull out. If it is loose, get a new bottom plug from the stockroom. Finish assembling the chromatography column as depicted in Figure 9.1. Be sure to clamp the column securely and vertically.

Grasp the valve with one hand and turn it with the other. Close the valve and fill the column with hexanes to the bottom of the plastic funnel. Weigh out approximately 4.5 g of activity grade III alumina in a small beaker and slowly sprinkle the dry alumina into the hexanes in the column while you tap the column with a pen or pencil. It may be necessary to drain off some of the solvent to keep it from flowing over the top. This amount of alumina should fill the column to a height of about 10 cm. It is extremely important *not* to let the column run dry at any time. This will allow air to enter the column, which will result in uneven bands and poor separation.

After all of the alumina has been added to the column, open the stopcock and continue to tap the column as you allow the solvent to drain slowly until the solvent just barely covers the surface of the alumina, collecting the solvent in an Erlenmeyer flask.

Adding the Sample

It is important to use a minimum amount of solvent when dissolving the sample. If too much is used, poor separation will result.

The solvent is drained just to the surface of the alumina, which should be perfectly flat. Dissolve the crude mixture of fluorene and fluorenone in 10 drops of dichloromethane and 10 drops of toluene and add this with a pipette to the surface of the alumina. Be sure to add the sample as a solution; should any sample crystallize, add 1 more drop of dichloromethane. (This is done so that the sample to be added to the column is in the most concentrated solution possible.) Drain some liquid from the column until the dichloromethane-toluene solution just barely covers the surface of the alumina. Then add a few drops of hexanes and drain out some solvent until the liquid just covers the alumina. Repeat until the sample is seen as a narrow band at the top of the column. Carefully add a 4–5 mm layer of sand and then fill the column with hexanes.

Collect 3-mL fractions in a combination of small vials, 10-mL Erlenmeyer flasks, 13 × 100 mm test tubes, vials, and small beakers. You will probably collect close to ten 3-mL fractions. While the chromatography is running, you will be determining the amount of fluorene or fluorenone in each fraction by TLC. Once you determine which fractions contain which compound, you will combine the "like" fractions and evaporate the solvent.

After you collect each 3-mL fraction in a flask, apply it to a TLC plate by spotting it 2 or 3 times on the plate in the same location. Four or five fractions can be applied and analyzed at the same time using one TLC plate. Allow the solvent to completely evaporate from the spot and examine the plate under a UV lamp to determine if there is any material present, as evidenced by a dark blue spot. The TLC plates can be developed using 20% dichloromethane in hexanes. You will probably collect a few fractions that contain little or no material; these fractions are likely to be the first or the middle of the series.

After spotting and developing each fraction on the TLC plate, combine like fractions and immediately start to boil off the solvent on the sand bath *in the hood* using a boiling stick broken in half. Tilt the flasks and vials on their side as much as possible to allow the heavy vapors to escape. As soon as all the liquid seems to have boiled off, set the flask *on its side* on the bench top in the hood to allow the last traces of heavy solvent vapors to escape. After the flask has cooled to room temperature, if that fraction contains any material, crystals may appear. If an oily, gooey residue is present, you may have to scratch it with a glass stirring rod to induce crystallization. With good organizational effort, you can do the TLC analysis and evaporate off the solvent at about the same rate at which you collect fractions; thus you can follow the progress of the chromatography simply by noting the amount of material in each flask, vial, or test tube. If, after solvent removal and cooling, the flasks are perfectly clean on careful inspection, they can be used to collect subsequent fractions.

If the yellow band has not moved one third of the way down the column after two fractions have been collected, you can speed up the elution by replacing the hexanes solvent at the top of the column with 20% dichloromethane in hexanes. Once the first component has completely eluted, you can speed up the elution of the second component by using 50% dichloromethane in hexanes. Decide when the product has been completely eluted from the column by using visual cues and TLC. Using a few drops of dichloromethane, wash all the fractions that contain fluorene as determined by TLC analysis into a tared container. Do the same for the fluorenone fractions. Evaporate the dichloromethane and obtain dry weights for the product and the recovered fluorene.

Mark all compound spots on all TLC plates with a pencil. Tape your developed TLC plates in your notebook with wide transparent tape. Calculate the theoretical yield and the percent yield of your pure fluorenone both with and without taking into account the amount of fluorene starting material recovered. Calculate the percent recovery of fluorene.

Cleaning Up. When you are done with the column, pour out the excess solvent into the proper waste container, pull out the bottom, and leave the wet column propped in a beaker in your desk hood. The column will dry out by the next lab, and the dry, used alumina can then be easily emptied out into a waste bin.

2. Chromium(VI) Oxidation of Fluorene to Fluorenone

Chromium(VI) oxidations are less favored today because of environmental concerns based on chromium's toxicity.

The 9-position of fluorene is unusually reactive for a hydrocarbon. The protons on this carbon atom are acidic by virtue of being doubly benzylic, and, consequently, this carbon can be oxidized by several reagents, including elemental oxygen. In this experiment, the very powerful and versatile oxidizing agent chromium(VI), in the form of chromium trioxide, is used to carry out the oxidation. Chromium(VI) in a variety of other forms is used for about a dozen oxidation reactions in this text. The *dust* of chromium(VI) salts is reported to be a carcinogen, so avoid breathing it.

Fluorene
mp 114°C
MW 166.22

Fluorenone
mp 83°C
MW 180.21

Microscale Procedure

> **IN THIS EXPERIMENT** the hydrocarbon fluorene is oxidized to the ketone fluorenone by sodium dichromate in acetic acid with heating. The mixture is diluted with water, and the crude product is isolated by filtration; then—in a standard procedure—it is dissolved in ether. The ether is dried and evaporated to give a mixture of fluorene and fluorenone, which is separated by column chromatography.

CAUTION: Sodium dichromate is toxic. The dust is corrosive to nasal passages and skin and is a suspected carcinogen. Hot acetic acid is very corrosive to skin. Handle sodium dichromate and acetic acid in the hood; always wear gloves.

Online Study Center

Photo: Column Chromatography; Video: Column Chromatography

The sample can also be applied to the column using the technique described on page 196.

In a reaction tube dissolve 50 mg of fluorene in 0.25 mL of acetic acid by heating and add this hot solution to a solution of 0.15 g of sodium dichromate dihydrate in 0.5 mL of acetic acid. Heat the reaction mixture to 80°C for 15 min in a hot water bath; then cool it and add 1.5 mL of water. Stir the mixture for 2 min; then filter it on a Hirsch funnel. Wash the product well with water and press out as much water as possible. Return the product to the reaction tube, add 2 mL of ether, and add anhydrous calcium chloride pellets until it no longer clumps together. Cork and shake the tube and allow the product to dry for 5–10 min before evaporating the ether in another tared reaction tube. Use ether to wash off the drying agent and to complete the transfer of product. Use this ether solution to spot a TLC plate. This crude mixture of fluorene and fluorenone will be separated by column chromatography.

Column Chromatography of the Fluorene-Fluorenone Mixture

Prepare a microscale chromatographic column exactly as described at the beginning of this chapter (*see* Fig. 9.1 on page 197). Use alumina as the adsorbent. Dissolve the crude mixture of fluorene and fluorenone in a mixture of 10 drops of dichloromethane and 10 drops of toluene and add this to the surface of the alumina. Be sure to add the sample as a solution; should any sample crystallize, add 1 more drop of dichloromethane. Run the hexanes down to the surface of the alumina, add a few drops more of hexanes, and repeat the process until the sample is seen as a narrow band at the top of the column. Carefully add a 3-mm layer of sand, fill the column with hexanes, and collect 5-mL fractions in tared 10-mL Erlenmeyer flasks. Sample each flask for TLC (*see* Fig. 9.5 on page 199) and evaporate each to dryness. Final drying can be done under vacuum using the technique shown in Figure 9.6. You are to decide when all of the product has been eluted from the column. The TLC plates can be developed using 20% dichloromethane in hexanes. Combine fractions that are identical and determine the melting points of the two substances.

Cleaning Up. The filtrate probably contains unreacted dichromate. To destroy it, add 3 *M* sulfuric acid until the pH is 1; then complete the reduction by adding solid sodium thiosulfate until the solution becomes cloudy and blue colored. Neutralize with sodium carbonate and then filter the flocculent precipitate of $Cr(OH)_3$ through Celite in a Büchner funnel. If regulations allow, the filtrate can be diluted with water and flushed down the drain; otherwise, it should be disposed of in a designated waste container. The precipitate and Celite should be placed in the heavy metal hazardous waste container.

Macroscale Procedure

In a 250-mL Erlenmeyer flask dissolve 5.0 g of practical grade fluorene in 25 mL of acetic acid by heating on a steam bath with occasional swirling. In a 125-mL Erlenmeyer flask dissolve 15 g of sodium dichromate dihydrate in 50 mL of acetic acid by swirling and heating on a hot plate. Adjust the temperature of the dichromate solution to 80°C, transfer the thermometer, and adjust the fluorene–acetic acid solution to 80°C; then, *in the hood*, pour in the dichromate solution. Note the time and the temperature of the solution and heat on a steam bath for 30 min. Observe the maximum and final temperature; then cool the solution and add 150 mL of water. Swirl the mixture for a full 2 min to coagulate the product and promote rapid filtration; collect the yellow solid in an 8.5-cm Büchner funnel using vacuum filtration (if filtration is slow, empty the funnel and flask into a beaker and stir vigorously for a few minutes). Wash the filter cake well with water and then suck the filter cake as dry as possible. Either let the product dry overnight or dry it quickly as follows: Put the moist solid into a 50-mL Erlenmeyer flask, add ether (20 mL) and swirl to dissolve, and add anhydrous calcium chloride (10 g) to scavenge the water. Decant the ethereal solution through a cone of anhydrous calcium chloride in a funnel into a 125-mL Erlenmeyer flask; then rinse the flask and funnel with ether. Evaporate on a steam bath under an aspirator, heat until all the ether is removed, and pour the hot oil into a 50-mL beaker to cool and solidify. Scrape out the yellow solid. The yield should be about 4.0 g.

Separation of Fluorene and Fluorenone

Prepare a column of 12.5 g of alumina, run out excess solvent, and pour onto the column a solution of 0.5 g of fluorene-fluorenone mixture. Elute at first with hexanes and use tared 50-mL flasks as receivers. The yellow color of fluorenone provides one index of the course of the fractionation, and the appearance of solid around the delivery tip provides another. Frequently wash the solid on the tip into the receiver with ether. When you believe that one component has been eluted completely, change to another receiver until you judge that the second component is beginning to appear. Then, when you are sure the second component is being eluted, change to a 1:1 hexanes and ether mixture and continue until the column is exhausted. It is possible to collect practically all of the two components in the two receiving flasks, with only a negligible intermediate fraction. After evaporation of the solvent, evacuate each flask under vacuum (Fig. 9.7) and determine the weight and melting point of the products. A convenient method for evaporating fractions is to use a rotary evaporator (Fig. 9.8).

Cleaning Up. All organic material from this experiment can go in the organic solvents container. If local regulations allow, evaporate any residual solvent from

Handle dichromate and acetic acid in laboratory hood; always wear gloves.

One-half hour unattended heating

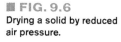
■ **FIG. 9.6**
Drying a solid by reduced air pressure.

■ **FIG. 9.7**

An aspirator tube in use. A boiling stick may be necessary to promote even boiling.

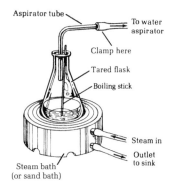

Aspirator tube

To water aspirator

Clamp here

Tared flask

Boiling stick

Steam in

Outlet to sink

Steam bath (or sand bath)

■ **FIG. 9.8**

A rotary evaporator. The rate of evaporation with this apparatus is very fast due to the thin film of liquid spread over the entire inner surface of the rotating flask, which is heated under vacuum. Foaming and bumping are also greatly reduced.

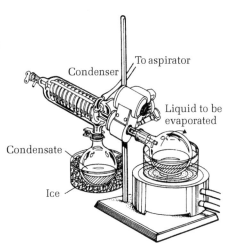

To aspirator

Condenser

Liquid to be evaporated

Condensate

Ice

the alumina in the hood and place the dried solid in the nonhazardous waste container. Otherwise, place the wet drying agent in a waste container designated for this purpose.

 ## 3. Acetylation of Cholesterol

Cholesterol is a solid alcohol; the average human body contains about 200 g distributed in brain, spinal cord, and nerve tissue and occasionally clogs the arteries and the gall bladder.

In the following experiment, cholesterol is dissolved in acetic acid and allowed to react with acetic anhydride to form the ester cholesteryl acetate. The reaction does not take place rapidly and consequently does not go to completion under the conditions of this experiment. Thus, when the reaction is over, both unreacted cholesterol and the product, cholesteryl acetate, are present. Separating these by fractional crystallization would be extremely difficult; but because they differ in polarity (the hydroxyl group of cholesterol is the more strongly adsorbed on alumina), they are easily separated by column chromatography. Both molecules are colorless and hence cannot be detected visually. Each fraction should be sampled by TLC. In that way not only the presence but also the purity of each fraction can be assessed. It is also possible to put 1 drop of each fraction on a watch glass and evaporate it to see if the fraction contains product. Solid will also appear on the tip of the column while a compound is being eluted.

 Microscale Procedure

IN THIS EXPERIMENT cholesterol is refluxed with acetic acid and acetic anhydride. In the standard procedure the mixture is diluted with water and extracted with ether, and the ether is dried and evaporated. The resulting mixture of cholesterol and cholesteryl acetate is separated by column chromatography on silica gel, eluting with hexanes and then hexanes and ether mixtures. Cholesteryl acetate comes off first, followed by cholesterol.

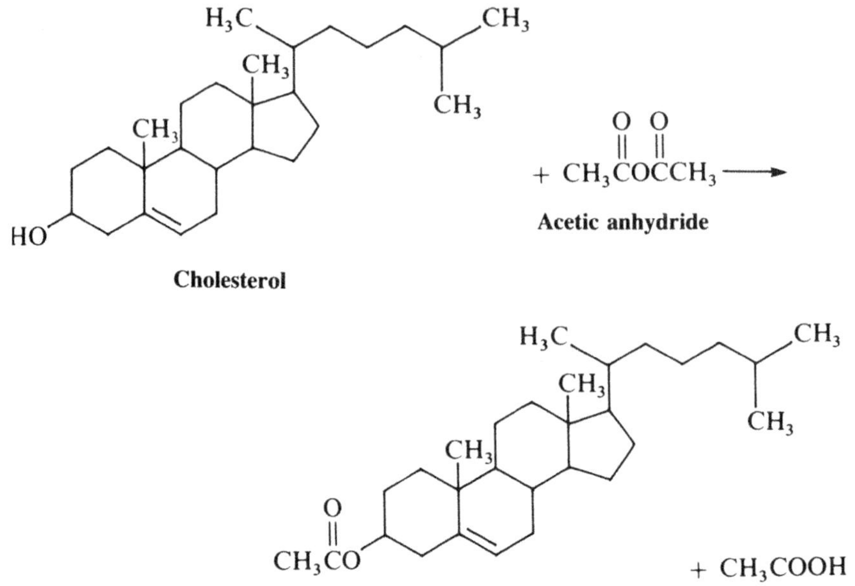

Cholesterol   **Acetic anhydride**

Cholesteryl acetate   + CH$_3$COOH

In a reaction tube, add 0.5 mL of acetic acid to 50 mg of cholesterol. The initial thin slurry may set into a stiff paste of the molecular complex consisting of one molecule of cholesterol and one molecule of acetic acid. Add 0.10 mL of acetic anhydride and a boiling chip and gently reflux the reaction mixture on a hot sand bath for no more than 30 min (*see* Fig. 1.1 on page 2).

While the reaction is taking place, prepare the microscale chromatography column as described previously, using silica gel as the adsorbent. Refer to Figure 9.1 on page 197 and the associated procedure earlier in the chapter. Cool the mixture, add 2 mL of water, and extract the product with three 2-mL portions of ether that are placed in the 15-mL centrifuge tube. Wash the ether extracts in the tube with two 2-mL portions of water and one 2.5-mL portion of 3 *M* sodium hydroxide (these three washes remove the acetic acid) and dry the ether by shaking it with 2.5 mL of saturated sodium chloride solution. Then complete the drying by adding enough anhydrous calcium chloride pellets to the solution so that the drying agent does not clump together.

Shake the ether solution with the drying agent for 10 min; then transfer it in portions to a tared reaction tube and evaporate to dryness. Use 1 drop of this ether solution to spot a TLC plate for later analysis. If the crude material weighs more than the theoretical weight, you will know that it is not dry or that it contains acetic acid, which can be detected by its odor. Dissolve this crude cholesteryl acetate in the minimum quantity of ether and apply it to the top of the chromatography column.

To prevent the solution from dribbling from the pipette, use a pipette pump to make the transfer. It also could be applied as a dry powder adsorbed on silica gel, as in Experiment 1. Elute the column with hexanes, collecting two 5-mL fractions in tared 10-mL Erlenmeyer flasks or other suitable containers. Add a boiling stick to each flask and evaporate the solvent under an aspirator tube on a steam bath or a sand bath (*see* Fig. 9.7). If the flask appears empty, it can be used to collect later fractions. Lower the solvent layer to the top of the sand and elute with 15 mL of a 70:30 mixture of hexanes and ether, collecting five 3-mL fractions. Evaporate the solvent under an aspirator tube (Fig. 9.7). The last traces of solvent can be removed using reduced pressure, as shown in Figure 9.6 on page 206. Follow the 70:30 mixture with 10 mL of a 50:50 mixture of hexanes and ether, collecting four 2-mL fractions. Save any flask that has any visible residue. Analyze the original mixture and each fraction by TLC on silica gel plates using a 1:1 mixture of hexanes and ether to develop the plates and either UV light or iodine vapor to visualize the spots.

Cholesteryl acetate (mp 115°C) and cholesterol (mp 149°C) should appear, respectively, in early and late fractions with a few empty fractions (no residue) in between. If so, combine consecutive fractions of early and late material and determine the weights and melting points. Calculate the percentage of the acetylated material compared to the total recovered and calculate the percentage yield from cholesterol.

Cleaning Up. After neutralization, acetic acid, the aqueous layers, and the saturated sodium chloride layers from the extraction can be flushed down the drain with water. Ether, hexanes, and TLC solvents should be placed in the organic solvents container. If local regulations allow, evaporate any residual solvent from the drying agents and the chromatography packing in the hood and place the dried solid in the nonhazardous waste container. Otherwise, place the wet materials in waste containers designated for this purpose.

Macroscale Procedure

Carry out procedure in laboratory hood.

Cover 0.5 g of cholesterol with 5 mL of acetic acid in a small Erlenmeyer flask; swirl and note that the initially thin slurry soon sets to a stiff paste of the molecular compound $C_{27}H_{45}OH \cdot CH_3CO_2H$. Add 1 mL of acetic anhydride and heat the mixture on a steam bath for any convenient period of time from 15 min to 1 h; record the actual heating period. While the reaction takes place, prepare the chromatographic column. Cool the reaction mixture, add 20 mL of water, and extract with two 25-mL portions of ether. Wash the combined ethereal extracts twice with 15-mL portions of water and once with 25 mL of 3 M sodium hydroxide and dry by shaking the ether extracts with 25 mL of saturated sodium chloride solution; then dry the ether over anhydrous calcium chloride pellets for 10 min in an Erlenmeyer flask, filter, and evaporate the ether. Save a few crystals of this material for TLC analysis. Dissolve the residue in 3–4 mL of ether, transfer the solution with a Pasteur pipette onto a column of 12.5 g of silica gel, and rinse the flask with another small portion of ether.[1] In order to apply the ether solution to the top of the sand and avoid having it coat the interior of the column, pipette the solution down a 6-mm-diameter glass tube that is resting on the top of the sand. Label a series of

---

1. Ideally, the material to be adsorbed is dissolved in hexanes (ligroin), the solvent of lowest elutant power. The present mixture is not soluble enough in hexanes, and so ether is used, but the volume is kept to a minimum.

■ **FIG. 9.9**
A bubbler for adding solvent automatically.

50-mL Erlenmeyer flasks as fractions 1 to 10. Open the pinchclamp, run the eluant solution into a 50-mL Erlenmeyer flask, and as soon as the solvent in the column has fallen to the level of the upper layer of sand, fill the column with a part of a measured 125 mL of a 70:30 mixture of hexanes and ether. When about 25 mL of eluant has collected in the flask (fraction 1), change to a fresh flask; add a boiling stone to the first flask and evaporate the solution to dryness on a steam bath under an aspirator tube (*see* Fig. 9.7 on page 207). Evacuation using an aspirator helps to remove last traces of ligroin (*see* Figure 9.6 on page 206). If fraction 1 is negative (no residue), use the flask for collecting further fractions. Continue adding the hexanes and ether mixture until the 125-mL portion is exhausted; then use 100 mL of a 1:1 hexanes and ether mixture. A convenient bubbler (Fig. 9.9) made from a 125-mL Erlenmeyer flask, a short piece of 10-mm-diameter glass tubing, and a cork will automatically add solvent. A separatory funnel with a stopper and a partially open stopcock serves the same purpose. Collect and evaporate successive 25-mL fractions of eluant. Save any flask that has any visible solid residue. The ideal method for the removal of solvents involves the use of a rotary evaporator (*see* Fig. 9.8 on page 207). Analyze the original mixture and each fraction by TLC on silica gel plates using a 1:1 mixture of hexanes-ether to develop the plates and either UV light or iodine vapor to visualize the spots.

Cholesteryl acetate (mp 115°C) and cholesterol (mp 149°C) should appear, respectively, in early and late fractions with a few empty fractions (no residue) in between. If so, combine consecutive fractions of early and of late material and determine the weights and melting points. Calculate the percentage of the acetylated material compared to the total recovered and compare your result with those of others in your class employing different reaction periods.

Cleaning Up. After neutralization, acetic acid, the aqueous layers, and the saturated sodium chloride layers from the extraction can be flushed down the drain with water. Ether, hexanes, and TLC solvents should be placed in the organic solvents container. Hexanes and other from the chromatography go into the organic solvents container. If local regulations allow, evaporate any residual solvent from the drying agents and silica gel in the hood and place the dried solid in the nonhazardous waste container. Otherwise, place the wet drying agent and silica gel in waste containers designated for this purpose.

4. Chromatography of a Mixture of Ferrocene and Acetylferrocene

Online Study Center

Photo: Column Chromatography;
Video: Column Chromatography

CAUTION: Acetylferrocene is toxic.

> **IN THIS EXPERIMENT** a mixture of two compounds is separated by chromatography on alumina. The first compound to come down the column is ferrocene (a yellow band). The solvent polarity is changed so that acetylferrocene is eluted as an orange band. The solvents are evaporated from the collection flasks and the compounds recrystallized. Both of these compounds are colored (*see* Chapter 49 for their preparation), so it is easy to follow the progress of the chromatographic separation.

Prepare the microscale alumina column exactly as described at the beginning of the chapter. Then add a dry slurry of 90 mg of a 50:50 mixture of acetylferrocene

(a)

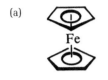

Ferrocene
MW 186.04
mp 172–174°C

(b)

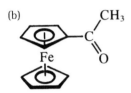

Acetylferrocene
MW 228.08
mp 85–86°C

■ **FIG. 9.10**
Evaporation of a low-boiling liquid under vacuum. Heat is supplied by the hand, the contents of the flask are swirled, and the vacuum is controlled with the thumb.

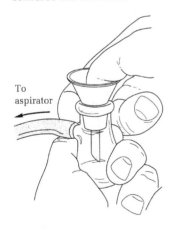

To aspirator

Isoprene

and ferrocene that has been adsorbed onto 300 mg of alumina, following the procedure for preparing and adding the sample given at the beginning of the chapter.

Carefully add hexanes to the column, open the valve (use both hands), and elute the two compounds. The first to be eluted, ferrocene, will be seen as a yellow band. Collect this in a 10-mL flask. Any crystalline material seen at the tip of the valve should be washed into the flask with a drop or two of ether. Without allowing the column to run dry, add a 50:50 mixture of hexanes and ether, and elute the acetylferrocene, which will be seen as an orange band. Collect it in a 10-mL flask. Spot a silica gel TLC plate with these two solutions. Evaporate the solvents from the two flasks and determine the weights of the residues. An easy way to evaporate the solvent is to place it in a tared 25-mL filter flask and heat the flask in your hand under vacuum while swirling the contents (Fig. 9.10).

Recrystallize the products from the minimum quantities of hot hexanes. Isolate the crystals, dry them, and determine their weights and melting points. Calculate the percent recovery of the crude and recrystallized products based on the 45 mg of each in the original mixture.

The TLC plate is eluted with a 30:1 mixture of toluene and absolute ethanol. Do you detect any contamination of one compound by the other?

Cleaning Up. If regulations allow, empty the chromatography column onto a piece of aluminum foil in the hood. After the solvent has evaporated, place the alumina and sand in the nonhazardous waste container. Otherwise, place the wet alumina and sand in a designated waste container. Evaporate the crystallization mother liquor to dryness and place the residue in the hazardous waste container.

5. Isolation of Lycopene and β-Carotene

Lycopene, the red pigment in tomatoes, is a C_{40}-carotenoid made up of eight five-carbon isoprene units. β-Carotene, the yellow pigment of the carrot, is an isomer of lycopene in which the double bonds at C_1—C_2 and C'_1—C'_2 are replaced by bonds extending from C_1 to C_6 and from C'_1 to C'_6 to form rings. The chromophore in each case is a system of 11 all-*trans* conjugated double bonds; the closing of the two rings renders β-carotene less highly pigmented than lycopene.

These colored hydrocarbons have been encountered in the TLC experiment (*see* Chapter 8). The isolation procedure described here affords sufficient carotene and lycopene to carry out analytical spectroscopy and some isomerization reactions. It might be of interest if some students isolate carotene from strained carrot baby food while others isolate lycopene from tomato paste.

Lycopene is responsible not only for the red color of tomatoes but also of red grapefruit and flamingos. If flamingos do not include foods containing lycopene in their diet, they will be white.

Until recently, lycopene was thought to have no utility until a study showed a lower incidence of prostate cancer among men who consumed such foods as spaghetti and pizza (but not tomato juice). It is theorized that the lycopene, which is insoluble in water, is dissolved in the fat of pasta sauce and pizza and thus absorbed through the intestine. Carotene does not have the same anticancer effect.

Lycopene (C₄₀H₅₆)
MW 536.85
mp 173°C, λ_{max}^{hexane} 475 nm

β-Carotene (C₄₀H₅₆)
mp 183°C, λ_{max}^{hexane} 451 nm

Lycopene is the predominant carotenoid in blood plasma and prostate tissue. It is not converted into vitamin A as carotene is, but it is a powerful antioxidant and is an efficient scavenger of singlet oxygen.

Carotenoids are highly sensitive to photochemical air oxidation; therefore, protect solutions and solids from undue exposure to light and heat and work as rapidly as possible. Do not heat solutions when evaporating solvents and, if possible, flush apparatus with nitrogen to exclude oxygen. Research workers isolate these compounds in dimly lit rooms and/or wrap all containers and chromatographic columns in aluminum foil and carry out extractions and crystallizations using solvents that have been deoxygenated.

Dehydration and Extraction of Tomato or Carrot Paste

IN THIS EXPERIMENT a vegetable paste is stirred with acetone to remove water, but not the coloring matter, from the paste. The mixture is filtered, the yellow filtrate discarded, and the material on the filter squeezed as dry as possible. This solid is then extracted three times with dichloromethane; the solution is dried over calcium chloride and evaporated at room temperature under vacuum to leave the crude carotenoids.

Add 5 g of tomato or carrot paste to a 15-mL centrifuge tube or 25 × 150-mm test tube; then add about 7 mL of acetone and stir the paste for several minutes until it is no longer gummy. This acetone treatment removes most of the water from the cellular mixture. Filter the mixture on a small Büchner funnel. Scrape out the tube with a spatula, let it drain thoroughly, and squeeze out as much liquid as possible from the solid residue in the funnel with a spatula. Discard the yellow filtrate. Then return the solid residue to the centrifuge tube and add 5 mL of dichloromethane to effect extraction. Cap the tube and shake the mixture vigorously. Filter the mixture on a Büchner funnel once more, repeat the extraction and filtration with two or three further 5-mL portions of dichloromethane, clean the tube thoroughly, and place the filtrates in it. Dry the solution over anhydrous calcium chloride pellets, filter the solution into a small flask, and evaporate the solution to dryness with a stream of nitrogen or under vacuum using a rotary evaporator (Fig. 9.8 on page 207) or the apparatus shown in Figure 9.10 on page 211, never heating the sample above 50°C.

Determine the weight of the crude material. It will be very small. If the residue is dry, as it should be, add just enough dichloromethane to dissolve the residue. Save 1 drop of this solution to carry out a TLC analysis (using dichloromethane as the eluent on silica gel plates; *see* Chapter 8). Then add 200 mg of alumina to the remaining dichloromethane solution and evaporate the mixture to dryness, again without heat.

Column Chromatography

The crude carotenoid is to be chromatographed on an 8-cm column of basic or neutral alumina, prepared with hexanes as solvent (see the detailed procedure at the beginning of this chapter). Run out excess solvent or remove it from the top of the chromatography column with a Pasteur pipette. Using the dry sample loading method described at the beginning of this chapter, add the 300 mg of alumina that has the crude carotenoids absorbed on it. Add a few drops of hexanes to wash down the inside of the chromatography column and to consolidate the carotenoid mixture at the top of the column. Elute the column with hexanes, discard the initial colorless eluate, and collect all yellow or orange eluates together. Place a drop of solution on a microscope slide and evaporate the remainder to dryness using a stream of nitrogen or a rotary evaporator (*see* Fig. 9.8 on page 207). Examination of the material spotted on the slide may reveal crystallinity. If you are using tomato paste, a small amount of yellow β-carotene will come off first, followed by lycopene. Collect the red lycopene separately by eluting with a mixture of 10% acetone in hexane and also evaporate that solution to dryness.

Finally, dissolve the samples obtained by evaporating the solvent in the least possible amount of dichloromethane and carry out TLC of the two products in order to ascertain their purity (*see* Chapter 8, Experiment 2). You may want to combine your purified products with those of several other students, evaporate the solution to dryness, dissolve the residue in deuterochloroform, $CDCl_3$, and determine the ^1H NMR spectrum (*see* Chapter 12). Also obtain an infrared spectrum and a visible spectrum (in hexane).

Note that β-carotene is in demand as a source of vitamin A and is manufactured by an efficient synthesis. Until very recently no use for lycopene had been found.

Cleaning Up. Place recovered and unused dichloromethane in the halogenated organic waste container; the solvents used for TLC in the organic solvents container. If local regulations allow, evaporate any residual solvent from the drying agents in the hood and place the dried solid in the nonhazardous waste container. Otherwise, place the wet drying agent in a waste container designated for this purpose. Used plant material and dry TLC plates can be discarded in the nonhazardous waste container.

For Further Investigation

The carotenoids of any leaf can be isolated in the manner described in this experiment. Grind the leaf material (about 10 g) in a mortar with some sand; then follow the above procedure. Waxy leaves do not work well. The carotenoids are present in the leaf during its entire life span, so a green leaf from a maple tree or euonymus shrub, also known as burning bush, known to turn bright red in the fall will show lycopene even when the leaf is green. In the fall the chlorophyll decomposes before the carotenoids, so the leaves appear in a variety of orange and red hues.

It is of interest to investigate the carotenoids of the tomato, of which there are some 80 varieties. The orange-colored tangerine tomato contains an isomer of lycopene. If a hexane solution of the prolycopene from this tomato is treated with a drop of a very dilute solution of iodine in hexane and then exposed to bright light, the solution will turn deep-orange in color, indicating that a *cis*-double bond has isomerized to the *trans* form. The product is, however, still not identical to natural lycopene.

Isomerization

Prepare a hexane solution of either carotene or lycopene and save a drop for TLC. Treat the solution with a very dilute solution of iodine in hexane, expose the resulting mixture to strong light for a few minutes, and then carry out TLC on the resulting solution. Also compare the visible spectrum before and after isomerization.

Iodine serves as a catalyst for the light-catalyzed isomerization of some of the *trans*-double bonds to an equilibrium mixture containing *cis*-isomers.

QUESTIONS

1. Predict the order of elution of a mixture of triphenylmethanol, biphenyl, benzoic acid, and methyl benzoate from an alumina column.
2. What would be the effect of collecting larger fractions when carrying out the experiments in this chapter?
3. What would have been the result if a large quantity of petroleum ether alone were used as the eluent in either of the experiments described?
4. Once the chromatographic column has been prepared, why is it important to allow the level of the liquid in the column to drop to the level of the alumina before applying the solution of the compound to be separated?
5. A chemist started to carry out column chromatography on a Friday afternoon, reached the point at which the two compounds being separated were about three-fourths of the way down the column, and then returned on Monday to find that the compounds came off the column as a mixture. Speculate the reason for this. The column had not run dry over the weekend.
6. Write a detailed mechanism for the formation of fluorenone from fluorene. Explain the purpose of the phase-transfer catalyst.
7. The primary cause of low yields in the isolation of lycopene is oxidation of the product during the procedure. Once crystalline, it is reasonably stable. Speculate on the primary products formed by photochemical air oxidation of lycopene.

Online Study Center

General Resources
Web Links

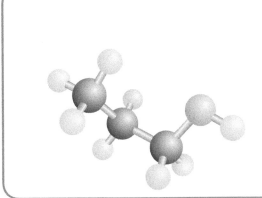

CHAPTER

10

Gas Chromatography: Analyzing Alkene Isomers

Online Study Center

This icon will direct you to techniques, equipment setups, and online resources at http://college.hmco.com/PIC/ williamsonMMOE5e.

PRELAB EXERCISE: If the dehydration of 2-methyl-2-butanol occurred on a purely statistical basis, what would be the relative proportions of 2-methyl-1-butene and 2-methyl-2-butene?

Gas chromatography (GC) is a rapid and sensitive method of separating and analyzing mixtures of gaseous or liquid compounds. The information it provides can tell you

1. if you have successfully synthesized your product;
2. whether your product contains unreacted starting material or other impurities;
3. whether your product is a mixture of isomers; and
4. the relative amounts of different materials or isomers in a mixture.

Gas chromatography is one of several *instrumental* analysis methods used by organic chemists. Unlike the apparatus for thin-layer (TLC) or column chromatography, the apparatus used for gas chromatography—a modern gas chromatograph like that shown in Figure 10.1—costs thousands of dollars and is therefore shared among many users. There are two major reasons for spending this much money for this instrument. First, very complex mixtures containing hundreds of components can be separated. Second, separated components can be detected in very small amounts, 10^{-6} to 10^{-15} g. The chromatogram in Figure 10.2 is a good example of the power of gas chromatography; it shows the separation of 1 mg of crude oil into the hundreds of compounds that are present: alkanes, alkenes, and aromatics, including many isomers. The exceptional sensitivity of GC instruments is the major reason they are often used in environmental and forensic chemistry labs, where the detection of trace amounts is necessary.

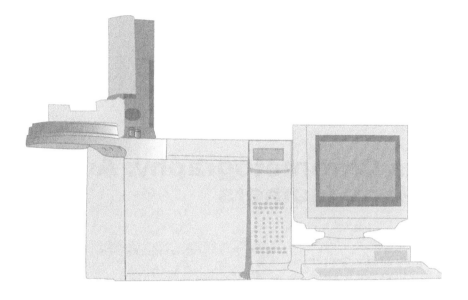

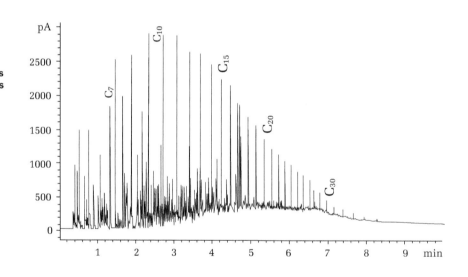

Polydimethylsiloxane

Carbowax

Gas chromatography involves the same principles that apply to all forms of chromatography, which were covered in Chapter 8. The mobile phase is a gas, usually helium. The stationary phase is often a fairly nonpolar polymer such as polydimethylsiloxane or poly(ethylene glycol) (Carbowax) that is stable at temperatures as high as 350°C. The most commonly used phase is polydimethylsiloxane in which 5% of the methyl groups have been replaced by phenyl groups.

There are two types of GC columns:

1. *Packed columns* in which the stationary phase consists of solid particles similar to the alumina or silica gel used in column chromatography but coated with a nonpolar polymer (e.g., polydimethylsiloxane or Carbowax) and packed into 3–6 mm diameter metal or glass tubes that are 1–3 m long and rolled into a compact coil.
2. *Capillary columns* are long, thin, flexible quartz tubes with a very thin coating of a nonpolar stationary phase polymer on the inside wall. They can be 15–60 m long with internal diameters of 0.1–0.5 mm. Their successful development came about because of the invention of fiber optics and the discovery that coating the outside of hollow quartz fibers with a polyimide polymer prevented them from being easily scratched and broken.

Packed columns can separate milligram quantities of materials, while capillary columns work best with microgram quantities or less. On the other hand, capillary columns have much better separating power than packed columns.

The GC instrument has the following components (Fig. 10.3). Helium at a pressure of 10–30 psi from a compressed gas cylinder flows through the heated injector port, the column (located in a temperature-controlled oven), and the heated detector at a flow rate of 10–60 mL/min for a packed column and about 1 mL/min for a capillary column. A microliter syringe is used to inject 1–25 μL of sample through a rubber septum into the hot (~250°C) injector, where any liquid compounds are instantly vaporized to the gaseous state. As the vaporized molecules are swept through the column, they interact with the stationary phase and are adsorbed and desorbed many times each second in a dynamic equilibrium. As discussed in Chapter 8, the equilibria between adsorbed and free states depends on each molecule's size (London forces), polarity, and ability to hydrogen

■ FIG. 10.3
A diagram of a gas chromatograph.

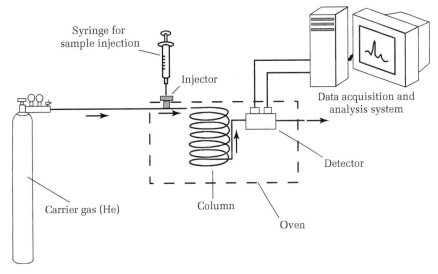

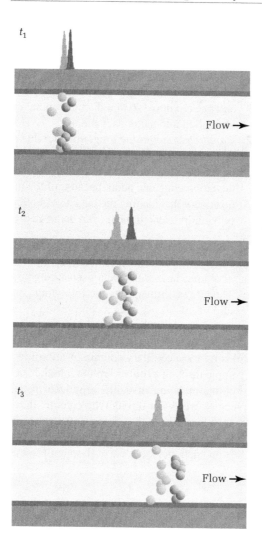

bond. Molecules spending more time absorbed to the stationary phase will not be
carried through the column as rapidly as those that are weakly bound and free. A
molecular view of this differential adsorption and flow is depicted in Figure 10.4.

The separated components pass one after the other into the detector. In column chromatography (*see* Chapter 9), human eyes are the detectors, either seeing
colored compounds flow out of the column or visualizing the presence of colorless compounds on a fluorescent TLC plate. Gas chromatographs have electronic
detectors that produce a signal voltage proportional to the number of molecules
passing through them at any instant in time. A record of this voltage versus time
is a gas chromatogram, with peak areas representing the amount of each individual component passing through the detector. Figure 10.5 is the chromatogram produced by the injection of a mixture of two compounds that were separated on the

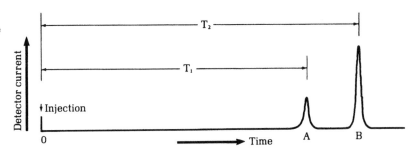

FIG. 10.5
A gas chromatogram of a mixture of two compounds, A and B.

column and detected to produce peaks A and B. The amount of each is proportional to the area under its peak, so we can say that there appears to be about twice as much B as A. If an integrating recorder or computer data acquisition system is used, the exact areas under all peaks are automatically determined and can be used to quantify the amount of each component in a mixture.

Most GC instruments use one of the following types of detectors:

- The *thermal conductivity detector* (TCD) consists of an electrically heated wire or thermistor. The temperature of the sensing element depends on the thermal conductivity of the gas flowing around it. Changes in thermal conductivity, such as when organic molecules displace some of the carrier gas, cause a temperature rise in the element that is sensed as a change in resistance. The TCD is the least sensitive (detecting micrograms per second, 10^{-6} g/s, of material) of the four detectors described here, but it is quite rugged. It will detect all types of molecules, not just those containing C—H bonds, and compounds passing through it are unchanged and can be collected.

- The *flame ionization detector* (FID) is at least a thousand times more sensitive than the TCD, easily detecting nanograms per second (10^{-9} g/s) of material. The molecules, however, must contain C—H bonds because the sample is actually burned to form ions. These ions carry a tiny current between two electrodes, which is greatly amplified to produce the signal output.

- The *electron capture detector* (ECD) contains a tiny amount of a radioactive substance, such as ^{63}Ni, that emits high-energy electrons (β^- particles). The sample molecules can capture these electrons, become charged, and carry a current between two electrodes, as in the FID detector. Halocarbons have exceptionally good cross sections for electron capture because of the electronegativity of the halogen atoms, and these types of compounds can be detected at picogram per second (10^{-12} g/s) levels. For this reason, ECD is used to measure levels of chlorocarbon pollutants, such as chloroform, DDT, and dioxin in the environment.

- A *mass spectrometer* detector is the only one of these four detectors that provides information about the structure of the molecules that pass into it. For this reason, combination gas chromatography–mass spectrometry (GC–MS) systems are often used in organic research, forensic science, and environmental analysis in spite of their high cost ($40,000 to $250,000). Sensitivity is at the nanogram to picogram per second level. Chapter 13 discusses mass spectrometry in detail.

The instant you inject your sample into the GC instrument, you normally press a button that starts a clock in the data recorder, either an electronic integrator or a computer data system. Then, as each eluting component produces a peak in the detector, the top of the peak is detected, and the elapsed time since injection, called the *retention time*, is recorded and stored. In Figure 10.5, the retention time of compound A is T_1 and of B is T_2. Which compound is more strongly absorbed by the stationary phase?

One powerful variable in gas chromatography that is not available in TLC or column chromatography is temperature. Because the GC column is in an oven whose temperature is programmable—that is, the temperature can be raised from near room temperature to as high as 350°C at a constant and reproducible rate—we are able to separate much more complex mixtures than if the oven could be set to only one temperature. At room temperature, a mixture containing lower-boiling, weakly adsorbed components (such as ether and dichloromethane) and higher-boiling, strongly adsorbed components (such as butanol and toluene) might take hours to separate because the latter would move significantly more slowly through the column at low temperature. If the column were kept at a high temperature, the higher-boiling compounds would come out in a shorter time, but the lower-boiling ones would not be retained at all and therefore would not be sep-

Temperature programming

arated. By programming the oven temperature, we can inject such a mixture at a low initial temperature so that weakly adsorbed components are separated. Then, as the temperature slowly rises, the more strongly retained components will move through the column faster and also have reasonable retention times. Temperature programming allows the separation of very complex mixtures containing both low-boiling small molecules and high-boiling large ones, as demonstrated by the GC analysis of crude petroleum in Figure 10.2 on page 216.

The General GC Analysis Procedure

Most GC separations are of gaseous and liquid mixtures. Although some solids will pass through a gas chromatograph at higher temperatures, it is best if all of the samples are distilled or vacuum transferred before injection into the GC instrument to ensure that only volatile compounds are present. If samples contain materials that cannot be vaporized and swept through the GC column, even at high temperatures, these will remain on the column and can ultimately ruin its performance. GC columns are expensive and can cost up to $400 to replace.

The guidelines given here are for GC analysis using a capillary column and an FID detector, which require very dilute solutions to avoid overwhelming the column, even if 90%–95% of the sample is split away so that it does not enter the column, a common practice in capillary chromatography. Samples analyzed with a packed column and a TCD detector can be much more concentrated or even undiluted (neat).

Sample Preparation

There are different ways of preparing a GC sample, depending on whether your product is a solid, a liquid (usually obtained by distillation), or a gaseous mixture; on the type of column (capillary or packed); and on the type of detector (FID, ECD, or TCD) being used.

Solid Samples

Low-melting, sublimable solids can be analyzed at higher column temperatures. Run a GC analysis on solids only if your experimental procedure explicitly says that you can do so. To prepare a solid sample for capillary column analysis, put a *small* crystal of the solid into a small vial and dissolve in 1 mL of dichloromethane.

Liquid Samples

Dilute solutions of liquid organic mixtures in dichloromethane are the most common form of GC samples. Place 1 or 2 drops of your distilled product into a small vial and add 1 mL of dichloromethane. Because GC columns can be damaged by moisture, it is suggested that enough *anhydrous* sodium sulfate be added to cover the bottom of the vial to a depth of 1–2 mm to ensure that all traces of water are removed. The sodium sulfate can be left in the sample because it usually will not go into the fine needle of the microliter syringe used to inject the sample. Injection of 1 μL of a sample prepared in this manner usually provides strong signals in an FID detector. In special cases, undiluted neat or pure liquids can be injected directly without dilution as noted, for example, the methylbutenes in Experiment 1. Also, if preparative gas chromatography is used to separate and collect the components of a mixture, neat liquids are usually injected.

Gaseous Samples

Gases are normally collected over water, as in Experiment 2. Do not remove the septum-capped collection tube from the beaker of water, but bring it and the beaker of water together to the gas chromatograph. A gas syringe is inserted through the rubber septum and 10–500 μL (depending on the type of the gas chromatograph) of the gas sample is withdrawn and injected into the unit.

Sample Analysis

Make sure that the required gases are flowing and that the instrument has been on at least 1 h so that all zones, the injector, the oven, and the detector have come to the required temperatures. Enter a temperature program: initial temperature, heating rate (in degrees per min), and final temperature, as required by the particular experiment. Fill the microliter syringe with the sample solution, normally 1 μL (Fig. 10.6a). A 10-μL capacity syringe with a plunger guide is recommended. Handle this expensive syringe carefully; it has a sharp needle, and the glass barrel is easily dropped and broken. Make sure that there are no bubbles in the syringe barrel when you draw up the sample. If bubbles are present, fill the syringe about

The steps in sample injection: (a) Fill the microliter syringe. (b) Insert the needle through the injector septum. (c) Depress the plunger. (d) Withdraw the syringe.

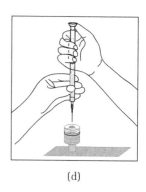

(a)   (b)   (c)   (d)

Do not touch the metal injector cap. It is very HOT.

halfway and depress the plunger quickly. Draw up the sample again slowly. Repeat as necessary to remove all bubbles. Inject the sample as shown in Figure 10.6c. To obtain a quality chromatogram, these steps should be done rapidly, and the syringe should be removed as soon as the plunger has been depressed and the sample injected (Fig. 10.6d). Then quickly press the start analysis button on the gas chromatograph, computer, or integrator to start data acquisition or mark the paper on the pen recorder.

Once all the compounds have eluted and the oven temperature program has finished, the run should be stopped. Plot your chromatogram using the computer or remove the chromatogram chart from the integrator or flat bed recorder. Record the GC analysis parameters: column diameter and length, column packing type, carrier gas and its flow rate, column temperature (initial, final, and heating rate if programmed), detector and injector temperatures, sample injection amount and solvent (if used), signal attenuation, and chart speed; keep these with your gas chromatogram. Write the structures of all identifiable peaks on the chromatogram. If you used dichloromethane to dilute your sample, the largest peak will be due to this substance and will have a short retention time.

Remember that retention times are not constants. If the programmed oven temperature rate is higher or lower than for a prior analysis, then the retention times for all of the peaks will be consistently shorter or consistently longer. Changes in helium flow and the aging of the column also affect retention times. Chemists primarily look for similarities in the pattern of eluting peaks, not for perfect matches in retention times. In general, for simple mixtures chromatographed on nonpolar stationary phases, the retention time order is the same as the order of increasing boiling points. For example, you would expect cyclohexene (bp 83°C) to have a shorter retention time than toluene (bp 110°C). You can often ascertain the identity of a certain peak by adding a small amount of a known standard to the sample and rerunning the GC analysis. The peak corresponding to the standard will increase significantly and can be identified. Often the chromatogram will show a number of small peaks that cannot be assigned to the product, the starting material, or the solvents; these are likely due to byproducts. As long as these impurities are minor, say, less than 10% of the total area (not including the solvent), they can probably be ignored. The ultimate method of identifying every peak is to analyze the sample on a GC–MS instrument.

Cleaning Up. Rinse out the syringe by filling it with dichloromethane and squirting it onto a tissue or paper towel at least four times. Clean and return any sample collection tubes, if used.

Collecting a Sample for an Infrared Spectrum

The small amount of sample analyzed in gas chromatography is an advantage in many cases, but it precludes isolating the separated components. Some specialized chromatographs can separate samples as large as 0.5 mL per injection and automatically collect each fraction in a separate container. At the other extreme, gas chromatographs equipped with FIDs can detect micrograms of substances, such as traces of pesticides in food or drugs in blood and urine. Clearly, a gas chromatogram gives little information about the chemical nature of the sample being detected. However, certain *preparative* chromatographs, like that shown in Figure 10.7, allow the collection of enough sample at the exit port to obtain an infrared spectrum. About 10–15 μL of a mixture (not diluted in solvent) is injected, and as the peak for the compound of interest appears, a 2-mm-diameter glass tube, 3 in. long and packed with glass wool, is inserted into the rubber septum at the exit port. The sample, if it is not too volatile, will condense in the cold glass tube. Subsequently, the sample is washed out with 1 or 2 drops of solvent, and an infrared spectrum is obtained. This process can be repeated to collect enough sample for obtaining an NMR spectrum (*see* Chapter 12), using a few drops of deuterochloroform ($CDCl_3$) to wash out the tube each time. See Figure 10.8 for another collection device.

■ **FIG. 10.7**

Some gas chromatographs with packed columns allow the collection of small amounts of separated compounds.

■ **FIG. 10.8**

A gas chromatographic collection device. Fill the container with ice or a dry ice–acetone mixture and attach to the outlet port of the gas chromatograph.[1]

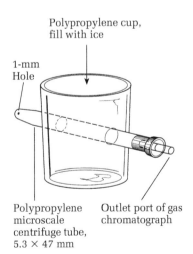

Polypropylene cup, fill with ice

1-mm Hole

Polypropylene microscale centrifuge tube, 5.3 × 47 mm

Outlet port of gas chromatograph

---

1. This apparatus is available from Kimble Kontes (Vineland, NJ).

EXPERIMENTS

 ### 1. 2-Methyl-1-Butene and 2-Methyl-2-Butene by Dehydration of an Alcohol[2]

The dilute sulfuric acid–catalyzed dehydration of 2-methyl-2-butanol (*t*-amyl alcohol) proceeds readily to give a mixture of alkenes that can be analyzed by gas chromatography. The mechanism of this reaction involves the intermediate formation of the relatively stable tertiary carbocation followed by the loss of a proton either from a primary carbon atom to give the terminal olefin, 2-methyl-1-butene, or from a secondary carbon to give 2-methyl-2-butene.

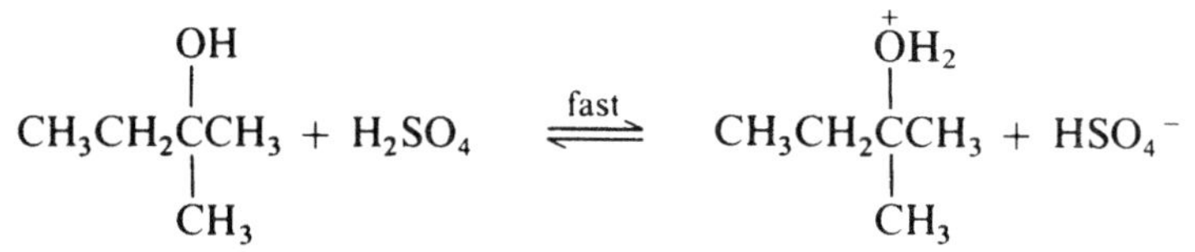

$$\underset{\substack{\text{OH} \\ | \\ \text{CH}_3\text{CH}_2\text{CCH}_3 \\ | \\ \text{CH}_3}}{} + \text{H}_2\text{SO}_4 \;\overset{\text{fast}}{\rightleftharpoons}\; \underset{\substack{\overset{+}{\text{O}}\text{H}_2 \\ | \\ \text{CH}_3\text{CH}_2\text{CCH}_3 \\ | \\ \text{CH}_3}}{} + \text{HSO}_4{}^-$$

2-Methyl-2-butanol
bp 102°C
den. 0.805
MW 88.15

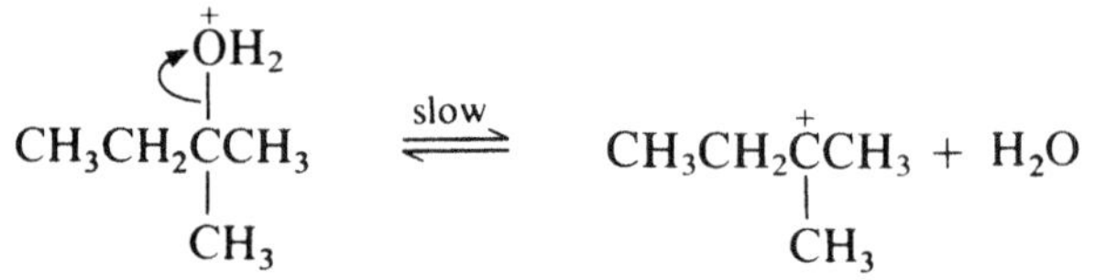

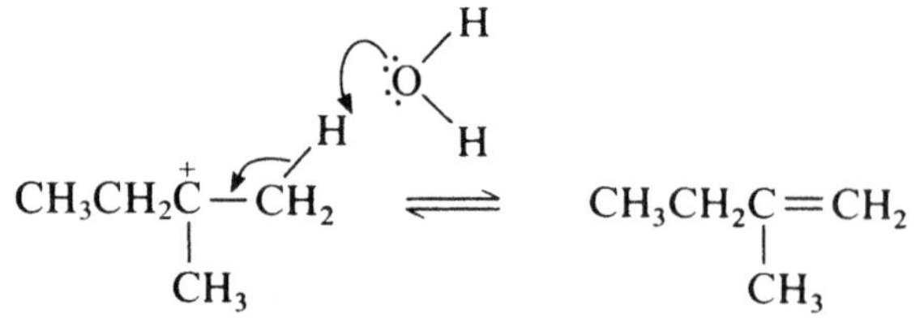

$$\underset{\substack{\overset{+}{\text{O}}\text{H}_2 \\ | \\ \text{CH}_3\text{CH}_2\text{CCH}_3 \\ | \\ \text{CH}_3}}{} \;\overset{\text{slow}}{\rightleftharpoons}\; \underset{\substack{ \\ | \\ \text{CH}_3\text{CH}_2\overset{+}{\text{C}}\text{CH}_3 \\ | \\ \text{CH}_3}}{} + \text{H}_2\text{O}$$

$$\underset{\substack{ \\ | \\ \text{CH}_3\text{CH}_2\overset{+}{\text{C}}\!-\!\text{CH}_2 \\ | \\ \text{CH}_3}}{} \;\rightleftharpoons\; \underset{\substack{ \\ | \\ \text{CH}_3\text{CH}_2\text{C}\!=\!\text{CH}_2 \\ | \\ \text{CH}_3}}{}$$

2-Methyl-1-butene
bp 31.16°C
den. 0.662
MW 70.14

2. Schimelpfenig, C. W. *J. Chem. Educ.* **1962**, *39*, 310.

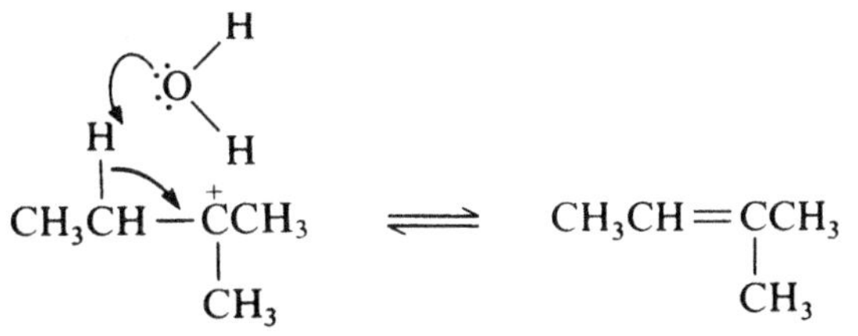

$$CH_3CH - \overset{+}{C}CH_3 \rightleftharpoons CH_3CH = CCH_3$$

2-Methyl-2-butene
bp 38.57°C
den. 0.662
MW 70.14

2-Methyl-2-butanol can also be dehydrated in high yield using iodine as a catalyst:

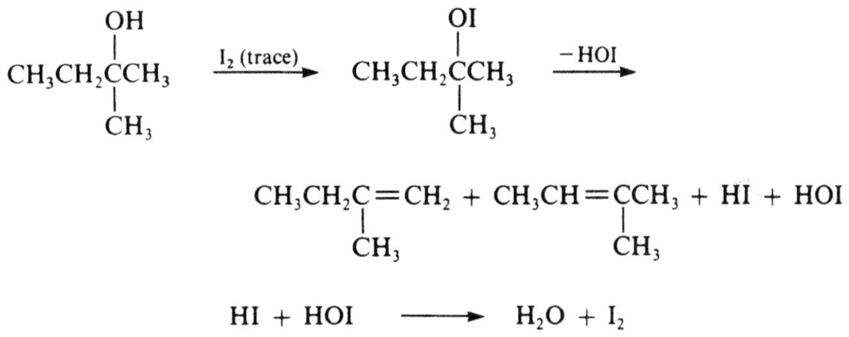

$$CH_3CH_2\underset{CH_3}{\overset{OH}{C}}CH_3 \xrightarrow{I_2 \text{ (trace)}} CH_3CH_2\underset{CH_3}{\overset{OI}{C}}CH_3 \xrightarrow{-HOI}$$

$$CH_3CH_2\underset{CH_3}{C}=CH_2 + CH_3CH=\underset{CH_3}{C}CH_3 + HI + HOI$$

$$HI + HOI \longrightarrow H_2O + I_2$$

Each step of this E_1 elimination reaction is reversible, and thus the reaction is driven to completion by removing one of the products, the alkene. Often these reactions produce several alkene isomers. The Saytzeff rule states that the more substituted alkene is the more stable and thus the one formed in larger amount. The *trans*-isomer is more stable than the *cis*-isomer. With this information it should be possible to deduce which peak on the gas chromatogram corresponds to a given alkene and to predict the ratios of the products.

In analyzing your results from this experiment, consider the fact that the carbocation can lose any of six primary hydrogen atoms but only two secondary hydrogen atoms to give the product olefins.

Microscale Procedure

IN THIS EXPERIMENT a tertiary alcohol is dehydrated to a mixture of alkenes in a reaction catalyzed by sulfuric acid. The products are collected by simple distillation into an ice-cold vial. Cold aqueous base is added to remove the acid, and the organic layer is dried with calcium chloride. The products are again distilled and then analyzed by gas chromatography.

■ FIG. 10.9

An apparatus for the dehydration of 2-methyl-2-butanol. The beaker can be clamped with a large three-prong clamp.

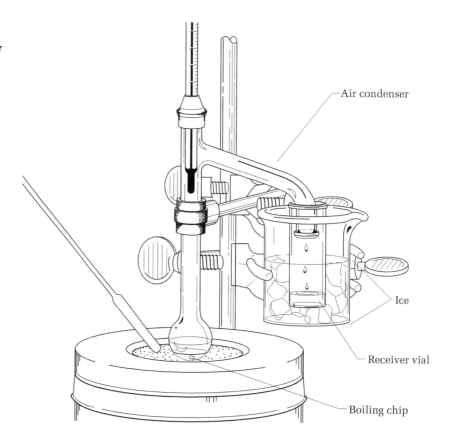

Air condenser

Ice

Receiver vial

Boiling chip

Handle sulfuric acid with care.

Keep the side arm down in the cold vial to avoid losing product.

Into a 5-mL long-necked, round-bottomed flask place 1 mL of water and then add dropwise with thorough mixing 0.5 mL of concentrated sulfuric acid. Cool the hot solution in an ice bath; then add to the cold solution 1.0 mL (0.80 g) of 2-methyl-2-butanol. Mix the reactants thoroughly, add a boiling stone, and set up the apparatus for simple distillation, as shown in Figure 10.9. Hold the flask in a towel while connecting the apparatus to guard against spills or flask breakage. Cool the receiver (a vial) in ice to avoid losing the highly volatile products. Warm the flask on a sand bath to start the reaction and distill the products over the temperature range 30°C–45°C. After all the products have distilled, the rate of distillation will decrease markedly. Cease distillation at this point. After this, the temperature registered on the thermometer will rise rapidly as water and sulfuric acid begin to distill.

To the distillate add 0.3 mL cold 3 *M* sodium hydroxide solution to neutralize sulfurous acid, mix well, draw off the aqueous layer, and dry the product over anhydrous calcium chloride pellets, adding the drying agent in small quantities until it no longer clumps together. Keep the vial cold. While the product is drying, rinse out the distillation apparatus with water, ethanol, and then a small amount of

Products are easily lost by evaporation. Store the mixture in a spark-proof refrigerator if GC analysis must be postponed.

acetone and dry it thoroughly by drawing air through it using an aspirator. If this is not done carefully, the products will become contaminated with acetone. Carefully transfer the dry mixture of butenes to the distilling apparatus using a Pasteur pipette, add a boiling chip to the now dry products, and distill into a tared, cold receiver. Collect the portion boiling up to 43°C. Weigh the product and calculate the yield, which is normally around 50% on this small scale compared to 84% on macroscale. Inject a few microliters of product into a gas chromatograph maintained at room temperature (packed column suggestion: 6-mm-diameter × 3-m column 10% SE-30 silicone rubber on Chromosorb-W).

For capillary column analysis, inject 1–10 μL of just the gaseous *vapors* above the liquid. Oven temperatures should be as low as room temperature (oven door open) to obtain baseline separation of the methylbutene isomers. In a few minutes two peaks should appear. Measure the relative areas under the two peaks. From your knowledge of the mechanisms of the dehydration of secondary alcohols, which olefin should predominate? Does this agree with the boiling points? (In general, the compound with the shorter retention time has the lower boiling point.)

Cleaning Up. The residue in the reaction tube should be diluted with water, neutralized with sodium carbonate, and then flushed down the drain.

Macroscale Procedure

Handle sulfuric acid with care.

Pour 18 mL of water into a 100-mL round-bottomed flask and cool in an ice-water bath while slowly pouring in 9 mL of concentrated sulfuric acid. Cool this 1:2 acid further with swirling while slowly pouring in 18 mL (15 g) of 2-methyl-2-butanol. Shake the mixture thoroughly and then mount the flask for fractional distillation over a flask heater, with the arrangement for ice cooling of the distillate seen in Figure 19.2 (on page 355). Cooling is needed because the olefin is volatile. Use a long condenser and an ample stream of cooling water. Heat the flask slowly with an electric flask heater until distillation of the hydrocarbon is complete. If the ice-cooled test tube will not hold 15 mL, be prepared to collect half of the distillate in another ice-cooled test tube. Transfer the distillate to a separatory funnel and shake it with about 5 mL of a 3 M sodium hydroxide solution to remove any traces of sulfurous acid. The aqueous solution sinks to the bottom and is drawn off. Dry the hydrocarbon layer by adding sufficient anhydrous calcium chloride pellets until the drying agent no longer clumps together. After about 5 min remove the drying agent by gravity filtration or careful decantation into a dry 25-mL round-bottomed flask and distill the dried product through a fractionating column (*see* Fig. 19.2 on page 355), taking the same precautions as before to avoid evaporation losses. Rinse the fractionating column with acetone and dry it with a stream of air to remove water from the first distillation. Collect in a tared bottle the portion boiling at 30°C–43°C. The yield reported in the literature is 84%; the average student yield is about 50%. Analyze the sample by injecting a few microliters of product into a gas chromatograph maintained at room temperature and equipped with a 6-mm-diameter × 3-m column packed with 10% SE-30 silicone rubber on Chromosorb-W or a similar inert packing. In a few minutes two peaks should appear. From your knowledge of the mechanisms of dehydration of secondary alcohols, which olefin should predominate? Does this agree with the boiling points? (In general, the compound with the shorter retention time has the lower boiling point.) Measure the relative areas under the two peaks.

Cleaning Up. The pot residue from the reaction is combined with the sodium hydroxide wash and neutralized with sodium carbonate. The pot residue from the distillation of the product is combined with the acetone used to wash out the apparatus and placed in the organic solvents container. If the calcium chloride pellets are dry, they can be placed in the nonhazardous solid waste container. If they are wet with organic solvents, they must be placed in the hazardous solid waste container for solvent-contaminated drying agent.

2. Computational Chemistry

The steric energies of the two isomeric butenes produced in Experiment 1 can be calculated using a molecular mechanics program, but the results are not valid because the bonding type is not the same. It is possible to compare the steric energies of *cis*- and *trans*-isomers of an alkene, but not a 1,1-disubstituted alkene with a 1,2-disubstituted alkene.

Compare your yields of the two isomers to their heats of formation, calculated using AM1 or a similar semiempirical calculation as described in Chapter 15. Do the calculated heats of formation correlate with the relative percentages of the isomers? What does this correlation or lack thereof tell you about the mechanism of the reaction?

3. 1-Butene and *cis*- and *trans*-2-Butene by Dehydration of an Alcohol[3]

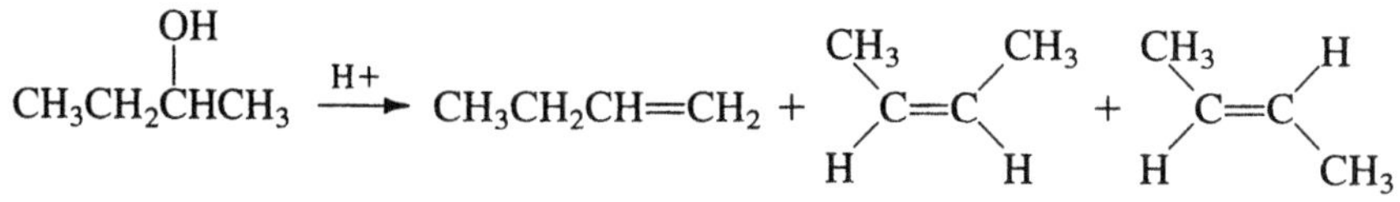

	2-Butanol	**1-Butene**	***cis*-2-Butene**	***trans*-2-Butene**
	MW 74.12	MW 56.11	MW 56.11	MW 56.11
	bp 98°C	bp −6.3°C	bp 3.7°C	bp 0.9°C

IN THIS EXPERIMENT 2-butanol is dehydrated with sulfuric acid to give a mixture of gaseous alkenes that are collected by the downward displacement of water. The gaseous mixture is analyzed by gas chromatography.

3. Helmkamp, G. K.; Johnson, H. W. Jr. *Selected Experiments in Organic Chemistry*, 3rd ed., Freeman: New York, 1983; 99.

■ **FIG. 10.10**

To thread a polyethylene tube through a septum, make a hole in the septum with a needle and then push a toothpick through the hole. Push the polyethylene tube firmly onto the toothpick; then pull and push on the toothpick. The tube will slide through the septum. Finally, pull the tube from the toothpick. A blunt syringe needle can also be used instead of a toothpick.

■ **FIG. 10.11**

An apparatus for the dehydration of 2-butanol. Butenes are collected by the downward displacement of water.

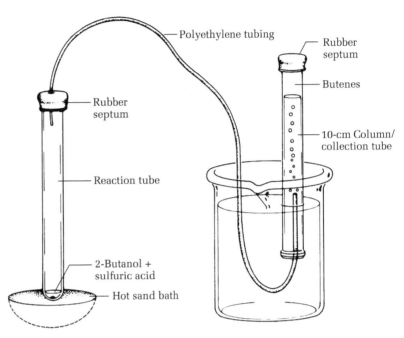

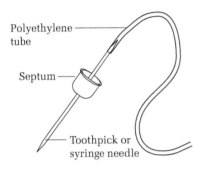

Online Study Center

Video: Butenes from Butanol; Photo: Placing a Polyethylene Tube through a Septum

Bend the polyethylene tubing to shape in hot sand or steam.

Online Study Center

General Resources
Additional Experiments

In a 10 × 100 mm reaction tube place 0.10 mL (81 mg) of 2-butanol and 0.05 mL (1 drop) of concentrated sulfuric acid. It is not necessary to measure the quantities of alcohol and acid exactly. Mix the reactants well and add a boiling chip. Stirring the reaction mixture is not necessary because it is homogeneous. Insert a polyethylene tube through a septum (Fig. 10.10), place this septum on the reaction tube, and lead the tubing into the mouth of the distilling column/collection tube, which is capped with a septum. The distilling column/collection tube is filled with water, capped with a finger, and inverted in a beaker of water and clamped for the collection of butene by the downward displacement of water (Fig. 10.11).

Lower the reaction tube into a hot (~100°C) sand bath and increase the heat slowly to complete the reaction. Collect a few milliliters of butene and *remove the polyethylene tube from the water bath*. If the polyethylene tube is not removed from the water bath, water will be sucked back into the reaction tube when it cools. Do not remove the septum-capped collection tube from the beaker of water, but bring it and the beaker of water together to the gas chromatograph. Use about 0.5 mL of the gaseous butene mixture to carry out the GC analysis with a TCD detector or a smaller amount if an FID detector is used.

Cleaning Up. The residue in the reaction tube may be diluted with water and flushed down the drain.

QUESTIONS

1. Write the structures of the three olefins produced by the dehydration of 3-methyl-3-pentanol.

2. When 2-methylpropene is bubbled into dilute sulfuric acid at room temperature, it appears to dissolve. What new substance has been formed?

3. A student wished to prepare ethylene gas by the dehydration of ethanol at 140°C, using sulfuric acid as the dehydrating agent. A low-boiling liquid was obtained instead of ethylene. What was the liquid, and how might the reaction conditions be changed to give ethylene?

4. What would be the effect on retention time of increasing the carrier gas flow rate?

5. What would be the effect on retention time of raising the column temperature?

6. What would be the effect of raising the temperature or increasing the carrier gas flow rate on the ability to resolve two closely spaced peaks?

7. If you were to use a column one-half the length of the one you actually used, how do you think the retention times of the butenes would be affected? How do you think the separation of the peaks would be affected? How do you think the width of each peak would be affected?

8. From your knowledge of the dehydration of tertiary alcohols, which olefin should predominate in the product of the dehydration of 2-methyl-2-butanol? Why?

9. What is the maximum volume (at standard temperature and pressure, STP) of the butene mixture that could be obtained by the dehydration of 81 mg of 2-butanol?

10. What other gases are in the collection tube besides the three butenes at the end of the reaction in Experiment 3?

11. Using a computational chemistry program (*see* Chapter 15), calculate the steric energies or heats of formation of the three isomeric butenes produced in Experiment 3. Do these energies correlate with the product distributions found? What does a correlation or lack thereof tell you about the mechanism of the reaction?

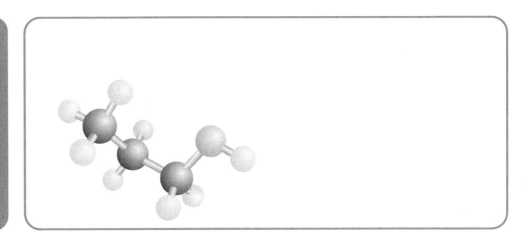

Infrared Spectroscopy

PRELAB EXERCISE: When an infrared (IR) spectrum is run, it is possible that the chart paper is not properly placed or that the spectrometer is not mechanically adjusted. Describe how you could calibrate an IR spectrum.

The types and molecular environment of functional groups in organic molecules can be identified by infrared (IR) spectroscopy. Like nuclear magnetic resonance (NMR) and ultraviolet (UV) spectroscopy, IR spectroscopy is nondestructive. Moreover, the small quantity of sample needed, the speed with which a spectrum can be obtained, the relatively low cost of the spectrometer, and the wide applicability of the method combine to make IR spectroscopy one of the most common structural elucidation tools used by organic chemists.

IR radiation consists of wavelengths that are longer than those of visible light. It is detected not with the eyes but by a feeling of warmth on the skin. When absorbed by molecules, radiation of these wavelengths (typically 2.5–5 μm) increases the amplitude of vibrations of the chemical bonds joining atoms.

2.5–25 μm equals 4000–400 cm^{-1}. Wavenumber(cm^{-1}) is proportional to frequency (c is the speed of light).

$$\overline{v}\,(\text{cm}^{-1}) = \frac{v}{c}$$

IR spectra are measured in units of frequency or wavelength. The wavelength is measured in micrometers[1] or microns, μm (1 μm = 1×10^{-6} m). The positions of absorption bands are measured in frequency units called wavenumbers v, which are expressed in reciprocal centimeters, cm^{-1}, corresponding to the number of cycles of the wave in each centimeter.

$$\text{cm}^{-1} = \frac{10{,}000}{\mu m}$$

Examine the scale carefully.

Unlike UV and NMR spectra, IR spectra are inverted, with the strongest absorptions at the bottom, and are not always presented on the same scale. Some

---

1. Although micrometers are also known as microns, the micron is not the official SI unit.

IR spectroscopy easily detects:

Hydroxyl groups   —OH

Amines   —NH$_2$

Nitriles   —C≡N

Nitro groups   —NO$_2$

IR spectroscopy is especially useful for detecting and distinguishing among all carbonyl-containing compounds:

$$
\begin{array}{ll}
\textit{Acids} & -\overset{\displaystyle O}{\overset{\|}{C}}-OH \\[2ex]
\textit{Amides} & -\overset{\displaystyle O}{\overset{\|}{C}}-NH_2 \\[2ex]
\textit{Anhydrides} & -\overset{\displaystyle O}{\overset{\|}{C}}-O-\overset{\displaystyle O}{\overset{\|}{C}}- \\[2ex]
\textit{Aldehydes} & -\overset{\displaystyle O}{\overset{\|}{C}}-H \\[2ex]
\textit{Ketones} & -\overset{\displaystyle O}{\overset{\|}{C}}- \\[2ex]
\textit{Esters} & -\overset{\displaystyle O}{\overset{\|}{C}}-O- \\[2ex]
\textit{Lactones} & -\overset{\displaystyle O}{\overset{\|}{C}}-O- \\
\end{array}
$$

spectrometers record the spectra on an ordinate linear in microns, but this compresses the low-wavelength region. Other spectrometers present the spectra on a scale linear in reciprocal centimeters, but linear on two different scales: one between 4000 and 2000 cm^{-1}, which spreads out the low-wavelength region; and the other a smaller one between 2000 and 667 cm^{-1}. Consequently, spectra of the same compound run on two different spectrometers will not always look the same.

To picture the molecular vibrations that interact with IR light, imagine a molecule as being made up of balls (atoms) connected by springs (bonds). The vibration can be described by Hooke's law from classical mechanics, which says that the frequency of a stretching vibration is directly proportional to the strength of the spring (bond) and inversely proportional to the masses connected by the spring. Thus we find C—H, N—H, and O—H bond-stretching vibrations are high frequency (short wavelength) compared to those of C—C and C—O because of the low mass of hydrogen compared to that of carbon or oxygen. The bonds connecting carbon to bromine and iodine, atoms of large mass, vibrate so slowly that they are beyond the range of most common IR spectrometers. A double bond can be regarded as a stiffer, stronger spring, so we find C=C and C=O vibrations at higher frequency than C—C and C—O stretching vibrations. And C≡C and C≡N stretch at even higher frequencies than C=C and C=O (but at lower frequencies than C—H, N—H, and O—H). These frequencies are in keeping with the bond strengths of single (~100 kcal/mol), double (~160 kcal/mol), and triple bonds (~220 kcal/mol).

These stretching vibrations are intense and particularly easy to analyze. A nonlinear molecule of n atoms can undergo $3n - 6$ possible modes of vibration, which means cyclohexane with 18 atoms can undergo 48 possible modes of vibration. A single CH$_2$ group can vibrate in six different ways because it is connected to two other atoms. Each vibrational mode produces a peak in the spectrum because it corresponds to the absorption of energy at a discrete frequency. These many modes of vibration create a complex spectrum that defies simple analysis, but even in very complex molecules, certain functional groups have characteristic frequencies that can easily be recognized. Within these functional groups are the above-mentioned atoms and bonds, C—H, N—H, O—H, C=C, C=O, C≡C, and C≡N. Their absorption frequencies are given in Table 11.1.

When the frequency of IR light is the same as the natural vibrational frequency of an interatomic bond, light will be absorbed by the molecule, and the amplitude of the bond vibration will increase. The intensity of IR absorption bands is proportional to the change in dipole moment that a bond undergoes when it stretches. Thus, the most intense bands (peaks) in an IR spectrum are often from C=O and C—O stretching vibrations, whereas the C≡C stretching band for a symmetrical acetylene is almost nonexistent because the molecule undergoes no net change of dipole moment when it stretches:

$$\overset{+}{\underset{/}{\diagdown}C=O} \longleftrightarrow \overset{+}{\underset{/}{\diagdown}C=O} \qquad H_3C-C\equiv C-CH_3 \longleftrightarrow H_2C-C\equiv C-CH_3$$

Change in dipole moment   No change in dipole moment

TABLE 11.1 • Characteristic IR Absorption Wavenumbers	
Functional Group	**Wavenumber (cm^{-1})**
O—H	3600–3400
N—H	3400–3200
C—H	3080–2760
C≡N	2260–2215
C≡C	2150–2100
C=O	1815–1650
C=C	1660–1600
C—O	1200–1050

The intensity of absorption is proportional to the change in dipole moment.

Unlike proton NMR spectroscopy, where the area of the peaks is strictly proportional to the number of hydrogen atoms causing the peaks, the intensities of IR peaks are not proportional to the numbers of atoms causing them. Given the chemical shifts and coupling constants, it is not too difficult to calculate a theoretical NMR spectrum that is an exact match to the experimental one. The calculation of all possible stretching and bending frequencies, for larger molecules, requires large amounts of time on a fast computer. Every peak or group of peaks in an NMR spectrum can be assigned to specific hydrogens in a molecule, but the assignment of the majority of peaks in an IR spectrum is usually not possible. Peaks to the right (longer wavelength) of 1250 cm^{-1} are the result of combinations of vibrations that are characteristic not of individual functional groups but of the molecule as a whole. This part of the spectrum is often referred to as the *fingerprint region* because it is uniquely characteristic of each molecule. Although two organic compounds can have the same melting points or boiling points and can have identical UV and NMR spectra, they cannot have identical IR spectra (except, as usual, for enantiomers). IR spectroscopy is thus the final arbiter in deciding whether two compounds are identical.

Analysis of IR Spectra

Only a few simple rules or equations govern IR spectroscopy. Because it is not practical to calculate theoretical spectra, the analysis is done almost entirely by correlation with other spectra. In printed form, these comparisons take the form of lengthy discussions, so detailed analysis of a spectrum is best done with a good reference book at hand.

In a modern analytical or research laboratory, a collection of many thousands of spectra is maintained on a computer. When the spectrum of an unknown compound is run, the analyst picks out five or six of the strongest peaks and asks the computer to list all the compounds that have peaks within a few reciprocal centimeters of the listed peaks. From the printout of a dozen or so compounds, it is often possible to pinpoint all the functional groups in the molecule being analyzed. There may be a perfect match of all peaks, in which case the unknown will have been identified.

For relatively simple molecules, a computer search is hardly necessary. Much information can be gained about the functional groups in a molecule from relatively few correlations.

To carry out an analysis, (1) pay most attention to the strongest absorptions; (2) pay more attention to peaks to the left (shorter wavelength) of 1250 cm^{-1}; and (3) pay as much attention to the absence of certain peaks as to the presence of others. The absence of characteristic peaks will definitely exclude certain functional groups. Be wary of weak O—H peaks because water is a common contaminant of many samples. Because potassium bromide is hygroscopic, water is often found in the spectra of KBr pellets.

The Step-by-Step Analysis of IR Spectra

IR spectra are analyzed as follows:

1. Is there a peak between 1820 cm^{-1} and 1625 cm^{-1}? If not, go to Step 2.
    (a) Is there a strong, wide O—H peak between 3200 cm^{-1} and 2500 cm^{-1}? If so, the compound is a carboxylic acid (Fig. 11.1, oleic acid). If not . . .
    (b) Is there a medium-to-weak N—H band between 3520 cm^{-1} and 3070 cm^{-1}? If there are two peaks in this region, the compound is a primary amide; if not, it is a secondary amide. If there is no peak in this region . . .
    (c) Are there two strong peaks, one in the region 1870 cm^{-1} to 1800 cm^{-1} and the other in the region 1800 cm^{-1} to 1740 cm^{-1}? If so, an acid anhydride is present. If not . . .
    (d) Is there a peak in the region of 2720 cm^{-1}? If so, is the carbonyl peak in the region 1715 cm^{-1} to 1680 cm^{-1}? If so, the compound is a conjugated aldehyde; if not, it is an isolated aldehyde (Fig. 11.2, benzaldehyde). However, if there is no peak near 2720 cm^{-1} . . .
    (e) Does the strong carbonyl peak fall in the region 1815 cm^{-1} to 1770 cm^{-1} and the compound give a positive Beilstein test? If so, it is an acid halide. If not . . .

▮ FIG. 11.1
The IR spectrum of oleic acid (thin film).

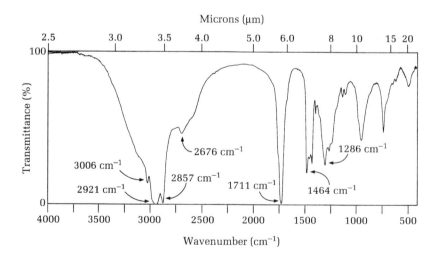

■ FIG. 11.2
The IR spectrum of benzaldehyde (thin film).

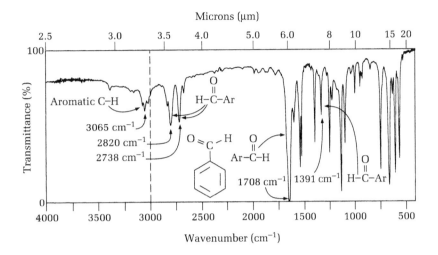

■ FIG. 11.3
The IR spectrum of n-butyl acetate (thin film).

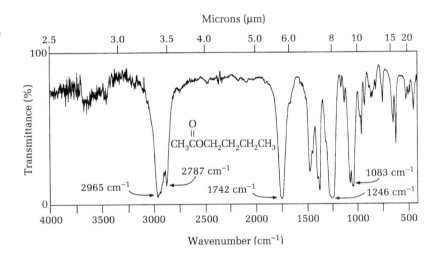

(f) Does the strong carbonyl peak fall in the range 1690 cm^{-1} to 1675 cm^{-1}? If so, the compound is a conjugated ketone. If not . . .

(g) Does the strong carbonyl peak fall in the range 1670 cm^{-1} to 1630 cm^{-1}? If so, the compound is a tertiary amide. If not . . .

(h) Does the spectrum have a strong, wide peak in the range 1310 cm^{-1} to 1100 cm^{-1}? If so, does the carbonyl peak fall in the range 1730 cm^{-1} to 1715 cm^{-1}? If so, the compound is a conjugated ester; if not, the ester is not conjugated (Fig. 11.3, n-butyl acetate). If there is no strong, wide peak in the range 1310 to 1100 cm^{-1}, then . . .

(i) The compound is an ordinary nonconjugated ketone.

■ FIG. 11.4
The IR spectrum of cyclohexanol
(thin film).

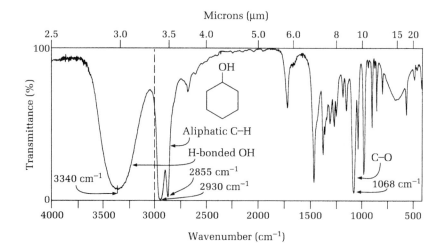

■ FIG. 11.5
The IR spectrum of benzonitrile
(thin film).

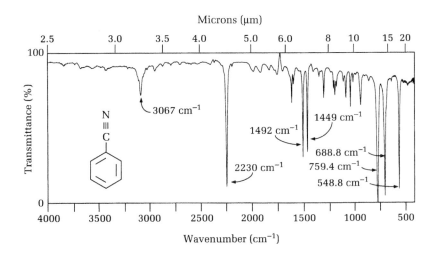

2. If the spectrum lacks a carbonyl peak in the range 1820 cm^{-1} to 1625 cm^{-1}, does it have a broad band in the region 3650 cm^{-1} to 3200 cm^{-1}? If so, does it also have a peak at about 1200 cm^{-1}, a C—H stretching peak to the left of 3000 cm^{-1}, and a peak in the region 1600 cm^{-1} to 1470 cm^{-1}? If so, the compound is a phenol. If the spectrum does not meet these latter three criteria, the compound is an alcohol (Fig. 11.4, cyclohexanol). However, if there is no broad band in the region 3650 cm^{-1} to 3200 cm^{-1}, then . . .

(a) Is there a broad band in the region 3500 cm^{-1} to 3300 cm^{-1}, does the compound smell like an amine, or does it contain nitrogen? If so, are there two peaks in this region? If so, the compound is a primary amine; if not, it is a secondary amine. However, if there is no broad band in the region 3500 cm^{-1} to 3300 cm^{-1}, then . . .

■ FIG. 11.6
The IR spectrum of
3-methylpentane (thin film).

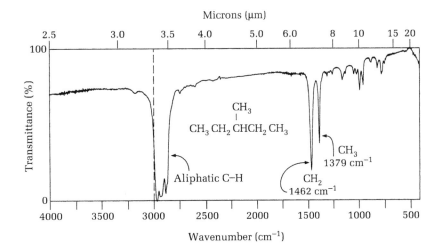

■ FIG. 11.7
The IR spectrum of *t*-butylbenzene
(thin film).

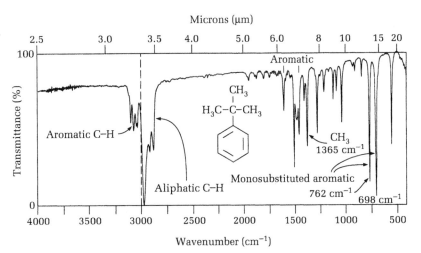

(b) Is there a sharp peak of medium-to-weak intensity at 2260 cm^{-1}
to 2100 cm^{-1}? If so, is there also a peak at 3320 cm^{-1} to 3310 cm^{-1}?
If so, then the compound is a terminal acetylene. If not, the compound
is most likely a nitrile (Fig. 11.5, benzonitrile), although it might be an
asymmetrically substituted acetylene. If there is no sharp peak of
medium-to-weak intensity at 2260 cm^{-1} to 2100 cm^{-1}, then . . .

(c) Are there strong peaks in the region 1600 cm^{-1} to 1540 cm^{-1} and
1380 cm^{-1} to 1300 cm^{-1}? If so, the molecule contains a nitro group.
If not . . .

(d) Is there a strong peak in the region 1270 cm^{-1} to 1060 cm^{-1}? If so, the
compound is an ether. If not . . .

(e) The compound is either a tertiary amine (odor?), a halogenated hydrocarbon (Beilstein test?), or just an ordinary hydrocarbon (Fig. 11.6,
3-methylpentane, and Fig. 11.7, *t*-butylbenzene).

The effect of ring size on the carbonyl frequencies of lactones and esters:

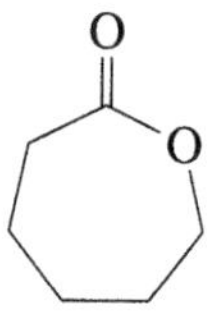

$1727\ \mathrm{cm}^{-1}$

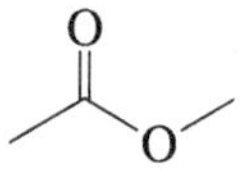

$1745\ \mathrm{cm}^{-1}$

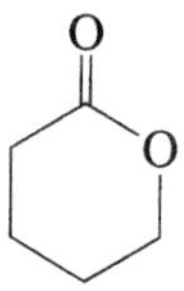

$1740\ \mathrm{cm}^{-1}$

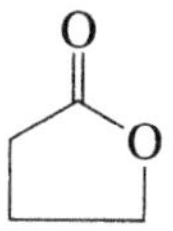

$1775\ \mathrm{cm}^{-1}$

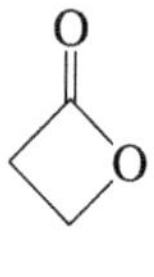

$1832\ \mathrm{cm}^{-1}$

Many comments can be added to this bare outline. For example, dilute solutions of alcohols will show a sharp peak at about $3600\ \mathrm{cm}^{-1}$ for a nonhydrogen-bonded O—H in addition to the usual broad hydrogen-bonded O—H peak.

Aromatic hydrogens give peaks just to the left of $3000\ \mathrm{cm}^{-1}$, whereas aliphatic hydrogens appear just to the right of $3000\ \mathrm{cm}^{-1}$. However, NMR spectroscopy is the best method for identifying aromatic hydrogens.

The carbonyl frequencies listed earlier refer to an open chain or an unstrained functional group in a nonconjugated system. If the carbonyl group is conjugated with a double bond or an aromatic ring, the peak will be displaced to the right by $30\ \mathrm{cm}^{-1}$. When the carbonyl group is in a ring smaller than six members or if there is oxygen substitution on the carbon adjacent to an aldehyde or ketone carbonyl, the peak will be moved to the left (refer to the margin notes about the effect of ring size and Table 11.2).

TABLE 11.2 • **Characteristic IR Carbonyl Stretching Peaks (Chloroform Solutions)**		
	Carbonyl-Containing Compounds	Wavenumber (cm^{-1})
RCR (ketone)	Aliphatic ketones	1725–1705
RCCl (acid chloride)	Acid chlorides	1815–1785
R—C=C—C—R (unsaturated)	α,β-Unsaturated ketones	1685–1666
Ar CR	Aryl ketones	1700–1680
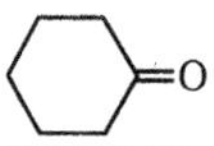 cyclohexanone =O	Cyclohexanones	1725–1705
—C—CH$_2$—C—	β-Diketones	1640–1540
RCH	Aliphatic aldehydes	1740–1720
R—C=C—CH	α,β-Unsaturated aldehydes	1705–1685
Ar CH	Aryl aldehydes	1715–1695

The effect of ring size on carbonyl frequency of ketones:

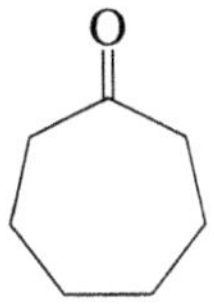

O (7-membered ring ketone)

1705 cm^{-1}

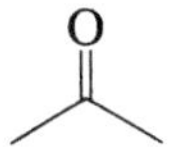

O (acetone)

1715 cm^{-1}

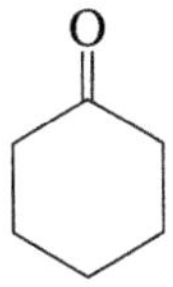

O (6-membered ring ketone)

1715 cm^{-1}

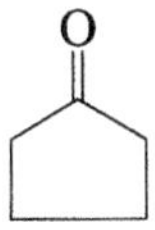

O (5-membered ring ketone)

1745 cm^{-1}

O (4-membered ring ketone)

1780 cm^{-1}

O (3-membered ring ketone)

1815 cm^{-1}

TABLE 11.2 • (Continued)		
	Carbonyl-Containing Compounds	**Wavenumber (cm^{-1})**
RCOOH	Aliphatic acids	1725–1700
R—C=C—COOH	α,β-Unsaturated acids	1700–1680
ArCOOH	Aryl acids	1700–1680
RCOR′ (O)	Aliphatic esters	1740
R—C=C—COR′ (O)	α,β-Unsaturated esters	1730–1715
Ar COR (O)	Aryl esters	1730–1715
HCOR (O)	Formate esters	1730–1715
$CH_2=CHOCCH_3$ (O), $C_6H_5OCCH_3$ (O)	Vinyl and phenyl acetate	1776
R—C(O)—O—R—C(O)	Acyclic anhydrides (two peaks)	1840–1800 1780–1740
RCNH$_2$ (O)	Primary amides	1694–1650
RCNHR′ (O)	Secondary amides	1700–1670
RCNR′$_2$ (O)	Tertiary amides	1670–1630

Methyl groups often give a peak near 1375 cm^{-1}, but NMR is a better method for detecting this group.

The pattern of substitution on an aromatic ring (mono-, *ortho-*, *meta-*, and *para*, di-, tri-, tetra-, and penta-) can be determined from C—H out-of-plane bending

■ **FIG. 11.8**
The band patterns of toluene and *o*-, *m*-, and *p*-xylene. These peaks are *very weak*. They are characteristic of aromatic substitution patterns in general, not just for these four molecules. Try to find these patterns in other spectra throughout this text.

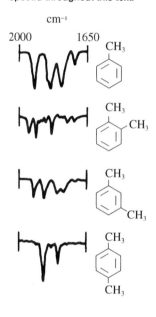

vibrations in the region 670–900 cm^{-1}. Much weaker peaks between 1650 cm^{-1} and 2000 cm^{-1} are illustrated in Figure 11.8.

Extensive correlation tables and discussions of characteristic group frequencies can be found in the specialized references listed at the end of this chapter.

The Fourier Transform Infrared (FTIR) Spectrometer

Like their NMR counterparts, Fourier transform IR instruments are computer based, can sum a number of scans to increase the signal-to-noise ratio, and allow enormous flexibility in the ways spectra can be analyzed and displayed because they are digitally oriented. In addition, they are faster and more accurate than double-beam spectrometers.

A schematic diagram of a single-beam FTIR spectrometer is given in Figure 11.9. IR radiation goes to a Michelson interferometer in which half the light beam passes through a partially coated mirror to a fixed mirror and half goes to a moving mirror. The combined beams pass through the sample and are then focused on the detector. The motion of the mirror gives a signal that varies sinusoidally. Depending on the mirror position, the different frequencies of light either reinforce or cancel each other, resulting in an interferogram. The mirror, the only moving part of the spectrometer, thus affects the light frequencies that encode the very high optical frequencies, transforming them to low-frequency signals that change as the mirror moves back and forth. The Fourier transform convert the digitized signal into a signal that is a function of frequency—creating a spectrum.

Like Fourier transform NMR spectroscopy, all IR frequencies are sampled simultaneously instead of being scanned successively, so a spectrum is obtained in seconds. Because there is no slit to sort out the wavelengths, high resolution is possible without losing signal strength. Extremely small samples or very dilute solutions can

■ **FIG. 11.9**
The optical path diagram of an FTIR spectrometer.

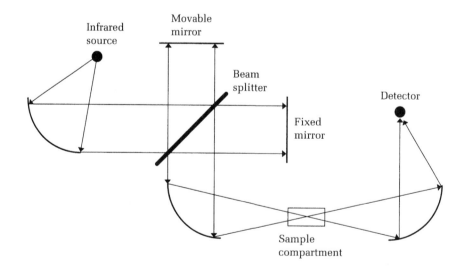

be examined because it is easy to sum hundreds of scans in the computer. It is even possible to take a spectrum of an object seen under a microscope. The spectrum of one compound can be subtracted from, say, a binary mixture spectrum to reveal the nature of the other component. Data can be smoothed, added, resized, and so on because of the digital nature of the output from the spectrometer.

Experimental Considerations

Glass, quartz, and plastics are opaque to IR radiation and absorb it. Therefore, sample holders must be made of other transparent materials. Metal halide salts such as sodium or silver chloride, calcium or barium fluoride, and potassium bromide are transparent to IR radiation; these are prepared as large polished crystals for use in cell holders, with NaCl and KBr being the most common. They are fragile and are easily attacked by moisture, so they must be handled gently by the edges and kept away from aqueous solutions. IR spectra can be determined on neat (undiluted) liquids, on solutions with an appropriate solvent, and on solids using mulls and KBr pellets, and by diffuse reflectance of the solid mixed with KBr. All IR cells or KBr powder must be stored in a desiccator when not in use.

> The sample, solvents, and equipment must be dry.

The Spectra of Neat Liquids

To run a spectrum of a neat (free of water!) liquid, remove a demountable cell (Fig. 11.10) from the desiccator, place a drop of the liquid between the salt plates, press the plates together gently to remove any air bubbles, and add the top rubber gasket and metal top plate. Next, put on all four of the nuts and *gently* tighten them to apply an even pressure to the top plate. Place the cell in the sample compartment (nearest the front of the spectrometer) and run the spectrum.

Although running a spectrum on a neat liquid is convenient and results in no extraneous bands to interpret, it is not possible to control the path length of the light through the liquid in a demountable cell. A low-viscosity liquid when squeezed between the salt plates may be so thin that the short path length gives peaks that are too weak. A viscous liquid, on the other hand, may give peaks that are too intense. A properly run spectrum will have the most intense peak with a transmittance of about 10%.

Another demountable cell is pictured in Figure 11.11. The plates are thin wafers of silver chloride, which is transparent to IR radiation. This cell has advantages over the salt cell in that the silver chloride disks are more resistant to breakage than NaCl plates, less expensive, and not affected by water. Because silver chloride is photosensitive, the wafers must be stored in a dark place to prevent them from turning black. Because one side of each wafer is recessed, the thickness of the sample can be varied according to the manner in which the cell is assembled. In general, the spectra of pure liquids are run as the thinnest possible films. Most of the spectra in this chapter have been obtained in this way. The disks are cleaned by rinsing them with an organic solvent such as acetone or ethanol and wiping dry with an absorbent paper towel.

> A fast, simple alternative: Place a drop of the compound on a round salt plate, add a top plate, and mount the two on the holder pictured in Figure 11.11.

FIG. 11.10
FIG. 11.10
An exploded view of a demountable salt cell for analyzing the IR spectra of neat liquids. The salt plates are fragile and expensive. Do not touch the surfaces. Use only dry solvents and samples.

FIG. 11.11
A demountable silver chloride or sodium chloride cell.

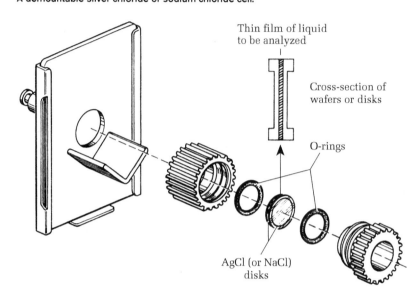

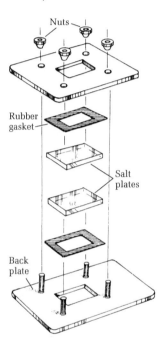

Solvents: $CHCl_3$, CS_2, CCl_4

CAUTION: Chloroform, carbon tetrachloride, and carbon disulfide are toxic. The first two are carcinogens. Use a laboratory hood to minimize exposure to vapors.

A very simple cell consists of two circular NaCl disks (1-in. diameter × 3/16-in. thick, 25 mm × 4 mm). The sample, 1 drop of a pure liquid or solution, is applied to the center of a disk with a polyethylene pipette (to avoid scratching the NaCl disk). The other disk is added, the sample is squeezed to a thin film, and the two disks are placed on the V-shaped holder shown in Figure 11.11.

The Spectra of Solutions

The most widely applicable method of running the spectra of solutions involves dissolving an amount of the liquid or solid sample in an appropriate solvent to give a 10% solution. Just as in NMR spectroscopy, the best solvents to use are carbon disulfide and carbon tetrachloride; because these compounds are not polar enough to dissolve many substances, chloroform is used as a compromise. Unlike NMR solvents, no solvent suitable in IR spectroscopy is entirely free of absorption bands in the frequency range of interest (Figs. 11.12, 11.13, and 11.14). In chloroform, for instance, no light passes through the cell between 650 cm^{-1} and 800 cm^{-1}. As can be seen from the figures, spectra obtained using carbon disulfide and chloroform cover the entire IR frequency range. Carbon tetrachloride would appear to be a good choice because it has few interfering peaks, but it is a poor solvent for polar compounds. In practice, a background spectrum is run with the same solvent in the same cell used to run the sample solution, and the data system subtracts the solvent peaks in background spectrum from the solution spectrum to yield an IR spectrum of the sample only.

■ FIG. 11.12
The IR spectrum of chloroform
(thin film).

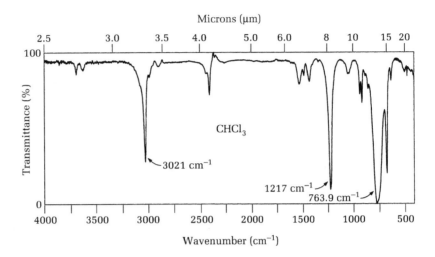

■ FIG. 11.13
The IR spectrum of carbon
disulfide (thin film).

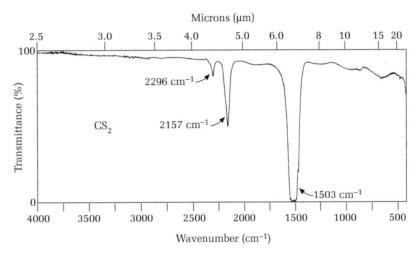

■ FIG. 11.14
The IR spectrum of carbon
tetrachloride (thin film).

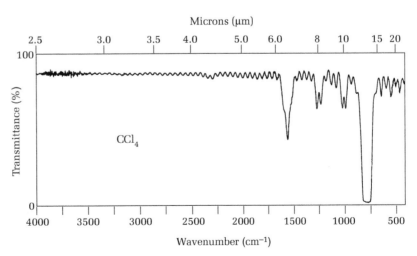

■ **FIG. 11.15**
A sealed IR sample cell.

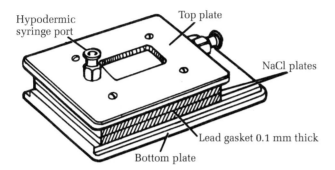

■ **FIG. 11.16**
Flushing an IR sample cell. The solvent used
to dissolve the sample is used in this process.

Three drops are needed to fill the cell.

Three large drops of solution will fill the usual sealed IR cell (Fig. 11.15). A 10% solution of a liquid sample can be approximated by diluting 1 drop of the liquid sample with 9 drops of the solvent. Because weights are more difficult to estimate, solid samples should be weighed to obtain a 10% solution.

The solvent and the sample must be dry. Do not touch or breathe on NaCl plates. Cells are very expensive.

First, obtain a background spectrum of the solvent. The IR cell is filled by inclining it slightly and placing about 3 drops of the solvent in the lower hypodermic port with a capillary dropper. The cell must be completely dry because the new sample will not enter if the cell contains solvent. The liquid can be seen rising between the salt plates by looking into the window. In the most common sealed cell, the salt plates are spaced 0.1 mm apart. Be sure that the cell is filled past the window and that no air bubbles are present. Then place a Teflon stopper lightly but firmly in the hypodermic port. Be particularly careful not to spill any of the sample on the outside of the cell windows. Place the cell in the instrument cell holder and acquire a background spectrum. Blow dry nitrogen or draw dry air through the cell to remove the solvent and dry the cell, fill with the sample solution prepared above, and acquire the spectrum.

After running the sample, force clean solvent through the cell using a syringe attached to the top port of the cell (Fig. 11.16). Dry using dry nitrogen or air and store in a desiccator.

Dispose of waste solvents in the container provided.

Cleaning Up. Discard halogenated liquids in the halogenated organic waste container. Other solutions should be placed in the organic solvents container.

Mulls

The sample must be finely ground.

Solids insoluble in the usual solvents can be run as mulls. In preparing a mull, the sample is ground to a particle size less than that of the wavelength of light going through the sample (2.5 μm) to avoid scattering light. About 15–20 mg of the

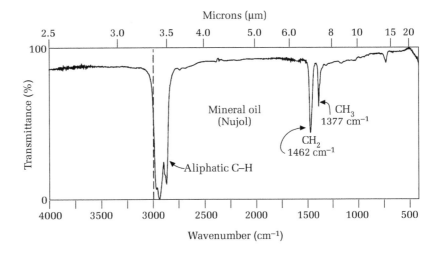

■ **FIG. 11.17**
The IR spectrum of Nujol (paraffin oil).

sample is ground for 3–10 min in an agate mortar until it is spread over the entire inner surface of the mortar and has a caked and glassy appearance. Then to make a mull, add 1 or 2 drops of paraffin oil (Nujol; Fig. 11.17) and grind the sample for 2–5 more minutes. The mull, which should have the consistency of thin margarine, is transferred to the bottom salt plate of a demountable cell (*see* Fig. 11.10 on page 242) using a rubber policeman, the top plate is added and rotated back and forth to distribute the sample evenly to eliminate all air pockets, and the spectrum is run. Because the bands from Nujol obscure certain frequency regions, running another mull using Fluorolube as the mulling agent will allow the entire IR spectral region to be covered. If the sample has not been ground sufficiently fine, there will be marked loss of transmittance at the short-wavelength end of the spectrum. After running the spectrum, the salt plates are wiped clean with a cloth saturated with an appropriate solvent.

A fast, simple alternative: Grind a few milligrams of the solid with 1 drop of tetrachloroethylene between two round salt plates. Mount on the holder as shown in Figure 11.11 on page 242.

Potassium Bromide Disk

The spectrum of a solid sample can also be run by incorporating the sample in a potassium bromide (KBr) disk. This procedure needs only one disk to cover the entire spectral range because KBr is completely transparent to IR radiation. Although very little sample is required, making the disk calls for special equipment and time to prepare it. Because KBr is hygroscopic, water is a problem.

Into a stainless steel capsule containing a ball bearing are weighed 1.5–2 mg of the compound and 200 mg of spectroscopic-grade KBr (previously dried and stored in a desiccator). The capsule is shaken for 2 min on a Wig-L-Bug (the device used by dentists to mix silver amalgam). The sample is evenly distributed over the face of a 13-mm die and subjected to a pressure of 14,000–16,000 psi for 3–6 min while under vacuum in a hydraulic press. A transparent disk is produced, which is removed from the die with tweezers and placed in a holder like that shown in Figure 11.11 on page 242 prior to running the spectrum.

■ **FIG. 11.18**
A simplified diagram of a diffuse
reflectance system.

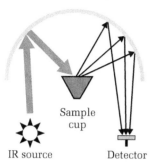

Sample
cup

IR source   Detector

Diffuse Reflectance

Reasonable IR spectra of solids can be obtained without pressing a KBr pellet if the IR instrument is equipped with diffuse reflectance optics. A mixture of the sample and KBr is put in a small sample cup, and this is placed in the diffuse reflectance optics system (Fig. 11.18). When the IR beam enters the sample, it can either be reflected off the surface of a particle or be transmitted through a particle. The IR beam that passes through a particle can either reflect off the next particle or be transmitted through the next particle. This transmission-reflectance event can occur many times in the sample, which increases the path length. Finally, any scattered IR energy that is not absorbed by the sample is collected by a spherical mirror that is focused onto a detector. A background spectrum of only pure KBr should be run before the sample to subtract out any absorbance that is not from the sample.

Weigh out 0.160–0.170 g of KBr quickly (this compound readily absorbs atmospheric humidity) and place it in a vial. Weigh out between 4 mg and 7 mg of your solid and add this to the vial. Mix and dump into a small, dry mortar and pestle. Grind for 1–2 min until a fine powder is obtained. Transfer to weighing paper and then to the cup of the diffuse reflectance sample holder. Place the holder in the reflectance attachment and acquire the spectrum.

Gas Phase IR Spectroscopy: The Williamson Gas Phase IR Cell

In the past it was difficult to obtain the IR spectra of gases. A commercial gas cell costs hundreds of dollars; the cell must be evacuated, and the dry gas carefully introduced—not a routine process. But we have found that a uniquely simple gas phase IR cell can be assembled at virtually no cost from a 105° microscale connecting adapter.[2]

■ **FIG. 11.19**
The Williamson Gas Phase IR Cell
utilizing the 105° connecting
adapter from the microscale kit
that accompanies this text.

Wad of cotton

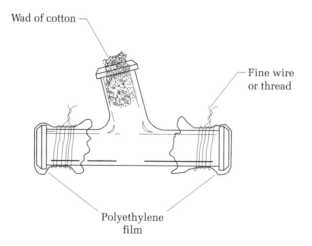

Fine wire
or thread

Polyethylene
film

---

2. The microscale connecting adapter is a available from Kimble Kontes (catalog no. 748001-1000).

■ FIG. 11.20
The Williamson Gas Phase IR Cell mounted in a cell holder.

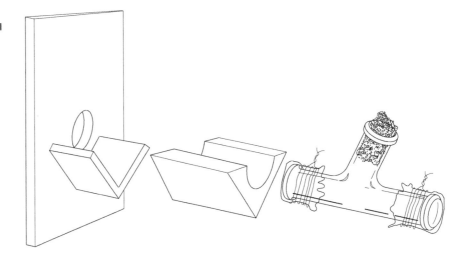

■ FIG. 11.21
The IR spectrum of polyethylene film.

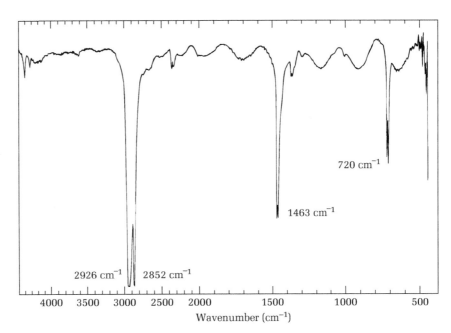

720 cm^{-1}

1463 cm^{-1}

2926 cm^{-1} 2852 cm^{-1}

4000   3500   3000   2500   2000   1500   1000   500

Wavenumber (cm^{-1})

    To make the Williamson gas phase IR cell, attach a thin piece of polyethyene film from, for example, a sandwich bag to each end of the adapter with a small rubber band, fine wire, or thread. Place a loose wad of cotton in the side arm and mount the IR cell in the beam of the spectrometer (Fig. 11.19 and Fig. 11.20).

    The adapter can be supported on a V-block as shown in Figure 11.20 or, temporarily, on a block of modeling clay. Be sure the beam, usually indicated by a red laser, strikes both ends of the cell. Run a background spectrum on the cell. The spectrum of polyethylene (Fig. 11.21) is very simple, so interfering bands will be few. This spectrum is automatically subtracted when you put a sample in the cell and run a spectrum.

A more permanent cell can be constructed by attaching two thin plates of silver chloride to the ends of the adapter with epoxy glue. These silver chloride plates can be made in the press used for making KBr discs, or they can be purchased.[3]

Obtaining a Spectrum

To run a gas phase spectrum, remove the cell from the spectrometer and spray a very short burst of a gas from an aerosol can onto the cotton. The nonvolatile components of the spray will stick to the cotton, and the propellant gas will diffuse into the cell. Run the spectrum in the usual way. The concentration of the gas is about the same as a thin film of liquid run between salt plates. If the concentration of the gas is too high, remove the cell from the spectrometer, remove the cotton, and "pour out" half the invisible gas by holding the cell in a vertical position.

You may be fortunate in having in your school's library a copy of Volume 3 of *The Aldrich Library of FT-IR Spectra: Vapor Phase* or an online database of IR spectra on your IR instrument data system. Either resource contains hundreds of reference spectra that will help you analyze the IR spectra of propellants in aerosol cans and any other gases you may encounter. Otherwise you may want to compare your spectra to the few in this text and available on the website for this text and assemble your own library of gas phase spectra. It is possible to make approximate calculations of IR vibrational frequencies using a semi-empirical calculation at the AM1 level (*see* Chapter 15).

Throughout the remainder of this book, representative IR spectra of starting materials and products are presented, and the important bands in each spectrum are identified.

■ FIG. 11.22
The IR spectrum of carbon dioxide (gas phase).

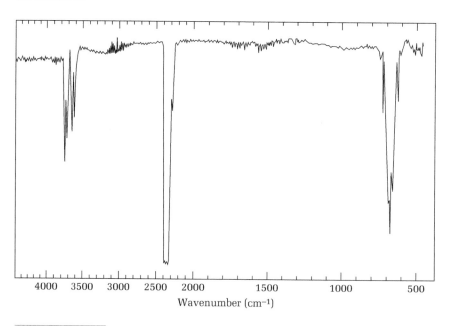

Wavenumber (cm⁻¹)

---

3. The silver chloride disks can be purchased from Fisher Scientific (catalog no. 14-385-860).

■ **FIG. 11.23**
The IR spectrum of methane
(gas phase).

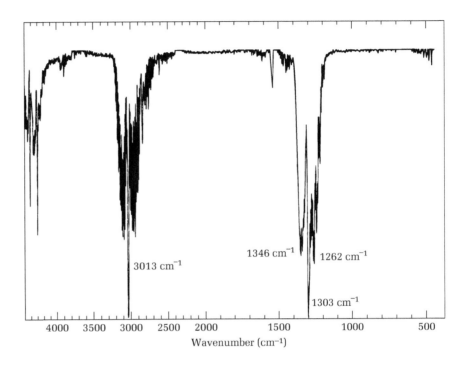

For Further Investigation

The Gas Phase IR Spectra of Aerosol Propellants

Acquire gas phase spectra from a number of substances from aerosol cans. Some cans will be labeled with the name of the propellant gas; others will not. Analyze the spectra, bearing in mind that the gas will be a small molecule such as carbon dioxide (Fig. 11.22), methane (Fig. 11.23), nitrous oxide (Fig. 11.24), propane, butane, isobutane, or a fluorocarbon such as 1,1,1,2-tetrafluoroethane. A very old aerosol can may contain an illegal chlorofluorocarbon, now outlawed because these substances destroy the ozone layer. Some examples of aerosol cans include hairsprays, lubricants (WD-40, Teflon), Freeze-It, inhalants for asthma, cigarette lighters, butane torches, and natural gas from the lab.

Other Gas Phase Spectral Research

You can deliberately generate hydrogen chloride (or deuterium chloride) and sulfur dioxide and then analyze the gas that is evolved from a thionyl chloride reaction. Can you detect carbon monoxide in automobile exhaust? How about marsh gas that you see bubbling out of ponds? With perhaps dozens of scans, can you record the spectra of vanillin over vanilla beans, eugenol vapor over cloves, anisaldehyde over anise seeds? What is the nature of the volatile components of chili peppers, orange peel, nail polish, and nail polish remover? Could you use this method to determine if someone has been drinking alcohol?

■ **FIG. 11.24**
The IR spectrum of nitrous oxide (gas phase).

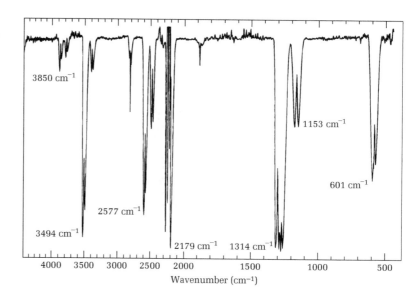

EXPERIMENT

Unknown Carbonyl Compound

The sample, solvents, and apparatus must be dry.

Run the IR spectrum of an unknown carbonyl compound obtained from your laboratory instructor. Be particularly careful that all apparatus and solvents are completely free of water, which will damage the NaCl plates. Determine the frequency of the carbonyl peak in the unknown *and* list the possible types of compounds that could correspond to this frequency (Table 11.2).

REFERENCES

1. Colthrup, Norman B., Lawrence H. Daly, and Stephen E. Wiberley, *Introduction to Infrared and Raman Spectroscopy*, 3rd ed. New York: Academic Press, 1990.
2. Cooper, James W., *Spectroscopic Techniques for Organic Chemists*. New York: John Wiley, 1980.
3. Lin-Vein, Daimay, Norma B. Colthrup, William G. Fateley, and Jeannette G. Grasselli, *The Handbook of Infrared and Raman Characteristic Frequencies of Organic Molecules*. Boston: Academic Press, 1991.
4. Pouchert, Charles J. *The Aldrich Library of FT-IR Spectra*, 2nd ed. Milwaukee: Aldrich Chemical Co., 1997.
5. Roeges, Noel P. G., *A Guide to the Complete Interpretation of Infrared Spectra of Organic Structures*. New York: John Wiley, 1994.
6. Silverstein, Robert M., Francis X. Webster, and David Kiemle, *Spectrometric Identification of Organic Compounds*, 7th ed.[4] New York: Wiley, 2005.

4. This book includes IR, UV, and NMR spectra.

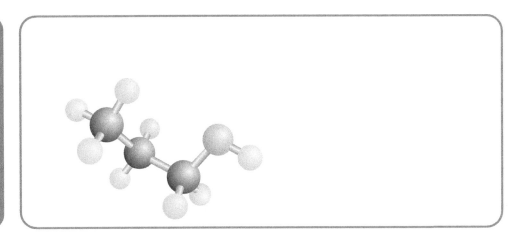

CHAPTER

12

Nuclear Magnetic Resonance Spectroscopy

Online Study Center

This icon will direct you to techniques, equipment setups, and online resources at http://college.hmco.com/PIC/williamson MMOE5e.

PRELAB EXERCISE: Outline the preliminary solubility experiments you would carry out using inexpensive solvents before preparing a solution of an unknown compound for nuclear magnetic resonance (NMR) spectroscopy using expensive deuterated solvents.

^{1}H NMR: Determination of the number, kind, and relative locations of hydrogen atoms (protons) in a molecule

Most organic chemists would agree that the most powerful instrumental method for revealing the structure of organic molecules is nuclear magnetic resonance (NMR) spectroscopy. The 2002 Nobel Prize in Chemistry was awarded, in part, to Kurt Wüthrich for advances in NMR spectroscopy that allowed the determination of the three-dimensional structure of biological macromolecules in solution. Because of such capabilities, introductory organic chemistry courses devote considerable time to the study of NMR. This concise chapter assumes that you have had some prior exposure to the concepts discussed; it focuses on the practical aspects of sample preparation, data acquisition, and interpretation of NMR spectra to elucidate and confirm organic structures.

Let us briefly review the theory of NMR. Certain nuclei such as ^{1}H, ^{13}C, ^{15}N, ^{18}O, ^{19}F, and ^{31}P are tiny magnets that can assume two energy states when placed in a magnetic field: aligned (lower energy) and opposed (higher energy). Like the energy states in electronic spectra, the difference in energy between these states is quantized, and energy is absorbed and emitted only at certain radio frequencies. These frequencies depend on the type of nucleus and the magnetic field strength, which is constant for a given instrument. More importantly, however, is the fact that slight differences in electronegativity and bonding state (sp, sp^2, and sp^3) of surrounding atoms cause small—parts per million (ppm)—variations in the magnetic field felt by each nucleus, causing them to have absorption signals at slightly different frequencies. These small variations, called *chemical shifts*, are plotted versus signal intensity to produce the NMR spectrum. The interpretation of these signals and other spectral features such as splitting

Chemical shift, δ (ppm)

251

patterns and peak areas, as described in the following sections, facilitates organic structure elucidation.

Interpretation of ^{1}H NMR Spectra

There are two approaches to interpreting proton NMR spectra:

1. The *structure from the spectrum* approach is the strategy of using the information in the NMR spectrum to draw the structure of the molecule based on reference tables and rules. This approach is used if the compound's structure is unknown. In this case, the NMR spectrum alone is often insufficient to "solve" the complete structure and must be combined with knowledge about the compound's source (synthetic reaction or natural product) and complementary spectral data (infrared, ultraviolet, and/or mass spectrometric).

2. The *spectrum from the structure* approach might be thought of as the reverse of the first approach and is commonly used when verifying products of known reactions using known starting materials. A hypothetical NMR spectrum is created, based on the known compound's structure (or related structural possibilities), using the same reference tables and rules as in the first approach. Computer programs are available that allow the calculation of a hypothetical NMR spectrum for any molecule simply by entering its structure. (Refer to the supplemental information for this chapter on this book's website.) Even better, there may be a published spectrum of the compound available in a collection of reference spectra. Either way, the sample spectrum should be compared to the hypothetical or reference spectrum, and if the two are closely matched, one can be confident the correct compound has been obtained.

H_3C I
  \\ /
  CH_2

Ethyl iodide

NMR spectra contain considerably more structural information than do infrared spectra, and this additional information should be used in interpreting them. The following four most informative features are the principal ones to look for when interpreting spectra. We will use ethyl iodide as a specific illustration of each feature.

1. The Number of Signals Due to Equivalent Hydrogen Nuclei or Protons

The three hydrogen atoms of the methyl group are said to be chemically and magnetically equivalent. Replacing any of these hydrogens with another atom will give the same compound. Similarly, the two methylene protons are magnetically and chemically equivalent. This is because there is very fast rotation about the carbon-carbon bond. Individual hydrogens or each set of *equivalent* hydrogens will experience slightly different magnetic fields depending on their chemical environment and, therefore, will produce a signal at different places in the spectrum. Thus, we would predict that ethyl iodide would produce two signals, one for the

CH_3 CH_3
\Si/
CH_3 CH_3

Tetramethylsilane

three equivalent CH_3 protons and the other from the equivalent protons on the CH_2 bonded to the electronegative iodine. The NMR spectra of ethyl iodide, regardless of whether it is obtained at low field strength (Fig. 12.1) or high field strength (Fig. 12.2), shows three signals, or peaks, at 0.00 ppm, 1.83 ppm, and 3.20 ppm, with the latter two split into patterns of lines. The peak at 0.00 ppm is due to the addition of a reference compound, tetramethylsilane (TMS), used as the zero reference point to assure that the instrument frequency assignments are accurate. Thus, there are only two sets of signal peaks from ethyl iodide as predicted.

■ FIG. 12.1
The ^1H NMR spectrum of ethyl iodide (60 MHz). The stair-step like line is the integral. In the integral mode of operation, the recorder pen moves from left to right and moves vertically a distance proportional to the areas of the peaks over which it passes. Hence the relative areas of the quartet of peaks at 3.20 ppm and the triplet of peaks at 1.83 ppm are given by the relative heights of the integrals (4 cm is to 6 cm as 2 is to 3). The relative numbers of hydrogen atoms are proportional to the peak areas (2 H and 3 H).

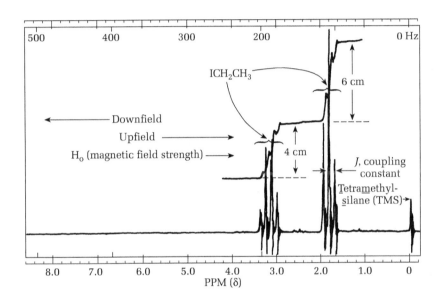

■ FIG. 12.2
The ^1H NMR spectrum of ethyl iodide (250 MHz). Compare this spectrum to Figure 12.1, run at 60 MHz. The peaks at 1.83 ppm and 3.20 ppm have been expanded and plotted on the left side of the spectrum.

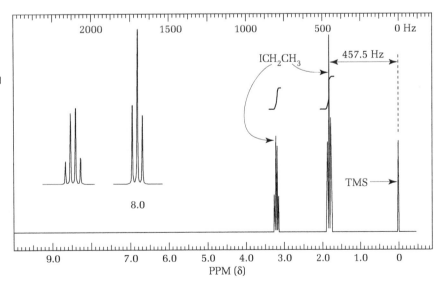

2. The Position of Each Signal on the Horizontal Axis

Each signal's position on the horizontal axis is called the *chemical shift*, and it indicates the effective magnetic field around a single proton or group of equivalent protons as affected by nearby bonds and atoms. This *chemical environment* depends on structure, that is, the bond types (sp^3, sp^2, sp, or aromatic) and the electronegativities of the atoms one, two, and three bonds away from the particular protons. Chemical shifts, symbolized by δ, are measured in dimensionless parts per million (ppm) to the left of, or *downfield* from, the reference TMS peak, according to the equation

$$\delta(\text{ppm}) = \frac{\text{shift of peak downfield from TMS (in Hz)}}{\text{spectrometer frequency (in MHz)}}$$

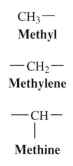

CH_3—
Methyl

—CH_2—
Methylene

—CH—
|
Methine

Table 12.1 gives the chemical shifts for protons in different chemical environments, spanning the region from 0 ppm to 12.5 ppm. Each signal in the spectrum of ethyl iodide can be assigned to a particular structural feature using this and similar tables. According to Table 12.1, the hydrogens on CH_2 groups attached to both an alkyl group and iodine in RCH_2I (where R can be methyl, methylene, or methine) should yield a signal between ~2.3 ppm and 3.2 ppm. We see a signal at 3.20 ppm in the spectrum of ethyl iodide, which is in this range. The table shows a range of 0.7 ppm to 1.1 ppm for a methyl group attached to a methylene or methine R group (RCH_3); R is –CH_2I for ethyl iodide. In actuality, the other signal in the spectrum of methyl iodide is not in this range; it is at 1.83 ppm, which is further downfield than predicted. This type of inconsistency occurs occasionally and demonstrates that reference tables are only generally representative of the majority of molecules that have been examined. It appears that in this particular case, the electronegative iodine, even though it is three bonds away, still strongly affects the magnetic field and shifts the peak for the methyl protons downfield. An examination of the other features of the spectrum will show that this peak assignment is correct.

3. Splitting Patterns and Coupling Constants

Patterns and coupling constants in the spectrum help define which groups are next to each other in a molecule's carbon skeleton. In an open-chain molecule with free rotation about the bonds and no chiral centers, protons couple with each other over three chemical bonds to give characteristic patterns of lines. If one or more equivalent protons couple to *one* adjacent proton, then the coupling hydrogen(s) appears as a *doublet* of equal intensity lines separated by the *coupling constant* (J) measured in Hertz (Hz). If the coupling proton or protons couple equally to *two* protons three chemical bonds away, they appear as a *triplet* of lines with relative intensities of 1:2:1, again separated by J. Finally, a *quartet* of lines in the ratio of 1:3:3:1 arises when one or more protons couple to the *three* protons on a methyl group. In general, chemically equivalent protons give a pattern of lines containing one more line than the number of protons being coupled to, and the intensities of the peaks follow the binomial expansion, conveniently represented by Pascal's triangle. Thus the methylene group of ethyl iodide appears as a quartet of lines at 3.20 ppm with relative intensities of 1:3:3:1 because the two methylene protons

Pascal's triangle

```
        1              singlet, s
      1   1            doublet, d
    1   2   1          triplet, t
  1   3   3   1        quartet, q
1   4   6   4   1      pentet
1  5  10  10  5  1     sextet
```

TABLE 12.1 • **Proton Chemical Shifts**

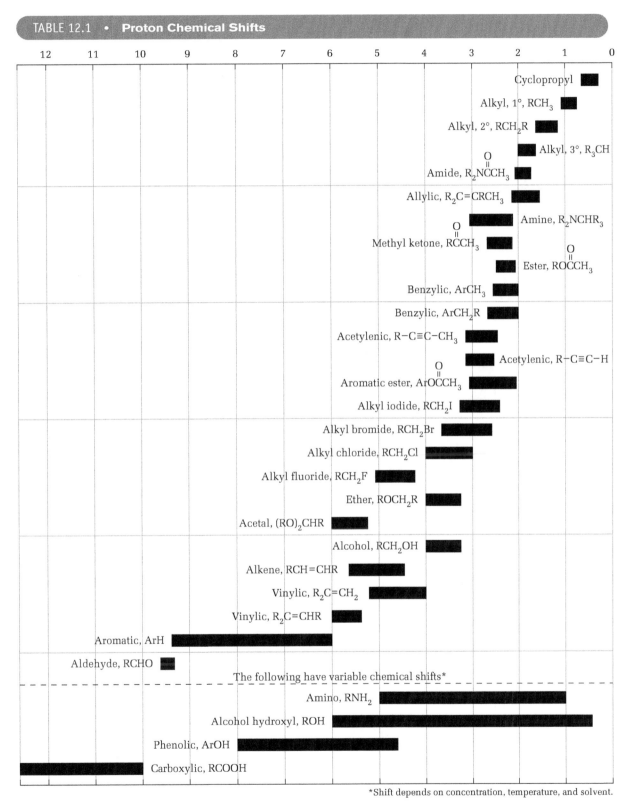

*Shift depends on concentration, temperature, and solvent.

are coupled to the three equivalent methyl protons (*see* Fig. 12.1 and Fig. 12.2). The methyl peak is split into a 1:2:1 triplet centered at 1.83 ppm because the methyl protons are coupled to the two adjacent methylene protons. This quartet-triplet combination is a strong indication of an ethyl group in any molecule. The *J* indicated in Figure 12.1 is the same 7.6 Hz between all lines in the quartet and triplet because the two groups are adjacent; therefore, the protons are coupled. In the 250 MHz spectrum, Figure 12.2, the splitting patterns are compressed and harder to distinguish because there are more Hz per ppm, but the expansion of these peaks on the left of the spectrum clearly shows a quartet and triplet.

Other characteristic proton coupling constants are given in Table 12.2. In alkenes *trans* coupling is larger than *cis* coupling, and both are much larger than

TABLE 12.2 • Spin-Spin Coupling Constants for Various Geometries

Fragment	J (Hz)	Fragment	J (Hz)
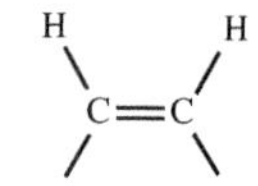	7–12	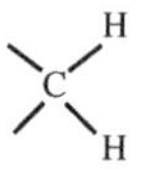	12–15
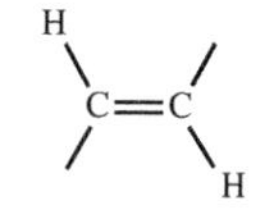	13–18	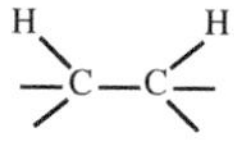	0–10
	0.5–3		6–9
			1–3
			0–1
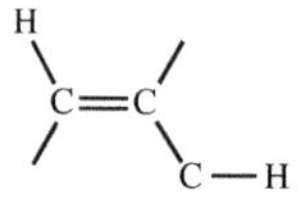	0.5–2.5	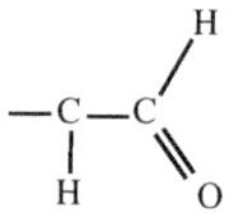	1–3
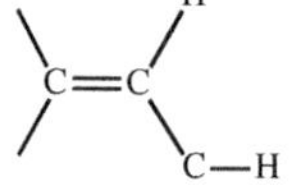	4–10	$CH_3 - CH_2 -$	6.5–7.5
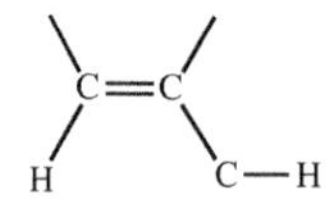	0–3	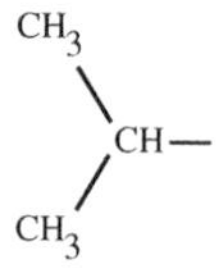	5.5–7
	9–13		5–9
			2–4

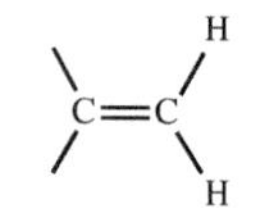

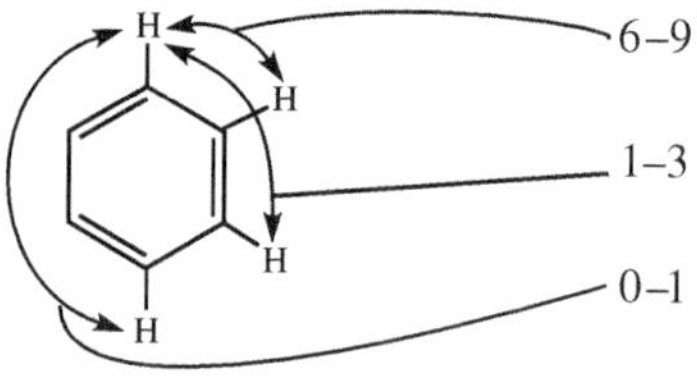

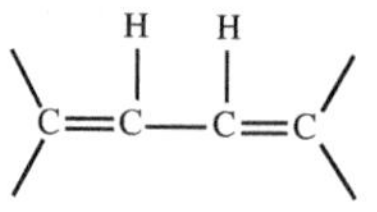

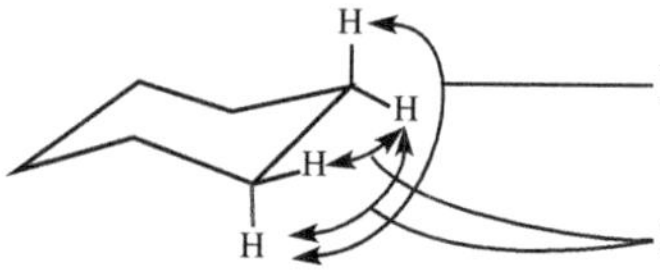

geminal coupling (coupling that takes place between two groups on the same carbon). *Ortho*, *meta*, and *para* couplings in aromatic rings range from 0 Hz to 9 Hz. The couplings in a rigid system of saturated bonds are strongly dependent on the dihedral angle between the coupling protons, as seen in cyclohexane.

4. The Integral

The relative numbers of distinctive hydrogen atoms (protons) in the molecule of ethyl iodide are determined from the *integral*, the stair-step line over the peaks shown in Figures 12.1 and 12.2. The height of the step is proportional to the area under the NMR signal for each group of equivalent protons. In NMR spectroscopy (contrasted with infrared spectroscopy, for instance) the area for each signal, including all splitting peaks, is directly proportional to the number of hydrogen atoms causing that signal. The methyl protons give a triplet of peaks, and the methylene protons give a quartet of peaks. Both ethyl iodide spectra in Figures 12.1 and 12.2 have a ratio of 3 to 2 for the area of the methyl triplet relative to the area of the methylene quartet, which is further evidence for the presence of an ethyl group. Integrators are part of all NMR spectrometers; running the integral takes little more time than running the spectrum. Most spectrometers print out a numerical value for the integral (with many more significant figures than are justified!).

In actuality, when chemists interpret NMR spectra, they use a combination of the *structure from the spectrum* and the *spectrum from the structure* approaches. A typical mental analysis of the NMR spectrum of 4-iodotoluene (Fig. 12.3) might go like this:

> This molecule should produce three signals: a singlet for the methyl protons, a doublet for the two equivalent aromatic protons *ortho* to the methyl group, and a doublet for the other two equivalent aromatic protons *ortho* to the iodo group. The methyl group is attached to an aromatic

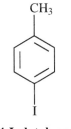

4-Iodotoluene

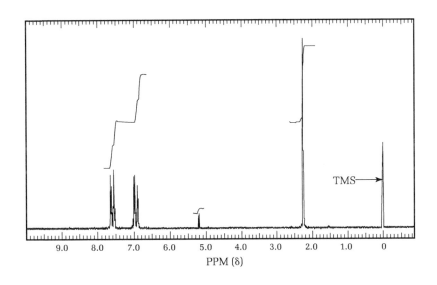

■ **FIG. 12.3**
The ¹H NMR spectrum of 4-iodotoluene (90 MHz in CDCl₃).

ring, and according to the table of proton chemical shifts, this should be detected as a singlet peak between 1.8 ppm and 2.8 ppm. The NMR spectrum of my product shows a singlet at 2.28 ppm, which is consistent with this. The aromatic protons show up as doublets in the right range, 6 ppm to 9 ppm according to the table. The spectrum has doublets at 6.95 ppm and 7.60 ppm. It also shows a singlet at 5.2 ppm and, according to the table, this implies that my product contains hydrogen on an alkene or in an acetal. Bad news! From other information I know, my product has neither of these functional groups. But wait! If I look at spectra of common sample impurities, I see that dichloromethane, the solvent I extracted the product with, appears at 5.25 ppm as a singlet. So this peak is not from my product and can be ignored . . .

This back-and-forth analysis correlating structure and NMR data, including coupling constants and integrals, is continued until all NMR peaks, splitting patterns, and integrals can be accounted for, and the identity of the product is verified.

When NMR data are reported in the literature, it is usually in a concise numerical form. For example, the NMR data for ethyl iodide derived from its spectrum (Fig. 12.2) would be reported as ^{1}H NMR (CDCl$_3$): 1.83 (3H, t, J = 7.6 Hz), 3.20 (2H, q, J = 7.6 Hz), where CDCl$_3$ (deuterochloroform) is the solvent, 1.83 and 3.20 are the chemical shifts in ppm, 3H and 2H are integrals, t = triplet, q = quartet, and J is the coupling constant in Hz.

A more complex spectrum

Some NMR spectra are not as easily analyzed as the spectrum for ethyl iodide. Consider the spectra shown in Figure 12.4. The proton spectrum of this unsaturated chloroester has been run at 500 MHz. Each chemically and magnetically nonequivalent proton is well resolved so that all of the couplings can be seen. Only the quartet of peaks at 4.1 ppm (relative area 2) and the triplet at 1.2 ppm (relative area 3) follow the simple first-order coupling rules outlined earlier. This quartet/triplet pattern is very characteristic of the commonly encountered ethyl group.

Because of the chiral carbon (marked with an asterisk), the protons on carbons 5 and 7 are diastereotopic, have different chemical shifts, and couple with each other and with adjacent protons to give the patterns seen in the spectrum. Many of the peaks can be assigned to specific hydrogens based simply on their chemical shifts. The coupling patterns then confirm these assignments.

Carbon-13 Spectroscopy

The element carbon consists of 98.9% carbon atoms with mass 12 (^{12}C) and spin 0 (NMR inactive) and only 1.1% carbon atoms with mass 13 (^{13}C) and spin 1/2 (NMR active). Carbon, with such a low concentration of spin 1/2 nuclei, gives a very small signal when run under the same conditions as used for a proton spectrum. Carbon resonates at 75 MHz in a spectrometer where protons resonate at 300 MHz. Because only 1 in 100 carbon atoms has mass 13, the chances of a molecule having two ^{13}C atoms adjacent to one another are small. Consequently, coupling of one carbon with another is not observed.

Each ^{13}C atom couples to hydrogen atoms over one, two, and three bonds. Because the coupling constants are large, there is a high probability of peak overlap. To simplify the spectra as well as to increase the signal-to-noise ratio, a

The 500-MHz ^{1}H and 75-MHz ^{13}C NMR spectra of ethyl (3-chloromethyl)-4-pentenoate.[1] *I. ^{13}C DEPT[2] spectrum.* The CH$_3$ and CH peaks are upright, and the CH$_2$ peaks are inverted. The quaternary carbon (the carbonyl carbon) does not appear. *II. The normal 75-MHz noise-decoupled ^{13}C spectrum.* Note the small size of the carbonyl peak. *III. Expansions of each group of proton NMR peaks.* Protons E and F as well as protons H and I are not equivalent to each other. These pairs of protons are diastereotopic because they are on a carbon adjacent to a chiral carbon atom. The frequencies of all peaks are found on this book's website. *IV. The integral.* The height of the integral is proportional to the number of protons under it. *V. The 500-MHz ^{1}H spectrum.*

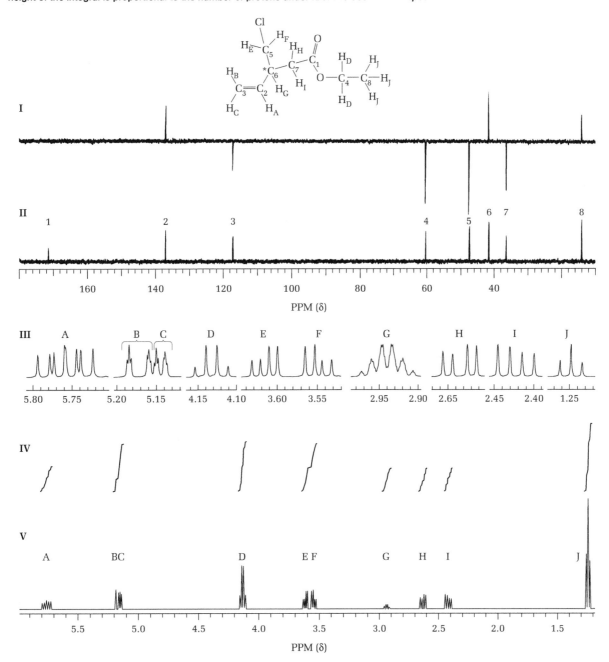

1. Spectra courtesy of Professor Scott Virgil.

2. DEPT: *d*istortionless *e*nhancement by *p*olarization *t*ransfer

TABLE 12.3 • **Carbon Chemical Shifts**

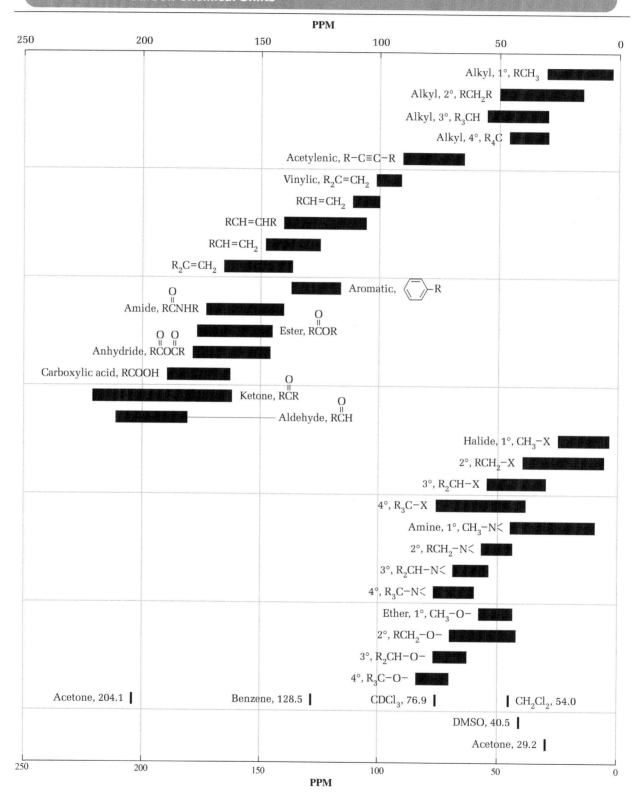

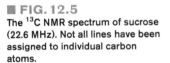

^{13}C spectra: Broadband noise decoupling gives a single line for each carbon.

special technique is routinely used in obtaining ^{13}C spectra: *broadband noise decoupling*. Decoupling has the effect of collapsing all multiplets (quartets, triplets, etc.) into a single peak. Further, the energy put into decoupling the protons can be looked on as appearing in the carbon spectrum in the form of an enhanced peak. This nuclear Overhauser enhancement (NOE) effect makes the peak appear three times larger than it would otherwise be. The result of decoupling is that every chemically and magnetically distinct carbon atom will appear as a single sharp line in the spectrum. Because the NOE effect is somewhat variable and does not affect carbons bearing no protons, one cannot do carbon counting from peak integrals the way hydrogen counting is done from proton NMR spectra.

Carbon Chemical Shifts

The range of carbon chemical shifts is 200 ppm compared to the 10-ppm range for protons. It is not common to have an accidental overlap of carbon peaks. In the spectrum for the unsaturated chloroester in Figure 12.4 (spectrum II), there are eight sharp peaks corresponding to the eight carbon atoms in the molecule.

The generalities governing carbon chemical shifts are very similar to those governing proton shifts, as seen by comparing Table 12.1 to Table 12.3. Most of the downfield peaks are due to those carbon atoms near electron-withdrawing groups. In Fig. 12.4 (spectrum II) the furthest downfield peak is that from the carbonyl carbon of the ester. The attached electronegative oxygens make the peak appear at about 172 ppm. The peak is smaller than any other in the spectrum because it does not have an attached proton and thus does not benefit from a NOE effect.

The ^{13}C spectrum of sucrose in Figure 12.5 displays a single line for each carbon atom; in Figure 20.3 (on page 364) a single line is seen for each of the 27 carbon atoms in cholesterol.

■ **FIG. 12.5**
The ^{13}C NMR spectrum of sucrose (22.6 MHz). Not all lines have been assigned to individual carbon atoms.

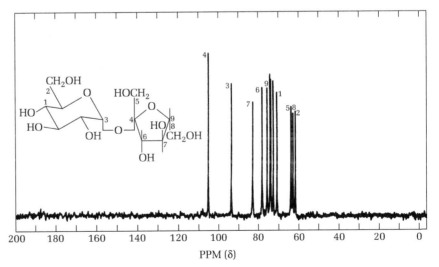

Fourier Transform Spectroscopy

Even with broadband decoupling, the signal from ^{13}C spectra can get lost in the random noise produced by the spectrometer. Similarly, dilute solutions of samples give very noisy proton spectra. To increase the signal-to-noise ratio, the data for a number of spectra are averaged in the spectrometer computer. Because noise is random and the signal coherent, the signal will increase in size, and the noise will decrease as many spectra are averaged together. The improvement in the signal-to-noise ratio is proportional to the square root of the number of scans.

Spectra are accumulated rapidly by applying a very short pulse of radio-frequency energy to the sample and then storing the resulting free induction decay signal in digital form in the computer. The free induction decay signal, which takes seconds to acquire, contains frequency information about all the signals in the spectrum. In a few minutes several hundred free induction decay signals can be obtained and averaged in the computer. The free induction decay signal is converted to a spectrum of conventional appearance by carrying out a Fourier transform (FT) computation on the signal using the spectrometer's computer.

Because the spectral information is in digital form, a large number of operations can be performed on it. The spectrum can be smoothed at the expense of resolution, or the resolution can be enhanced at the expense of noise. As indicated earlier, the spectrum can be numerically integrated and the position of each peak, in hertz or ppm, printed above the peak. The spectra can be expanded both vertically and horizontally to clarify complex couplings, and all the parameters used to acquire the spectrum can be printed as well.

Protons with Variable Chemical Shifts

Labile proton(s)	δ (ppm)
ROH	0.5–6
ArOH	4.5–8
RCOOH	10–12
Enols	10–17
RNH$_2$	1–5
Amides	5–6.5

Molecules that have protons attached to oxygen and nitrogen form hydrogen bonds with each other and with protic solvents. As a result, the chemical shifts of these protons depend on solvent, temperature, and concentration and can appear almost anywhere in a spectrum. They will exchange with each other, a process catalyzed by small concentrations of acid or base. If this exchange is rapid (a common occurrence in alcohols, for example), only a single sharp line for the hydroxyl proton will be seen. At intermediate rates of exchange, the line becomes broad; at slow rates of exchange, it couples to other protons within three chemical bonds.

Protons bound to nitrogen often give a broad line because of the nonuniform distribution of charge on the nucleus of ^{14}N. Obviously, NMR spectroscopy is a poor means for detecting hydroxyl, amine, and amide protons; infrared spectroscopy, on the other hand, is an excellent means of doing this. Because labile protons can undergo rapid exchange with each other, they can also exchange with deuterium. If 1 drop of D_2O is added to a $CDCl_3$ solution of an NMR sample of a compound that contains such protons and gently mixed, the protons will exchange for deuterium atoms. The water will, of course, float on top of the denser chloroform. The peak for the labile proton will disappear, thus simplifying the spectrum and allowing for assignment of the peak.

■ FIG. 12.6

The 60-MHz ^{1}H NMR spectrum of 2-methyl-3-pentanol (0.4 M in CS$_2$) with various amounts of shift reagent present. (A) No shift reagent present. All methyl peaks are superimposed. The peak for the proton adjacent to the hydroxyl group is downfield from the others because it is adjacent to the electonegative oxygen atom. (B) 2-Methyl-3-pentanol + Eu(dpm)$_3$. The mole ratio of Eu(dpm)$_3$ to alcohol is 0.05. The hydroxyl proton peak at 1.6 ppm in spectrum A appears at 6.2 ppm in spectrum B because it is closest to the Eu in the complex formed between Eu(dpm)$_3$ and the alcohol. The next closest proton, the one on the hydroxyl-bearing carbon atom, gives a peak at 4.4 ppm. Peaks due to the three different methyl groups at 1.1 ppm to 1.5 ppm begin to differentiate. (C) The mole ratio of Eu(dpm)$_3$ to alcohol is 0.1. The hydroxyl proton does not appear in this spectrum because its chemical shift is greater than 8.6 ppm with this much shift reagent present. (D) The mole ratio of Eu(dpm)$_3$ to alcohol is 0.25. Further differentiation of methyl peaks (2.3 ppm to 3.0 ppm) is evident. (E) The mole ratio of Eu(dpm)$_3$ to alcohol is 0.5. Separate groups of peaks begin to appear in the region 4.2 ppm to 6.2 ppm. (F) The mole ratio of Eu(dpm)$_3$ to alcohol is 0.7. Three groups of peaks (at 6.8 ppm, 7.7 ppm, and 8.1 ppm) due to the protons on C-2 and C-4 are evident, and three different methyls are now apparent. (G) The mole ratio of Eu(dpm)$_3$ to alcohol is 0.9. Only the methyl peaks appear on the spectrum. The two doublets come from the methyls attached to C-2, and the triplet comes from the terminal methyl at C-5.

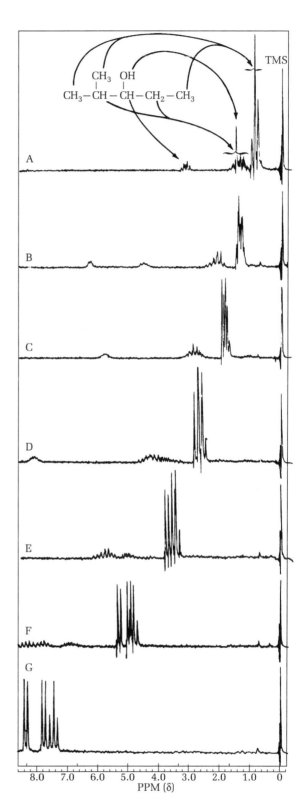

Tris(dipivaloylmethanato)europium(III), Eu(dpm)$_3$
MW 701.78, mp 188–189°C

Shift Reagents

Some spectra can be simplified by adding shift reagents to the sample. The resulting spectrum is similar to a spectrum one might obtain in a high-field spectrometer.

Adding a few milligrams of a hexacoordinate complex of the lanthanide europium called tris(dipivaloylmethanato)europium(III), or Eu(dpm)$_3$, to an NMR sample that contains a Lewis base center (an amine or basic oxygen, such as a hydroxyl group) has a dramatic effect on the spectrum. This reagent (soluble in CDCl$_3$) causes large shifts in the positions of peaks arising from the protons near the metal atom in this molecule and is therefore referred to as a *shift reagent*. It produces the shifts by complexing with the unshared electrons of the hydroxyl oxygen, the amine nitrogen, or other Lewis base center.

With no shift reagent present, the ^{1}H NMR spectrum of 2-methyl-3-pentanol (Fig. 12.6a) is not readily analyzed. Adding 10-mg portions of the shift reagent to the sample causes very large downfield shifts of peaks, owing to protons near the coordination site (Figs. 12.6b–12.6g).[3] The two protons on C-4 and the two methyls on C-2 are diastereotopic and thus are magnetically nonequivalent because they are adjacent to a chiral center, C-3; each gives a separate set of peaks. When enough shift reagent is added, this spectrum can be analyzed by inspection (Fig. 12.7). Quantitative information about molecular geometry can be obtained from such shifted spectra. The shift induced by the shift reagent is related to the distance and the angle of the proton with respect to the europium atom.

Chiral shift reagents (derivatives of camphor) will cause differential shifts of the protons or carbon atoms in enantiomers. The separated peaks can be integrated to determine enantiomeric purity. A chiral shift reagent is used in Chapter 65 to determine the enantiomeric purity of a chiral alcohol made by the enzymatic reduction of a ketone.

Chiral shift reagents

■ **FIG. 12.7**
The ^{1}H NMR spectrum of 2-methyl-3-pentanol containing Eu(dpm)$_3$ at 60 MHz. The mole ratio of Eu(dpm)$_3$ to alcohol is 1.0. Compare this spectrum with those shown in Figure 12.6. Protons nearest the hydroxyl group are shifted most. Methyl groups are recorded at reduced spectrum amplitude. Note the large chemical shift difference between the two diastereotopic protons on C-4.

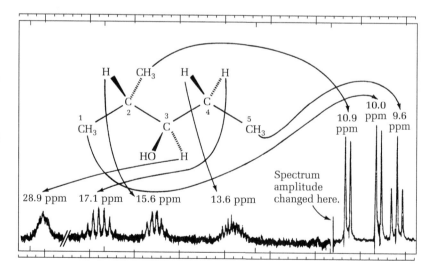

3. Williamson, K. L.; Clutter, D. R.; Emch, R.; Alexander, M.; Burroughs, A. E.; Chua, C.; Bogel, M. E. *J. Am. Chem. Soc.* **1974,** *96,* 1471.

EXPERIMENTS

1. Running an NMR Spectrum

Sample Preparation

A typical ^{1}H NMR sample consists of about 10–50 mg of sample dissolved in about 1.0 mL of $CDCl_3$ that contains a very small percentage of TMS, the reference compound. Because the tiniest particles of dust, especially metal particles, can adversely affect resolution, it is best to filter every sample routinely through a wad of cotton or glass wool in a Pasteur pipette (Fig. 12.8).

If very high resolution spectra (all lines very sharp) are desired, oxygen, a paramagnetic impurity, must be removed by bubbling a fine stream of pure nitrogen through the sample for 60 s. Routine samples do not require this treatment.

It is most convenient if all the tubes in a laboratory are filled to exactly the same height with the $CDCl_3$ solution; this will greatly facilitate tuning of the spectrometer. High-quality NMR tubes give the best spectra, which are free of spinning side bands. A good tube should roll at a slow, even rate down a very slightly inclined piece of plate glass.

Chemical shifts of protons and carbon are measured relative to the sharp peak in the TMS (taken as 0.0 ppm). Because $CDCl_3$ is not 100% pure, there will always be a very small peak at 7.27 ppm in the proton spectrum that arises from the tiny amount of residual ordinary chloroform, $CHCl_3$, in the solvent (Fig. 12.10). In the carbon spectrum three lines of equal intensity appear at 77 ppm due to coupling of the deuterium atom with the carbon in $CDCl_3$.

If the sample does not dissolve in $CHCl_3$, a number of other deuterated solvents are available, including deuteroacetone (CD_3COCD_3), deuterodimethyl-sulfoxide (DMSO-d_6), and deuterobenzene (C_6D_6). All of these solvents are expensive, so the solubility of the sample should first be checked in nondeuterated solvents. For highly polar samples, a mixture of the more expensive DMSO-d_6 with the less expensive $CDCl_3$ is often satisfactory. If it is necessary to use D_2O as the solvent, then a special water-soluble reference, sodium 2,2-dimethyl-2-silapentane-5-sulfonate [$(CH_3)_3Si(CH_2)_3SO_3^-Na^+$ (DSS)], must be used. The protons on the three methyl groups bound to the silicon in this salt absorb at 0.0 ppm.

The usual NMR sample has a volume of about 1.0 mL, even though the volume sensed by the spectrometer receiver coils (referred to as the *active volume*) is much smaller. To average the magnetic fields produced by the spectrometer within the sample, the tube is spun by an air turbine at about 30 revolutions per second while taking the spectrum. Spinning that is too rapid or an insufficient amount of solution will cause the vortex produced by the spinning to penetrate the active volume (Fig. 12.9), giving erratic and nonreproducible spectra.

Cleaning Up. Place halogenated solvents and compounds in the halogenated organic waste container. All other substances go into the organic solvents container.

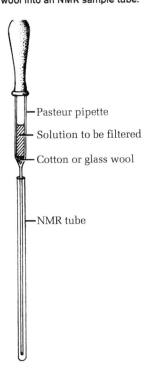

■ FIG. 12.8
Microfiltration for NMR samples. The solution to be filtered is placed in the top of the Pasteur pipette, the rubber bulb is put in place, and pressure is applied to force the sample through the cotton or glass wool into an NMR sample tube.

Pasteur pipette

Solution to be filtered

Cotton or glass wool

NMR tube

■ FIG. 12.9
The effect of too rapid spinning
or insufficient sample. The active
volume is the only part of the
sample detected by the
spectrometer.

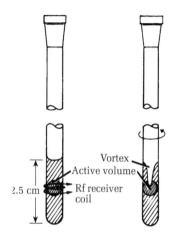

Adjusting the Spectrometer

To be certain the spectrometer is correctly adjusted and working properly, record the spectrum of the standard sample of chloroform and TMS usually found with the instrument. A 5% ethylbenzene in *d*-chloroform (CDCl$_3$) is also a useful performance check sample.

Small peaks symmetrically placed on each side of a principal peak are artifacts called *spinning side bands*. They are recognized as such by changing the spin speed, which causes the spinning side bands to change positions. They can be minimized by proper adjustment of the homogeneity controls.

Because Fourier transform spectrometers lock on the resonance of deuterium to achieve field/frequency stabilization, all samples must be dissolved in a solvent containing deuterium. Confirmation that the lock is obtained is registered on a meter or oscilloscope. The Z and Z^2 controls are used to maximize the field homogeneity and achieve the highest resolution. This adjustment is needed for virtually every sample, but, as noted earlier, if all sample tubes in a laboratory are filled to exactly the same height, these adjustments will be minimized. Many modern instruments have automatic tuning capabilities.

Common Contaminant Peaks

Almost every NMR spectrum contains spurious peaks that arise from a number of sources: residual protium (^{1}H) in deuterated (^{2}H) NMR solvents, dirty NMR tubes, and impure or inadequately dried products. NMR spectra of common contaminants are shown in Figure 12.10. The deuterated NMR solvents used are usually only about 99% deuterated and the residual protium present will produce detectable signals, particularly if sample signals are weak relative to the solvent. This type of residual peak is shown for CDCl$_3$, DMSO-d_6, and D$_2$O in Figure 12.10. Due to exchange with H$_2$O from glass surfaces and the atmosphere, D$_2$O is especially prone to contain a lot of HOD. If a product is crystallized from a solvent or isolated by removal of a solvent such as water, ethanol, dichloromethane, or ether, any solvent that remains will produce interfering signals, as shown in this figure. Atmospheric moisture absorbed by samples or solutions also gives rise to water peaks of varying intensities. NMR tubes are often rinsed out with acetone and even after prolonged drying, traces of acetone will remain and show up at about 2.2 ppm. All chemical shifts vary over a considerable range, depending on the solvent. For example, dichloromethane appears at 4.5 ppm in *d*-benzene, and 5.8 ppm in DMSO-d_6. The H$_2$O peak varies over an even wider range.

Two-Dimensional NMR Spectroscopy

The fast computers associated with FT spectrometers allow for a series of precisely timed pulses and data accumulations to give a large data matrix that can be subjected to Fourier transformation in two dimensions to produce a two-dimensional (2D) NMR spectrum.

The ^1H NMR spectra of common sample impurities and deuterated solvents. Note how the chemical shift due to water as an impurity varies over a wide range.

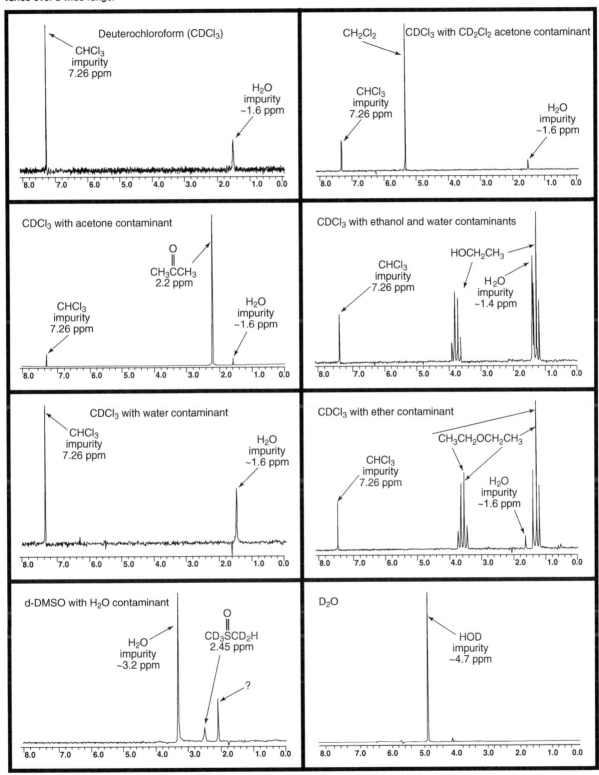

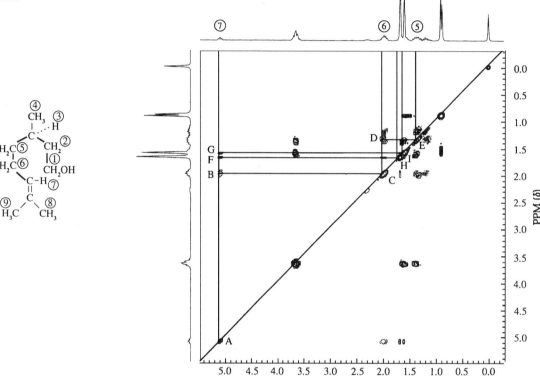

One of the most common and useful of these is the COSY (correlated spectroscopy) spectrum (Fig. 12.11). In this spectrum, two ordinary one-dimensional (1D) spectra are correlated with each other through spin-spin coupling. The 2D spectrum is a topographic representation, where spots represent peaks. The 1D spectra at the top and side are projections of these peaks. Along the diagonal of the 2D spectrum is a spot for each group of peaks in the molecule.

Figure 12.11 shows the 2D spectrum of citronellol. (Fig. 12.12 is the 1D spectrum of this compound.) Consider spot A on the diagonal of the 2D spectrum. From a table of chemical shifts, we know that this is a vinyl proton, 7, the single proton on the double bond. From the structure of citronellol, we can expect this proton to have a small coupling over four bonds to the protons on methyls 8 and 9, and a stronger coupling to the protons on carbon-6. In the absence of a 2D spectrum, it is not obvious which group of peaks belongs to carbon-6, but the spot B correlates with spot C on the diagonal, which is directly below the peak labeled 6 on the 2D spectrum.

The diagonal spot C, which we have just assigned to carbon-6, correlates through the off-diagonal spot D with the diagonal spot E, which lies just below the

■ FIG. 12.12
The 1D ^{1}H NMR spectrum
of citronellol (250 MHz).

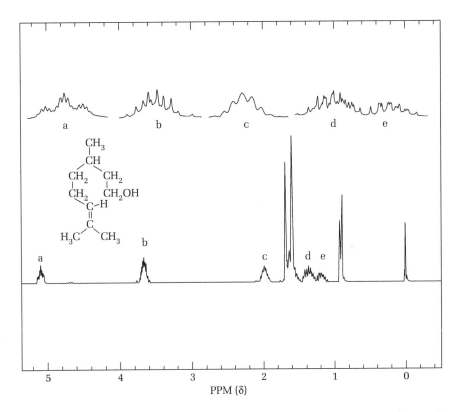

group of peaks labeled 5. Spot A on the diagonal also correlates through spots F and G with spots H and I on the diagonal, which lie directly below methyl peaks 8 and 9. In this way we can determine the complete connectivity pattern of the molecule, seeing which protons are coupled to other protons.

Other 2D experiments allow proton spectra on one axis to be correlated with carbon spectra on the other axis. NOESY (nuclear Overhauser effect spectroscopy) experiments give cross-peaks for protons that are near to each other in space but not spin coupled to each other.

2. Identification of an Unknown Alcohol or Amine by ^{1}H NMR

Shift reagents are expensive.

Prepare a solution of the unknown in $CDCl_3$ and run a spectrum in the normal way. To the unknown solution add about 5 mg of $Eu(dpm)_3$ (the shift reagent), shake thoroughly to dissolve, and run another spectrum. Continue adding $Eu(dpm)_3$ in 5-mg to 10-mg portions until the spectrum is shifted enough for easy analysis. Integrate peaks and groups of peaks if in doubt about their relative areas. To protect the $Eu(dpm)_3$ from moisture, store it in a desiccator.

Cleaning Up. All samples containing shift reagents go into a hazardous waste container for heavy metals.

QUESTIONS

1. Propose a structure(s) consistent with the proton NMR spectrum in Figure 12.13. Numbers adjacent to groups of peaks refer to relative peak areas. Account for missing lines.

2. Propose a structure(s) consistent with the proton NMR spectrum in Figure 12.14. Numbers adjacent to groups of peaks refer to relative peak areas.

3. Propose a structure(s) consistent with the proton NMR spectrum in Figure 12.15.

4. Propose structures for (a), (b), and (c), consistent with the ^1H and ^{13}C NMR spectra in Figures 12.16, 12.17, and 12.18. These are isomeric alcohols with the empirical formula $C_4H_{10}O$.

FIG. 12.13
The ^1H NMR spectrum (60 MHz) for question 1.

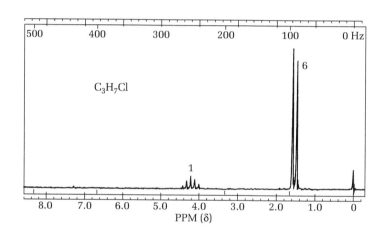

FIG. 12.14
The ^1H NMR spectrum (60 MHz) for question 2.

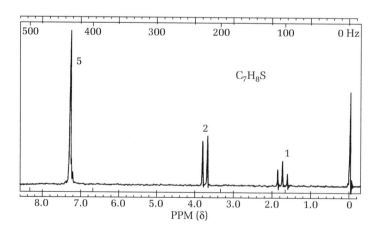

■ FIG. 12.15
The ¹H NMR spectrum (60 MHz)
for question 3.

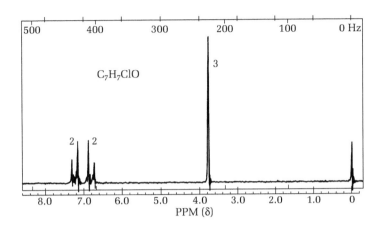

■ FIG. 12.16
The ¹H NMR spectrum (60 MHz)
of compound (a), $C_4H_{10}O$, for
question 4.

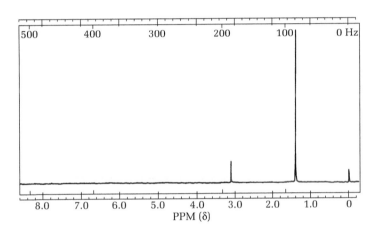

■ FIG. 12.17
The ¹³C NMR spectrum
(22.6 MHz) of compound (b),
$C_4H_{10}O$, for question 4.

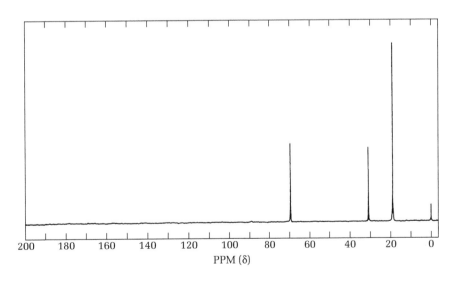

◼ FIG. 12.18
The ^{13}C NMR spectrum (22.6 MHz) of compound (c), $C_4H_{10}O$, for question 4.

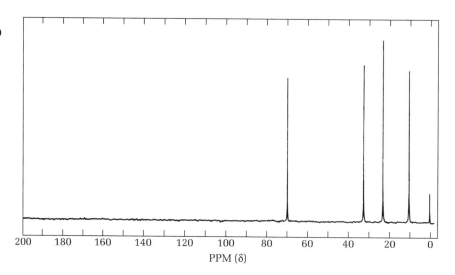

PPM (δ)

REFERENCES

Online Study Center

General Resources
Additional Experiments, Web Links

1. Abraham, Raymond J., J. Fisher, and P. Loftus, *Introduction to NMR Spectroscopy*. Chichester, UK: John Wiley & Sons, 1988.

2. Croasmun, William R., and Robert M. K. Carlson, *Two-Dimensional NMR Spectroscopy: Applications for Chemists and Biochemists*, 2nd ed. New York: VCH Publishers, 1994.

3. Friebolin, Horst, *Basic One- and Two-Dimensional NMR Spectroscopy*, 3d ed. New York: VCH, 1998.

4. Gadian, David G., *NMR and Its Applications to Living Systems*, Oxford, UK: Oxford University Press, 1996.

5. Pavia, Donald L., Gary M. Lampman, and George S. Kriz, *Introduction to Spectroscopy*, 3rd ed.[4] Brooks/Cole Thomson Learning, 2001.

6. Sanders, Jeremy K. M., and Brian K. Hunter, *Modern NMR Spectroscopy: A Guide for Chemists*, 2d ed. Oxford, UK: Oxford University Press, 1993.

7. Silverstein, Robert M., Francis X. Webster, and David Kiemle, *Spectrometric Identification of Organic Compounds*, 7th ed.[4] New York: Wiley, 2005.

8. Young, Ian R., Dave M. Grant, and Ryan K. Harris, eds., *Methods in Biomedical Magnetic Resonance Imaging and Spectroscopy*, 2 vol. New York: John Wiley & Sons, 2000.

---

4. This book includes IR, UV, and NMR spectra.

CHAPTER 13

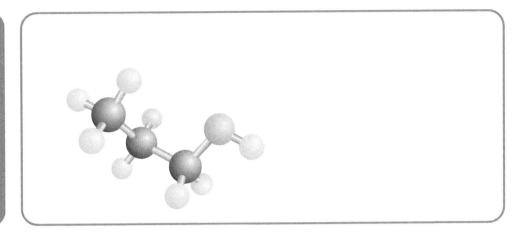

Mass Spectrometry

Online Study Center

This icon will direct you to techniques, equipment setups, and online resources at http://college.hmco.com/PIC/williamson MMOE5e.

PRELAB EXERCISE: Calculate the molecular weight of bromobutane (in grams per mole) and the exact masses of its molecular ions (in daltons) as measured by mass spectrometry.

Although its origins date back to the early part of the 20th century, mass spectrometry (MS) has experienced a renaissance in the past 2 decades due to major improvements in ion formation and analysis. Because mass spectrometry is capable of providing composition and structure information for a broad range of compounds at sensitivities much greater than those of other techniques, more money is invested worldwide in mass spectrometers than in any other type of instrumentation. In the chemical and life sciences, mass spectrometry is used to identify and quantify compounds present in complex organic mixtures; to identify structures of biomolecules such as carbohydrates, nucleic acids, and steroids; to sequence proteins and oligosaccharides; to determine how drugs are used by the body; to monitor the breath of patients anesthetized during surgery; and to check fermentation processes for the biotechnology industry. In environmental science, mass spectrometry is used to detect environmental pollutants such as dioxins in fish and humans and to determine gene damage from environmental causes. In forensic science, mass spectrometry is used to confirm and quantify drugs of abuse and steroid use by athletes and to identify accelerants used in arson. Mass spectrometry is used to determine the age and origins of specimens in geochemistry and archaeology; to locate oil deposits by measuring petroleum precursors in rock; to report the composition of molecular species found in space; to perform ultrasensitive, multielement inorganic analyses; and to establish the elemental composition of semiconductor materials.

Notice that the infrared (IR), nuclear magnetic resonance (NMR), and ultraviolet (UV) methods are called *spectroscopy*, but MS is mass *spectrometry*. This difference helps remind us that mass spectrometry does not involve the absorption of specific electromagnetic energies resulting in spectra of frequency versus intensity; mass spectrometry involves the production and detection of different masses to yield spectra recorded as ion mass versus ion abundance.

The Mass Spectrometer

Mass spectrometers have the following components, connected as shown in Figure 13.1.

1. Ion Source

Neutral molecules in the gas state move randomly, and it is impossible to move them in any particular direction without using a pump to create a pressure gradient. However, if molecules can be positively or negatively charged to form ions, these ions can be attracted and repelled by charged metal surfaces so that they can be moved or accelerated in any chosen direction. The ionization of molecules is done in numerous ways, but the most common way is shown in Figure 13.2. Sample molecules in the gaseous state are introduced into an *ion source*, a small metal enclosure containing a hot filament, like that in a light bulb, on one side. Electrons are emitted from the filament and attracted to a small metal collector electrode on the other side by a slight positive voltage, usually +70 eV, relative to the filament. These energetic electrons can collide with the gaseous sample molecules and ionize them by knocking an electron out of their electron clouds, producing *molecular ions*, most having a single positive charge, but a few with two positive charges. Some negative ions are formed by the capture of an electron, but these are much less abundant. Therefore, most electron ionization spectra are of positive ions.

■ **FIG. 13.1**
The components of a mass spectrometer.

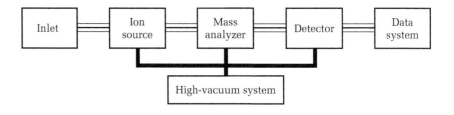

■ **FIG. 13.2**
A view into the electron ionization source enclosure. Sample molecules, CO_2 in this illustration, are ionized by high-energy electron impacts and together with fragment ions are accelerated into the mass analyzer as a continuous beam of pulsed ion packets.

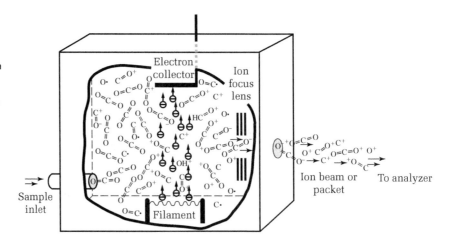

The ionization process is illustrated in Figure 13.2 and in the following text using a sample of CO_2. Electron ionization produces $O{=}C{=}O^{+\cdot}$ molecular ions written using the cation radical symbol $^{+\cdot}$ because they are odd electron species.

Ionization:

$$O{=}C{=}O \quad + \quad e^{\ominus} \quad \longrightarrow \quad O{=}C{=}O^{+\cdot} \quad + \quad 2e^{\ominus}$$

Fragmentation:

$$O{=}C{=}O^{+\cdot} \quad \longrightarrow \quad O\cdot \quad + \quad C{=}O^{+} \quad \text{or} \quad O^{+} \quad + \quad C{=}O\cdot$$

$$C{=}O^{+} \quad \longrightarrow \quad C^{+\cdot} \quad + \quad O^{\cdot} \quad \text{or} \quad C\cdot \quad + \quad O^{+\cdot}$$

The energy of the bombarding electrons (about 70 eV) is generally much greater than that of the bonds holding the molecule together. Thus, when high-energy electrons interact with the molecule, not only does ionization occur, but bonds are broken and *fragmentation* occurs, giving rise to *fragment ions* and *fragment neutrals*. The $C{=}O$ bonds of CO_2 can be broken to produce either a $C{=}O^{+}$ ion and a neutral O radical or an O^{+} ion and a neutral $C{=}O$ radical. If both $C{=}O$ bonds are broken, a C^{+} ion can be produced.

The electron ionization source, as well as the other parts of a mass spectrometer, must be maintained at very low pressure (high vacuum; $\sim 10^{-8}$ atm), and only very small amounts of sample can be introduced; if the pressure is too high, the ions would collide with unionized molecules or other ions instead of being accelerated from the ion source and into the ion mass analyzer. Also, the electron filament would burn out if an atmosphere of air were present. Therefore, sample inlets are typically a fine capillary or a tiny pinhole to restrict sample flow into the source chamber. Solids are introduced on a metal rod that slides through a sealed vacuum lock. Molecules that are not ionized and neutral fragments are pumped away.

There are several other ionization methods that have the advantage of producing intense molecular ions or protonated molecular ions, which are just as informative. For samples that are gaseous or that can be vaporized in a vacuum, chemical ionization and field ionization are often used. These will not be discussed here, but the references at the end of this chapter provide details. High molecular weight nonvolatile compounds can be ionized by electrospray ionization (ESI) or matrix-assisted laser desorption ionization (MALDI), both of which are discussed in a later section on bioanalytical mass spectrometry.

2. Mass Analyzer

Positive ions are propelled into the analyzer by maintaining the ion source at a more positive electrical potential relative to the analyzer and by focusing the ions with voltages applied to an electronic lens system located between the source and the analyzer. The beam of ions passes through the *mass analyzer* where the ions are separated according to mass. Actually, ions are separated based on the mass to charge ratio (m/z). For example, for CO_2^{+} ions $z = 1$, so $m/z = 44/1$, or 44; for CO_2^{++} ions

■ FIG. 13.3
A time-of-flight mass
spectrometer. A packet of
molecular and fragment ions
is pulsed out of the ion source.
Lower mass ions arrive at the
detector first, with higher mass
ions following.

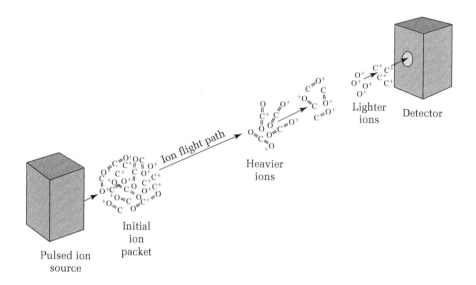

$z = 2$, so $m/z = 44/2$, or 22. Doubly charged ions are rare, so $z = 1$ in almost all cases, and $m/1 = m$; therefore, for practical purposes, ions are separated according to their mass, and the ensuing discussion will mainly use mass instead of m/z.

In the earliest mass spectrometers, the ions of different masses were separated by passing the ion beam through a magnetic field. Today, however, two types of mass analyzers are commonly used. The simplest in theory is the *time-of-flight (TOF) analyzer* (Fig. 13.3). The ion beam is pulsed through an electronic gate in little packets, and the start time of each packet of ions is recorded. The packet of ions travels down a long tube, typically 1–2 m in length. Lower mass, lighter ions travel faster and arrive at the detector to produce signals with short times, while heavier ions produce signals with longer times. The development of high-speed electronics now allows the detection of ions that differ in mass by only 1 part in 10,000, which corresponds to a difference in flight times of a few microseconds (10^{-6} s) or less.

Quadrupole mass analyzers are commonly used in GC–MS systems (discussed in a later section of this chapter). The continuous beam of ions from the source passes between four parallel metal rods or poles (Fig. 13.4). Depending on the magnitude and frequency of the direct current (DC) and alternating current (AC) voltages on the rods, only ions of a certain *resonant* mass will pass between the rods and into the detector. Ions of other masses are nonresonant and will be deflected to the side and miss the detector. The mass range is scanned by increasing the rod voltage and AC frequency so that ions of increasing masses hit the detector.

3. Detector

After separation by mass, the ions pass into the *detector* where they impact a metal electrode called a dynode that has a high negative voltage. This releases a number of electrons (~5–10) that are drawn toward a slightly more positive

■ **FIG. 13.4**
A quadrupole mass spectrometer.
Ions oscillate as they move
through the electrical fields on the
four rods. Changing these fields
allows a specific resonant ion to
pass to the detector. Ions of other
masses are deflected.

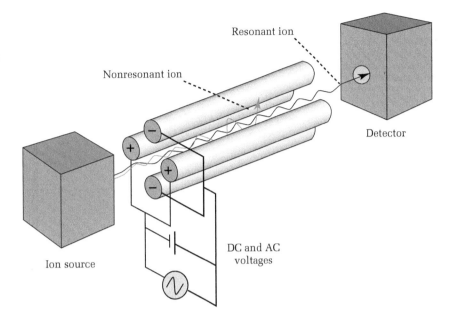

■ **FIG. 13.4**
A quadrupole mass spectrometer.
Ions oscillate as they move
through the electrical fields on the
four rods. Changing these fields
allows a specific resonant ion to
pass to the detector. Ions of other
masses are deflected.

■ **FIG. 13.5**
A diagram of an electron multiplier
detector.

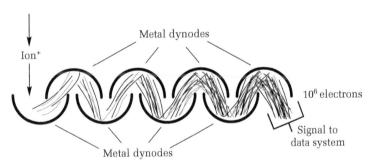

dynode; each of these electrons releases a number of electrons, which accelerate toward a third dynode, releasing even more electrons (Fig. 13.5). These *electron multipliers* have enough dynode stages that a single positive ion impact produces a cascade of a million or more electrons at the final electrode. Single ions can be detected, thus making mass spectrometry a very sensitive method.

The signal output voltage is proportional to the number of ions hitting the detector at a given time (TOF analysis) or AC and DC voltages (quadrupole analysis) and is continuously digitized and stored in a computer. The computer has files of mass versus calibrated TOFs or quadrupole voltages and can therefore plot the mass spectrum as the *percent relative abundance* versus mass or more correctly m/z (see earlier discussion). All signal peaks are normalized with the most abundant ion as 100%.

The electron ionization spectrum for a sample of CO_2 is shown in Figure 13.6. The ionized CO_2 molecule (or molecular ion) appears at mass 44. The ion is singly

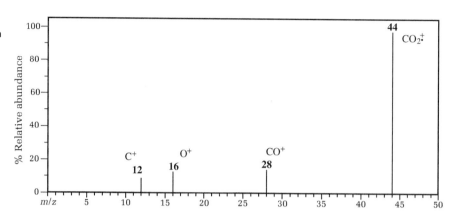

A dalton (Da) is a mass unit used
in mass spectrometry and is
defined as 1/12 of the mass of
^{12}C, which has the atomic weight
12.00000.

charged, and the *nominal ion mass* is 44 atomic mass units or daltons (Da): carbon =
12 Da and oxygen = 16 Da (in calculating nominal ion mass, atomic masses are
rounded to the nearest integer). Because the ionization process breaks up or fragments some of the CO$_2$ molecules, a fraction of the ions appear in the spectrum at
mass values less than the molecular ion. Cleavage of a carbon-oxygen bond in
the molecular ion, which produces ionized carbon monoxide (CO$^+$) and a neutral
oxygen radical (O·) or ionized atomic oxygen (an O$^+$ ion) and a neutral carbon
monoxide radical (C=O·), results in the fragment ions at masses 28 and 16; loss
of two neutral oxygen atoms results in an additional fragment at mass 12 for carbon (a C$^+$ ion). Note that only charged fragments, not neutral fragments, appear in
the spectrum.

Most mass spectrometers are set up so that they do not scan below $m/z = 35$
because there are very large background signals at masses 32 (O$_2$), 28 (N$_2$), and
18 (H$_2$O) due to residual air and moisture in the instrument, in spite of the high
vacuum used. Therefore, unlike Figure 13.6, the remaining spectra in this chapter
start at mass 35. Weak background signals at masses 40 and 44 from atmospheric
argon and carbon dioxide are also seen in many spectra but are usually ignored.

Interpretation of Electron Ionization Spectra

Molecular Ions

The mass of a molecular ion M$^+$, as measured by a mass spectrometer, is not the
same as the molecular weight conventionally used to weigh out mole quantities on
a balance. The analyzer of a mass spectrometer separates individual molecular ions
based on the masses of the isotopes in those ions. The important isotopes for organic mass spectrometry are listed in Table 13.1. Every organic compound contains both ^{12}C isotopes with mass 12.0000 Da and ^{13}C isotopes with mass 13.0034
Da, with a 1.1% probability of a given carbon being a ^{13}C. (Radioactive ^{14}C is at

TABLE 13.1 • The Natural Abundance of Isotopes (atoms per 10,000 atoms and percent abundance) and the Atomic Mass of Each Isotope								
	Abundance and Mass of the Most Abundant Isotope (given 10,000 atoms)				Abundance and Mass of Other Isotopes (given 10,000 atoms)			
	Isotope	No. of Atoms	Mass (Da)	% Abundance	Isotope	No. of Atoms	Mass (Da)	% Abundance	Average Mass (g/mol)
C	^{12}C	9890	12.0000	98.9	^{13}C	110	13.0034	1.1	12.0107
H	^{1}H	9999	1.0078	99.99	^{2}H	1	2.0141	0.01	1.00794
O	^{16}O	9985	15.9949	99.76	^{17}O	4	16.9991	0.04	15.9994
					^{18}O	20	17.9992	0.2	
N	^{14}N	9963	14.0031	99.63	^{15}N	37	15.0001	0.37	14.0067
Cl	^{35}Cl	7577	34.9689	75.8	^{37}Cl	2423	36.9659	24.2	35.453
Br	^{79}Br	5069	78.9183	50.7	^{81}Br	4931	80.9163	49.3	79.904
S	^{32}S	9493	31.9721	94.93	^{33}S	76	32.9715	0.76	32.065
					^{34}S	429	33.9679	4.29	

The elements ^{19}F, ^{31}P, and ^{127}I are monoisotopic.

■ FIG. 13.7
The mass spectrum of decane showing isotopic molecular ions, fragment ions, and neutral losses due to bond cleavages at a, b, c, and d.

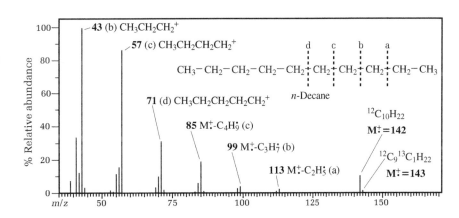

a much lower natural abundance.) About 89% of molecules in a sample of decane contain only ^{12}C and have the composition $^{12}C_{10}H_{22}$, and 11% of the molecules have one ^{13}C (10 carbons times 1.1% probability of each one being a ^{13}C) and the composition $^{12}C_9{}^{13}C_1H_{22}$. When 1 mol of decane is weighed out, huge numbers of carbon atoms ($\sim 60 \times 10^{23}$) are present, and therefore it is reasonable to use an average mass of the two isotopes, 12.0107 g/mol, to calculate the molecular weight of decane as 142.285 g/mol. But in a mass spectrometer, $^{12}C_{10}H_{22}{}^+$ ions are separated from $^{12}C_9{}^{13}C_1H_{22}{}^+$ ions, as seen in the mass spectrum of decane shown in Figure 13.7.

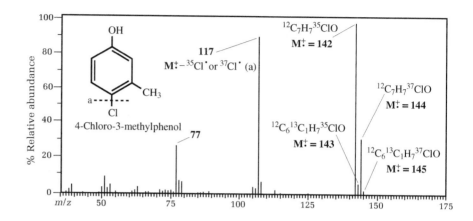

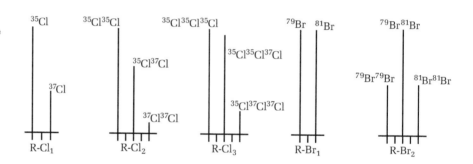

To calculate the exact masses of these ions, isotopic masses for ^{12}C (12.0000 Da) and ^{13}C (13.0034 Da) must be used. The $^{12}C_{10}H_{22}{}^{+}$ molecular ion has a mass of 142.1716 Da and that of $^{12}C_9{}^{13}C_1H_{22}{}^{+}$ has a mass of 143.1750 Da. The heavier isotopes of hydrogen (^2H and ^3H) and oxygen (^{17}O and ^{18}O) have abundances that are so low that they are difficult to detect and are usually ignored. For decane, the *nominal* mass of the most abundant molecular ion containing just ^{12}C, obtained by rounding the exact mass to the closest integer, is 142 Da, the same as the integer molecular weight of decane, 142 g/mol. For other isotopes, this may not be true. In the mass spectrum of 4-chloro-3-methylphenol (Fig. 13.8), there are four molecular ions with isotopic compositions shown due to the presence of not only ^{13}C but also the two isotopes of chlorine, ^{35}Cl and ^{37}Cl. The exact mass of the largest molecular ion, $^{12}C_7H_7O^{35}Cl$ is 142.0184 Da. This is quite different than the molecular weight of this compound, 142.5774 g/mol. After rounding, the nominal mass of the ion is 142 Da, but the integer molecular weight is 143 g/mol. This difference is the primary reason that daltons are used in mass spectrometry—to help differentiate exact masses measured by mass spectrometry from the conventional molecular weight in grams/mole used in lab weighing.

Both chlorine and bromine have isotopes separated by 2 Da, creating unique and easily recognized patterns in a mass spectrum. The ratio of peaks depends on

TABLE 13.2 • Six Compounds with Molecular Ions of Nominal Mass 142 That Can Be Differentiated by Accurate Mass Measurement		
Compound	Molecular Formula	Exact Mass of Molecular Ion
3-Fluoro-4-methoxyphenol	$C_7H_7FO_2$	142.0430
Octenoic acid	$C_8H_{14}O_2$	142.0993
Cyclohexylurea	$C_7H_{14}N_2O$	142.1106
3-Nonanone	$C_9H_{18}O$	142.1357
1,2-Diaminocyclooctane	$C_8H_{18}N_2$	142.1469
Decane	$C_{10}H_{22}$	142.1721

the number of chlorines or bromines in the molecule, as shown in Figure 13.9. The Cl_1 pattern is seen in the mass spectrum of the chlorophenol in Figure 13.8.

Many modern instruments are able to measure ion masses to within a few parts per million of accuracy. For example, even though all of the compounds listed in Table 13.2 (as well as hundreds of other possible compounds) have molecular ions at nominal mass 142, exact mass measurement would easily differentiate among them.

Conversely, given an exact mass measured by mass spectrometry, a computer can compare this with all possible exact masses calculated for all reasonable combinations of ^{12}C, ^1H, ^{14}N, ^{16}O, ^{35}Cl, and so forth and produce a list of compositions that come within a few parts per million of the measured mass. Compositions determined in this manner are required whenever a new compound is reported in a publication. (Note, however, that the composition $C_9H_2O_2$, for example, is not reasonable even though it has a mass of 142.)

Compounds containing C, H, O, F, Cl, Br, I, and S or an even number of nitrogens will have an even mass molecular ion. *The nitrogen rule* states that all molecules with an odd number of nitrogen atoms will have an odd mass molecular ion. Compare the molecular ions in the mass spectrum of a nitrogen-containing compound (Fig. 13.10) with the other spectra in this chapter. If we look at the simplest compounds of C, O, and N, the reason becomes apparent. C and O have even masses and form an even number of bonds, so H_2O has a molecular mass of 18 and CH_4 has a molecular mass of 16, both even. Nitrogen (^{14}N) has an even mass but forms an odd number of bonds, so NH_3 has an odd mass, 17. Organic compounds are simply derivatives of these simple molecules. The halogens form only single bonds (odd) but have odd masses, so replacement of a hydrogen by a halogen does not change the molecular mass from even to odd.

Fragmentation

With electron ionization, the percent relative abundance, or intensity, of the molecular ion varies dramatically with compound structure. Carbon dioxide (*see* Fig. 13.6 on page 278) and aromatic compounds (*see* Fig. 13.8) contain multiple

■ **FIG. 13.10**
The mass spectrum of methyl 3-nitrobenzoate.

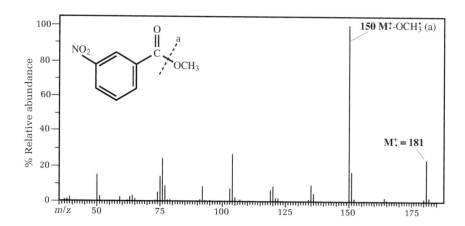

■ **FIG. 13.11**
The mass spectrum of 4,4-dimethyloctane, an isomer of decane.

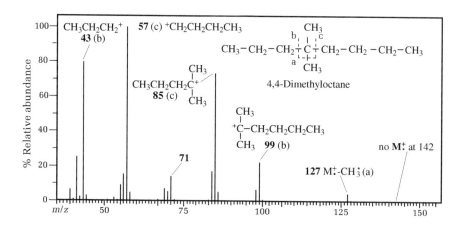

double bonds that are stronger than single bonds, and the π bonding systems can help stabilize the positive charge so that less bond fragmentation occurs and the molecular ion abundance is high. Decane has all single bonds and fragments more readily. The most abundant ions are fragments, and a weak molecular ion is observed. Although the mass of the molecular ion is considered the most important piece of information to be derived from a mass spectrum, the sad truth is that many compounds, like hydrocarbons, alcohols, and alkyl halides, do not show any molecular ion signal. For example, the mass spectrum of 4,4-dimethyloctane (Fig. 13.11) shows no signal at mass 142, which indicates that it fragments even more readily than decane. Nevertheless, the masses and mass differences of abundant fragment ions can provide useful structural information.

We have already discussed the fragment ions (detected) and fragment neutrals (not detected) generated by bond cleavage in CO_2. It is clear that these peak masses are calculated by the simple arithmetic of adding up the atomic isotopic masses of the atoms in the fragments formed by the cleavage of bonds in the molecule. Table 13.3 gives the masses and possible structures for several commonly observed fragment

TABLE 13.3 • Common Fragment Ion Masses and Structures

Fragment Ion Mass (Da)	Structure(s)
. . . 113, 99, 85, 71, 57, 43 Alkyl series	. . . $C_8H_{17}^+$, $C_7H_{15}^+$, $C_6H_{13}^+$, $C_5H_{11}^+$, $C_4H_9^+$, $C_3H_7^+$ Differ by CH_2 or 14 Da
. . . 111, 97, 83, 69, 55, 41 Alkenyl series	. . . $C_8H_{15}^+$, $C_7H_{13}^+$, $C_6H_{11}^+$, $C_5H_7^+$, $C_4H_7^+$, $C_3H_5^+$ Differ by CH_2 or 14 Da
. . . 113, 99, 85, 71, 57, 43 Alkanoyl series	(a) $\quad C_6H_{13}\overset{\overset{\textstyle O}{\|}}{C}{}^+,\ C_5H_{11}\overset{\overset{\textstyle O}{\|}}{C}{}^+,\ C_4H_9\overset{\overset{\textstyle O}{\|}}{C}{}^+,\ C_3H_7\overset{\overset{\textstyle O}{\|}}{C}{}^+,\ C_2H_5\overset{\overset{\textstyle O}{\|}}{C}{}^+$ Differ by CH_2 or 14 Da
105 plus mass of any group on ring	(b) (benzoyl)
91 plus mass of any group on ring	(c) (benzyl)
77 plus mass of any group on ring	(d) (phenyl)

ions. The abundance of a particular fragment ion depends on two factors: the ease with which bonds to it can be broken and the stability of the resulting ion. In the mass spectrum of decane (*see* Fig. 13.7 on page 279), we can see a series of fragment ions at masses 41, 57, 71, 85, 99, and 113, differing by increments of 14 Da, which is the mass of a CH_2 unit. Table 13.3 lists this alkyl series of ions. These arise from the cleavage of bonds at a, b, c, and d, as marked on the decane molecular ion, to form a charged fragment and an uncharged or neutral fragment. From the ion abundances, it is clear that the positive charge prefers to reside on fragments containing three or four carbons, with masses 43 and 57, so the neutral fragments formed by bond cleavage contain the other seven or six carbons.

This alkyl series of fragment ions is common in all alkanes or molecules with large alkyl groups and is observed in the mass spectrum 4,4-dimethyloctane (Fig. 13.11). However, there are several important differences in the percentage ion abundances in its spectrum when compared to that of decane (Fig. 13.7). These can be explained in terms of the increasing stability of carbocations as we go from primary to secondary to tertiary carbocations, the same stability observed in solution organic reactions. The structure of dimethyloctane promotes fragmentation at b and c to give the stable tertiary cations shown at masses 85 and 99. Both of these ions are much more abundant in this spectrum than in the spectrum of decane. Notice also the decreased intensity at mass 71, for the C_6H_{11} alkyl ion. Because there are no C_6 alkyl groups in 4,4-dimethyloctane, the formation of an ion of mass 71 requires the breaking of two C—C bonds and a hydrogen transfer, a much less favorable process. Finally, unlike decane, no molecular ion signal is detected, which is more evidence of the facile fragmentation of this molecule.

Strong peak intensities at masses 77, 91, and 105 are uniquely diagnostic for the presence of an aromatic ring (*see* Table 13.3). These ions are increased in mass by any groups attached to the aromatic ring. Chapter 28 describes the synthesis of methyl 3-nitrobenzoate. The mass spectrum of this product is shown in Figure 13.10. The most abundant ion at mass 150 corresponds to the benzoyl ion of mass 105 shown in Table 13.3, plus the weight of a nitro group, NO_2, mass 46, minus 1, the mass of a hydrogen that the NO_2 replaces.

Notice that the fragment ion masses of the alkanoyl series are the same as that of the alkyl series; therefore, it might seem impossible to distinguish carbonyl compounds like ketones, acids, and esters from alkanes. However, the presence of a C=O leads to some diagnostic fragmentations. For example, the mass spectrum of 3-nonanone in Figure 13.12 shows a very diagnostic fragment ion at mass 72, which is even. This ion arises by a fragmentation of the molecular ion involving a rearrangement called the McLafferty rearrangement:

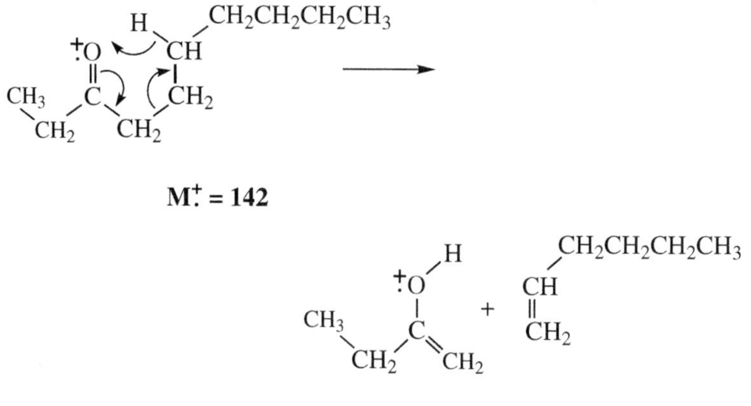

$$M^{+\cdot} = 142$$

Mass 72   **Neutral loss**
fragment ion   **fragment**

Electron pair bonds shift in the six-member intermediate, and a hydrogen is transferred from carbon 6 of the ketone to the carbonyl oxygen. The bond rearrangement releases the neutral alkene, and the positive charge remains on the oxygen. Fragmentation with a hydrogen rearrangement is designated with ~H on the spectrum. Such fragmentations are observed quite frequently.

Neutral Loss Masses

When a molecule breaks into two fragments, only one of them carries the positive charge of the parent molecular ion and can be detected. The uncharged piece, normally a radical or in certain cases a small molecule, is not detected. However, its

■ FIG. 13.12
The mass spectrum of
3-nonanone.

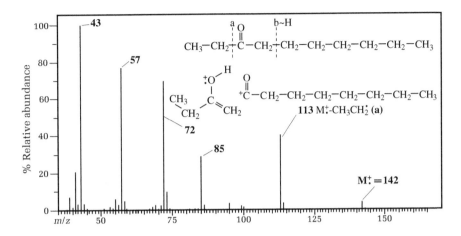

mass can be easily calculated by subtracting the mass of the charged fragment from the molecular ion mass. Like fragment ion masses, the neutral fragment difference masses provide structural information. And they are often easier to understand. Common neutral loss masses are shown in Table 13.4.

Fragment ions can be specified in either of two ways: the fragment structure with a positive charge attached or the M^+ minus the neutral fragment. Both ways are used in the spectra discussed here. The latter method is much more concise and readily understood for the higher mass fragments close to the M^+. For example, the fragment at mass 113 in the decane spectrum (Fig. 13.7) is much easier to designate as M^+ minus ethyl ($M^+ - C_2H_5$) than as $CH_3CH_2CH_2CH_2CH_2CH_2CH_2CH_2^+$. The same is true for fragments at masses 99 and 85. In the spectrum of the chlorophenol in Figure 13.8, the difference in mass between the largest M^+ at mass 142 and the major fragment ion at mass 117 is 35. From Table 13.4, we see this can be attributed to a neutral loss of chlorine and the fragment ion is specified as $M^+ -$ Cl, where Cl is either ^{35}Cl or ^{37}Cl depending on the isotopes in the fragmenting molecular ion.

One of the major differences between the mass spectra of decane and 4,4-dimethyloctane (Fig. 13.7 and Fig. 13.11) lies in the neutral loss of methyl. This fragmentation, specified here as $M^+ - CH_3$, is seen for the dimethyloctane because it leads to a stable tertiary carbocation fragment. This type of fragment is not possible with the linear alkane decane, and a neutral loss of methyl is not observed. The major fragmentation of the ester shown in Figure 13.10 involves a neutral loss with a mass difference of 31, which is attributable to cleavage of neutral OCH_3 radical. In Figure 13.12, the ion at mass 113 is specified both as its fragment ion structure and neutral loss designation.

Many neutral losses involve the elimination of an intact neutral molecule such as water, carbon monoxide, carbon dioxide, or acetic acid. Figure 13.13 shows the mass spectrum of the alcohol 2-nonanol, and, as is typical of alcohols, no molecular ion is seen because of the facile neutral loss of H_2O. The mass of the molecular ion can be deduced, however, by examining other ions observed at

TABLE 13.4 • Some Common Neutral Loss Masses and Possible Fragment Structures

Mass Difference (Da)	Neutral Radical or Molecule Lost
1	·H (hydrogen)
15	·CH$_3$ (methyl)
17	·OH (hydroxyl)
18	H$_2$O (water)
28	C≡O (carbon monoxide) or CH$_2$=CH$_2$ (ethene)
29	·CH$_2$CH$_3$ (ethyl) or ·$\overset{\text{O}}{\overset{\|}{\text{C}}}$—H (formyl)
31	·OCH$_3$ (methoxyl)
35 or 37	^{35}Cl· or ^{37}Cl· (chloro)
42	CH$_2$=C=O (from acetyl group)
43	CH$_3$—$\overset{\text{O}}{\overset{\|}{\text{C}}}$· (acetyl) or ·C$_3H_7$ (propyl)
44	CO$_2$ (carbon dioxide)
45	·OCH$_2$CH$_3$ (ethoxyl) or ·$\overset{\text{O}}{\overset{\|}{\text{C}}}$OH (carboxyl)
60	CH$_3$—C(=O)—OH (acetic acid)
79 or 81	^{79}Br· or ^{81}Br· (bromo)

higher molecular weight. In particular, the presence of two ions 3 Da apart (masses 126 and 129 in Fig. 13.13) indicates neutral losses of water and a methyl group from a molecular ion. For 2-nonanol, adding 15 to mass 129 for a methyl radical equals 144 and adding 18 to mass 126 for a water molecule also equals mass 144. This is strong evidence that the molecular ion mass is indeed 144. The fact that the alkene fragment ion formed by the loss of water has an even mass (except for a compound containing nitrogen) is another indicator of the loss of a molecule rather than a radical.

Each of the bond fragmentations designated a, b, and c in Figure 13.13 is called an *alpha cleavage* because each involves the cleavage of the bond

■ FIG. 13.13
The mass spectrum of 2-nonanol.

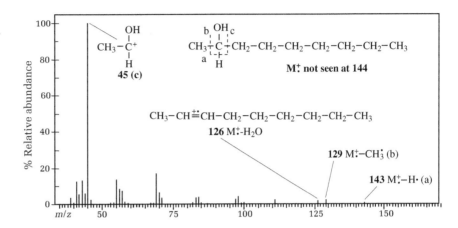

adjacent, or alpha, to the C—OH bond. The electron movement for each of the fragmentations to give ions at masses 143, 129 and 45 are shown below (R = $CH_2CH_2CH_2CH_2CH_2CH_2CH_3$):

Computer-Aided Mass Spectral Identification

Today, there are computer searchable databases of digitized mass spectra of over 200,000 different organic compounds. With less than 1 mg of sample, a mass spectrum can be acquired by a mass spectrometer and stored in a computer. This digitized spectrum can be compared with all the spectra in these collections, and those reference spectra that have ion masses and intensities similar to the sample spectrum can be displayed. If the sample and a reference library spectrum match closely enough, the chemist can feel confident that the sample has been correctly identified. This sort of analytical power is critical to the success of forensic and environmental chemistry laboratories. Be aware, however, that mass spectra of many structural isomers, for example, the three possible dimethylbenzenes, will be so similar that a secure identification is not possible. In this case, other spectral data such as IR and NMR must be acquired or known standards must be run using a GC–MS system to compare retention times as well as mass spectra.

Gas Chromatography–Mass Spectrometry (GC–MS)

The development of capillary column gas chromatography greatly simplified the direct coupling of a gas chromatograph with a mass spectrometer. The combination is one of the most powerful organic analysis techniques available, and thousands of GC–MS instruments are found in research, analytical, and teaching labs around the world. As described in Chapter 10, gas chromatography has the ability to separate complex mixtures containing hundreds of compounds (*see* Fig. 10.2 on page 216). However, many detectors, such as the flame ionization detector or the electron capture detector, while being very sensitive, do not provide any structural information. Replacing these with a mass spectrometer maintains the sensitivity but provides the rich structural information available from this instrument.

The typical mass analyzer is of the quadrupole type (*see* Fig. 13.4 on page 277), but TOF analyzers and other types are also used. The advantages of the TOF analyzer are rapid scanning (10 ms per spectrum or less) and accurate mass measurement. Quadrupole instruments acquire spectra at the rate of about one spectrum per second for a mass range of 35–600 Da. A typical GC–MS analysis using temperature programming of the GC oven to elute all the compounds in a mixture requires about 30 min. Every spectrum acquired by the mass spectrometer is digitized and stored by the data system; almost 2000 digitized spectra are acquired and stored in 30 min. A typical capillary gas chromatography peak is 3–10 s wide, so from three to six spectra are recorded as each compound enters the mass spectrometer ion source from the end of the GC column. The computer constructs a gas chromatogram of the separation by plotting the sum of all the ion signal intensities in each scan, called the total ion current (TIC), versus time. The instrument operator can click on any point on the chromatogram, typically at the top of each peak, to view the mass spectrum at that time in a separate window. These can be compared to a digital library of reference spectra as discussed in the previous section and identified if there is a good match. In this way complex mixtures of hundreds of compounds can be separated and structurally characterized.

Bioanalytical Mass Spectrometry

In 2002, John Fenn of the United States and Koichi Tanaka of Japan shared the Nobel Prize in Chemistry for their development of *soft ionization* methods for the analysis and structural identification of biological macromolecules such as proteins. Because the electron ionization method requires that the sample be in the gaseous vapor state in order to be ionized, most of the high molecular weight

The MALDI ionization process.

■ FIG. 13.15

The MALDI mass spectrum of polypropylene glycol (structure shown). Each peak differs by the mass of one monomer unit.

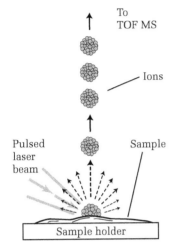

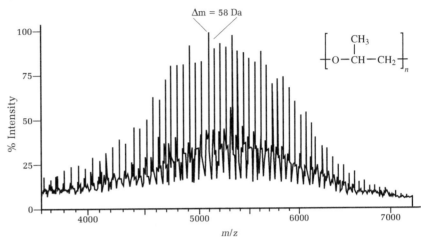

polar materials, such as polysaccharides, nucleic acids, and proteins, would decompose long before they could be heated to a high enough temperature to produce appreciable gaseous molecules. Therefore, for most of its history, mass spectrometry was limited to molecules of molecular weight 2000 Da or less, mostly less than 1000 Da. Then in the 1990s, two new methods were developed that allowed "soft" ionization, without heating, and thus the mass analysis of materials with molecular weights of over 1 million Da. Both of these new techniques—MALDI and ESI—have the advantage that fragmentation is minimal and primarily molecular ions are observed. Actually, the ions formed are *pseudomolecular ions* produced by the attachment of one or more protons or other positive ions such as sodium or potassium to the neutral molecule.

MALDI (*matrix-assisted laser desorption ionization*) uses laser pulses to sputter ions from a mixture of a high molecular weight sample embedded in a crystalline matrix of a low molecular weight compound, such as a cyanobenzoic acid (Fig. 13.14). The matrix facilitates the lifting of the large macromolecules from the surface into the gas phase, where they are mass analyzed, typically using TOF mass spectrometry. A MALDI mass spectrum of polypropylene glycol is shown in Figure 13.15.

ESI (electrospray ionization) is illustrated in Figures 13.16a and 13.16b. When a liquid solution of sample molecules flows from the tip of a fine metal capillary maintained at a high voltage toward a flat metal plate with a potential of thousands of volts between them (Fig. 13.16a), a fine spray of charged droplets forms. Flowing nitrogen gas aids the evaporation of the solvent, and the charged droplets shrink (Fig. 13.16b). At some point the charge becomes too concentrated, and repulsive forces break the droplet into finer droplets. The evaporation, shrinking, and droplet fracture continues until only charged sample molecules remain. Macromolecules typically pick up many charges; the larger the molecule, the

■ FIG. 13.16

(a) The electrospray ionization source. (b) The electrospray ionization process.

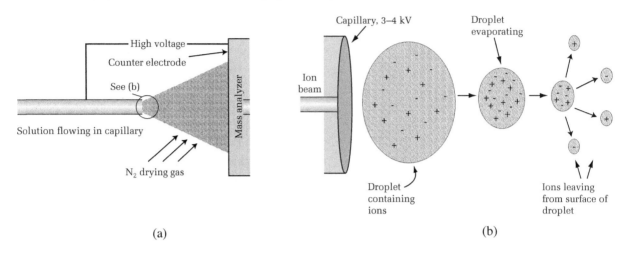

(a)   (b)

■ FIG. 13.17

The electrospray mass spectrum of hen egg lysozyme, showing z and m/z for seven charge states observed. A is the adduct ion of the molecule with 7 to 13 protons attached to give the multiply charged ion masses shown.

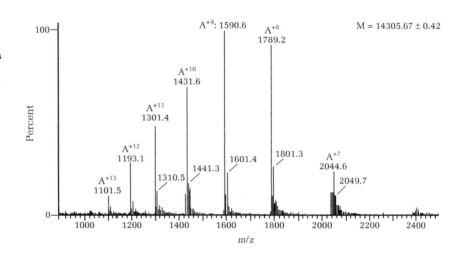

more the charges it can hold. A pin size hole in the plate leads into the high vacuum of a mass analyzer of the quadrupole, TOF, or some other type. The fact that molecules have multiple charges complicates the mass spectrum because each different charge state appears at a different m/z. The electrospray spectrum of the hen egg lysozyme is shown in Figure 13.17. Ions with seven different charge states are detected. Computer-aided calculations allow the determination of these charge states, and from this a molecular weight of 14,305.67 Da is calculated.

As in the case of GC–MS, the development of ESI has allowed the coupling of high performance liquid chromatography (HPLC) with mass spectrometry to give HPLC–MS. This has become a powerful tool for the analysis of complex biological mixtures, such as protein digests.

QUESTIONS

1. Show by calculation that if you weighed 10,000 molecules of decane consisting of the expected percentages of $^{12}C_{10}H_{22}$ and $^{12}C_9{}^{13}C_1H_{22}$ molecules, you would obtain an average molecular weight that is the same as that listed in reference books.

2. Give both the isotopic composition formula and the relative intensity for the two peaks marked with a question mark in the mass spectra in Figures 13.18 and 13.19.

FIG. 13.18
A mass spectrum for question 2.

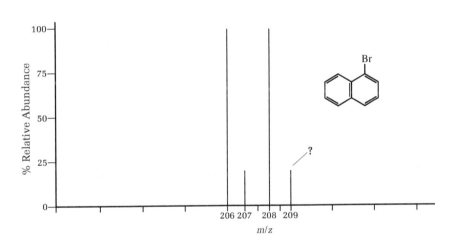

FIG. 13.19
A mass spectrum for question 2.

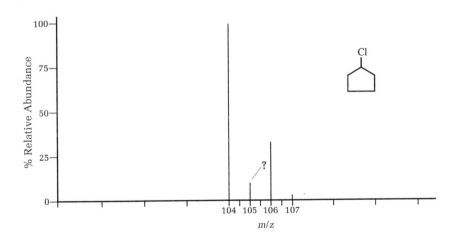

■ FIG. 13.20
A mass spectrum for question 3.

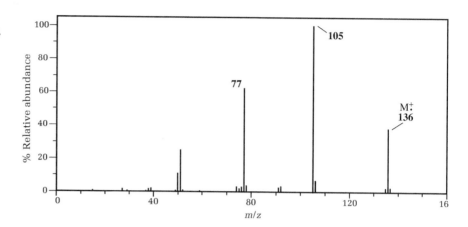

3. Give a structure for the compound with the empirical formula $C_8H_8O_2$ and the following spectral data:
   NMR: Singlet at 3.7 ppm (3 H); multiplet at 7.2 ppm (5 H)
   IR: 2850 cm^{-1} (sharp), 1720 cm^{-1}, 1600 cm^{-1}, 1503 cm^{-1}
   MS: as shown in Figure 13.20

REFERENCES

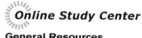

Online Study Center

General Resources
Web Links

1. Lee, Terrence A., *A Beginner's Guide to Mass Spectral Interpretation*. New York: John Wiley and Sons, 1998.
2. McLafferty, Fred W., and František Tureček, *Interpretation of Mass Spectra*, 4th ed. Sausalito, CA: University Science Books, 1993.
3. McMaster, Marvin, and Christopher McMaster, *GC/MS: A Practical User's Guide*. New York: Wiley-VCH, 1998.
4. Silverstein, Robert M., Francis X. Webster, and David Kiemle, *Spectrometric Identification of Organic Compounds*, 7th ed.[1] New York: Wiley, 2005.
5. Sparkman, O. David, *Mass Spectrometry Desk Reference*. Pittsburgh: Global View Publishing, 2000.

---

1. This book includes IR, UV, and NMR spectra.

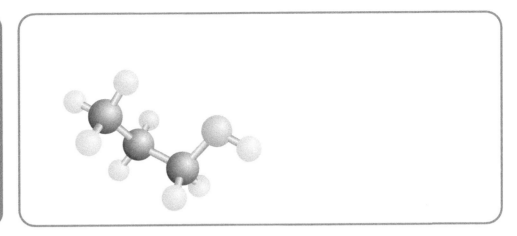

CHAPTER

15

Computational Chemistry

Online Study Center

This icon will direct you to techniques, equipment setups, and online resources at http://college.hmco.com/PIC/williamsonMMOE5e.

The high-speed digital computer has, as in so many other aspects of modern life, changed the way we do and visualize organic chemistry. The ability to predict the outcome, in a quantitative sense, of a chemical reaction and to visualize organic molecules in three dimensions was, until fairly recently, the province of the specialist. Calculations that once taxed the largest computer facilities at a university can now be carried out on a desktop computer. This computational approach to chemistry is being used extensively in drug design, protein mutagenesis, biomimetics, catalysis, studies of DNA-protein interactions, and the determination of structures of molecules using nuclear magnetic resonance (NMR) spectroscopy, to name just a few applications.

Graphical and Mathematical Models

Physical properties and chemical reactivity are a direct consequence of molecular structure, commonly represented by graphical models showing atomic geometry and electronic distribution. Models can be simple or complex; although all models are simplified representations of reality, it is not absolutely necessary that a particular model correspond exactly to reality. The level of complexity depends on how much information is needed to understand and predict chemical behavior. For example, all of the models for methanol shown in Figure 15.1 are correct, but certain models provide more information than others. What is essential is that models allow chemists to predict some measurable quantity with sufficient accuracy. For example, the simplest model of methanol in Figure 15.1, the formula CH_4O, provides all the information necessary to calculate the amount of CO_2 that will be released if one burns a certain quantity of methanol. However, this simple model is not adequate for predicting the hydrogen bonding that occurs between methanol molecules. To do this, a Lewis structure that shows the nonbonding pairs of electrons on the oxygen is a more appropriate model. Chemists often use Lewis structures and think about these representations in terms of the valence bond theory. A Lewis structure shows a molecule as a skeleton of atoms with their accompanying *valence shell electrons*. Valence shell electrons are shown as either

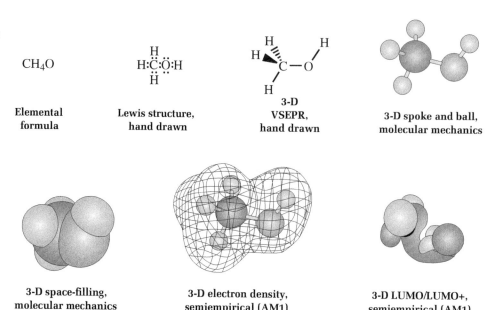

CH_4O

Elemental
formula

Lewis structure,
hand drawn

3-D
VSEPR,
hand drawn

3-D spoke and ball,
molecular mechanics

3-D space-filling,
molecular mechanics

3-D electron density,
semiempirical (AM1)

3-D LUMO/LUMO+,
semiempirical (AM1)

electron pair bonds connecting atoms or as nonbonded pairs localized on atoms. Lewis structures are simple to draw, and the information conveyed is usually sufficient, within the rules of *valence shell electron pair repulsion (VSEPR) theory*, to obtain an approximate prediction of the three-dimensional shape of a molecule. Lewis structures are also usually adequate to give a general picture of the electronic changes that occur during the course of a reaction and are used to write mechanisms involving electron "pushing," which is represented by arrows. Simple hand-drawn structures provide enough information and insight that most of the organic structure and reactivity can be understood using just these. However, hand-drawn structures do not provide detailed quantitative information about bond lengths, angles, and strengths; ionization potentials; electron affinities; or dipoles and molecular energies. There are many times when such information is useful in solving a problem.

By way of analogy, if only the skeletal structure of an animal were presented to an untrained person, that person would probably make mistakes when trying to predict the animal's complete shape and behavior. A trained anatomist, however, will more likely be able to make reasonable predictions based on past experience. Hand-drawn structures present a similar situation. Novices may have more difficulty making correct predictions, whereas a trained chemist is more likely to derive useful information about the structure and properties of a molecule based on its Lewis structure. When drawing accurate structures and predicting physical properties and chemical behavior as determined by molecular energies, however, both the expert and the novice alike encounter problems. Clearly, the development of more numerically accurate, mathematical models would help. They yield tables of numbers such as bond lengths, molecular energies, and orbital information that can be used directly or converted to graphical representations. These more accurate and detailed graphical structures can be rotated in any direction on a computer screen and would be hard to draw by hand. They help to visualize complex numerical data, providing much greater insight into molecular and reaction chemistry.

There are two major types of mathematical models. In the *molecular mechanics* model, molecules are made up of atoms and bonds (as opposed to nuclei and electrons), and the model uses the ball and spring equations developed in physics for classical mechanics. Atom positions are adjusted to best match known empirical parameters, such as bond lengths and angles derived from experimental measurement. In the *quantum chemical* model, molecules are considered to be positively charged atomic nuclei in a cloud of negative electrons. This model relies on the pure theory of quantum physics and mathematical equations to describe the attractive and repulsive interactions between the charges that hold the molecule together. It uses no experimental parameters, working only from theoretical principles, and is therefore also called the *ab initio* ("from the beginning") model.

There is also a third model, called the *semiempirical* model, which is a hybrid of the pure theory-based quantum chemical model and the experimentally parameterized molecular mechanics model. It helps overcome certain limitations associated with these two models and can be effectively applied to a broad range of problems.

You do not need to be a highly trained theoretician to run these calculations, but you should be critical and even skeptical of results produced by computational chemistry. One of the major limitations of all these computations is that they assume noninteracting molecules, in other words, gaseous molecules in a vacuum. Obviously, reactions run in solvents involve solvent-molecule interactions that can significantly change the experimentally measured energetics from those predicted by calculation. (There are models that can estimate solvent effects.) It is quite possible to run a series of calculations that produce results that are absolutely meaningless. Therefore, it is imperative that you use your knowledge of organic chemistry and repeatedly ask, "Are these results consistent with what is observed experimentally? Do they make sense using the traditional hand-drawn, arrow-pushing chemistry taught in textbooks?" If they don't, it probably means that you have chosen the wrong computational model to do the calculation. Computers just generate numbers; they cannot tell you whether the chosen method is valid for the problem being studied. People have to make these decisions to correctly solve real problems.

The biggest challenge in computational chemistry lies in understanding the capabilities and limitations of each model. These will be emphasized in the following sections. One also has to choose which computational method will provide useful information in a minimum amount of time. Ab initio calculations can take hours and even days to complete. If the same information can be derived from a semiempirical calculation in one-tenth the time, the fancier method is unnecessary and wasteful of computer resources.

Molecular Mechanics Models

The simplest, fastest, and easiest to understand calculations involve molecular mechanics, and the results from this approach are usually adequate for many of the modeling exercises in this text. Molecular mechanics is used to calculate the structures of molecules based not on a complete solution of the Schrödinger equation but on a mechanical model for molecules. This model regards molecules as being masses (atoms) connected by forces somewhat like springs (bonds) at certain preferred lengths and angles.

Every organic chemist routinely uses a set of models to examine the three-dimensional structures of molecules because it is difficult to visualize these from a two-dimensional drawing. Molecular mechanics programs allow one to determine quickly the best (lowest-energy) structure for a molecule. Experienced chemists have been doing this intuitively for years, but simple plastic or metal models can lead to incorrect structures in many cases, and, of course, one obtains no quantitative data regarding the stability of one structure compared to another.

Modern computational chemistry software makes it quite easy to draw an approximate structure of a molecule on the computer screen. After the rough structure is drawn and converted to a digital format, a molecular mechanics program such as SYBYL or MMFF automatically steps through a series of iterative calculations; such programs adjust every bond angle, bond length, dihedral angle, and van der Waals interaction to produce a new structure, often called the *equilibrium geometry*, having the lowest possible *strain energy*. The absolute value of the strain energy often differs from one program to another, and so it can be used only in a relative sense to determine the difference in energies between two possible structural geometries or conformations. Molecular mechanics assumes that all bond properties are the same in all molecules. For exceptional molecules such as cyclopropane and cyclobutane, it merely substitutes new parameters for these unusual rings. The experimental data on which many of the parameters for molecular mechanics rest come from x-ray diffraction studies of crystals, electron diffraction studies of gases, and microwave spectroscopy. Even if an ab initio quantum chemical calculation is ultimately planned, the starting structure should always be energy minimized using the faster molecular mechanics program before going to the next level.

Molecular mechanics calculations do *not* provide any information about the locations of electrons and other electronic properties such as ionization potentials, and they *cannot* be used to explore processes that involve bond breaking or bond making. This computational method cannot evaluate the energy of a transition state, and thus the activation energy from which reaction rates can be derived, nor can it calculate the heat of formation or of combustion. These sorts of calculations are the province of the quantum chemical computational models; unlike them, molecular mechanics calculations are empirical in nature and are not derived from first principles.

If a molecule has a single minimum-energy conformation and that conformation has been found, then the shape of the final molecule can be used to predict or confirm approximate vicinal (three-bond) proton-proton NMR coupling constants, which are dependent on the dihedral angle between the protons.

(Couplings are also dependent to a smaller degree on electronegativity of substituents, bond angles, etc.) Optical rotatory dispersion curves can also be predicted when molecular conformations are known.

In a typical empirical force field used in molecular mechanics, the steric energy of a molecule can be represented as the sum of five energies given by the following equation:

$$E_{\text{steric}} = E_{\text{str}} + E_{\text{bnd}} + E_{\text{tor}} + E_{\text{oop}} + E_{\text{vdW}}$$

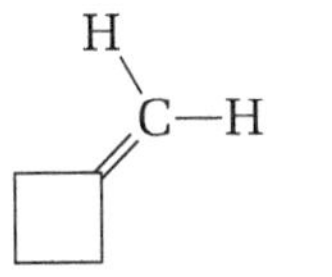

where E_{str} is the energy needed to stretch or compress a bond. Plastic student molecular models are realistic in that there is no way to adjust this value; in molecular mechanics calculations one finds that there is a big force constant (more correctly, potential constant) for this parameter. Most sets of molecular models are constructed so that modest bending of bond angles is possible, and similarly one finds that the force constant for angle bending (E_{bnd}) has an intermediate value. The force constant for torsional motion (E_{tor}) is very small as seen in the free rotation about bonds in most mechanical models. The out-of-plane bending term (E_{oop}) arises in molecules such as methylene cyclobutene where, without this term, the methylene group would want to be bent out of the plane to relieve angle strain. The energy due to van der Waals interactions (E_{vdW}) is very important. It takes two forms, either attractive or repulsive, depending on internuclear distance.

Molecular mechanics allows the calculation of the equilibrium geometry of a molecule (Fig. 15.2). The individual mathematical equations describing the mechanics of a molecule are not terribly complex. For a typical organic molecule containing 30–40 atoms, only a few seconds at most are needed by a modern desktop computer to carry out the millions of calculations and summations for the pair-wise interactions between atoms described by these equations.

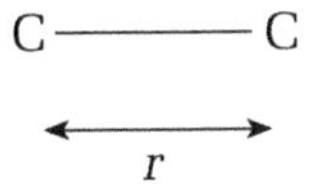

Let us examine in detail the five most important factors entering into molecular mechanics calculations. The bond between two atoms can be regarded just like a spring. It can be stretched, compressed, or bent and, provided these distortions are small, will follow Hooke's law from classical physics.

For bond stretching, the energy is given by

$$E_{\text{str}} = (1/2)k_r(r - r_0)^2$$

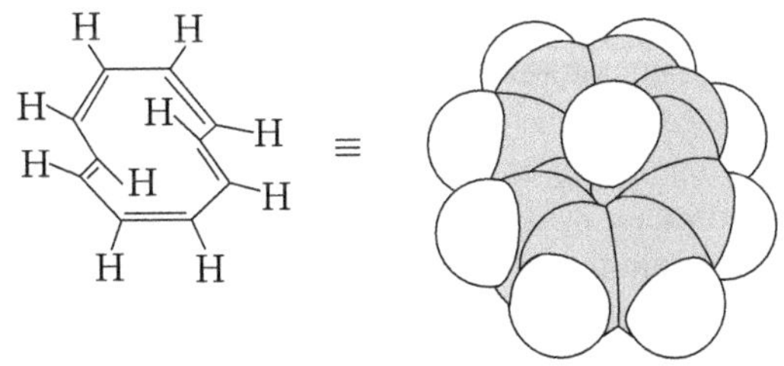

Methylene cyclobutene

All of these energies are empirical in nature, not calculated from first principles.

C ——————— C

r

■ FIG. 15.2
Cyclodecapentaene. From this energy-minimized structure, it is easy to see that the molecule cannot be flat and thus cannot benefit from aromatic stabilization.

where E_{str} is the bond stretching energy in kJ/mol, k_r is the bond stretching force constant in kJ/mol·Å^2, r is the bond length, and r_0 is the equilibrium bond length.

Similarly, the energy for bond angle bending (in-plane bending) is given by

$$E_{bnd} = (1/2)k_\theta(\theta - \theta_0)^2$$

where E_{bnd} is the bond bending energy in kJ/mol, k_θ is the angle bending force constant in kJ/mol·deg^2, θ is the angle between two adjacent bonds in degrees, and θ_0 is the equilibrium value for the angle between the two bonds in degrees.

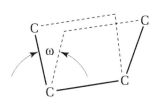

The energy for twisting a bond, E_{tor}, is given by

$$E_{tor} = (1/2)k_\omega[1 - \cos(j\omega)]$$

where E_{tor} is the torsional energy in kJ/mol, k_ω is the torsional force constant in kJ/mol·deg, ω is the torsion angle in degrees, and j is usually 2 or 3 depending on the symmetry of the bond (e.g., twofold in ethylene or threefold in ethane).

The energy term for out-of-plane bending (E_{oop}) is not used very often, but it is a necessary term. It arises in cases such as cyclobutanone, where the C—C—O bond angle is not the sp^2 bond angle of 120° but 133°. To relieve this bond angle strain, the oxygen could tip up out of the plane, but this would require a large amount of energy because of the distortion of the π bonding of the carbonyl. So the force constant for out-of-plane bending is different from that for in-plane bending (bond angle bending; E_{bnd}) discussed earlier.

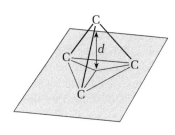

$$E_{oop} = (1/2)k_\delta(d^2)$$

where E_{oop} is the out-of-plane bending energy in kJ/mol, k_δ is the out-of-plane bending constant in kJ/mol·Å^2, and d is the height of the central atom in angstroms above the plane of its substituents.

Very important is the van der Waals energy term (E_{vdW}) given by

$$E_{vdW} = \varepsilon[(r_0/r)^{12} - 2(r_0/r)^6]$$

where E_{vdW} is the van der Waals energy in kJ/mol, ε is the van der Waals potential of the two interacting atoms in kJ/mol, r_0 is the equilibrium distance between the two atoms, and r is the interatomic distance in the molecule in angstroms.

This important function is best understood by referring to Figure 15.3, which shows a curve that describes the energy of two atoms as a function of their distance (r) from each other. At very short distances, the two atoms repel each other very strongly. This interaction raises the energy of the system as the 12th power of the interatomic distance. At a slightly longer distance, a weak attraction is described by an r^6 term, and then at large separation the interaction energy goes asymptotically to zero. The value of ε determines the depth of the potential well. At distance r_0, when the attractive forces overwhelm the repulsive forces, the system has minimum energy. It is r_0 that we call the *bond length*. The molecular mechanics van der Waals term is just an approximation of the ideal curve. It comes close enough and is easy to compute.

■ **FIG. 15.3**

The van der Waals energy function.

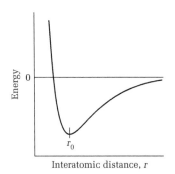

Carrying out molecular mechanics calculations on a personal computer or a small workstation is not difficult. One first draws the structure of the molecule on a computer screen in the same way or even using the same program that is used to draw organic structures for publication in reports and papers. This results in a wire model such as the hexagon we associate with cyclohexane with the proper x, y, z coordinates for each carbon. The program can be directed to add hydrogens at all the unfilled valences. Then the molecular mechanics program is told to minimize the structure, and in a few seconds a minimum energy structure is calculated. Usually, the molecule appears to wiggle on the screen as the minimum energy geometry is calculated.

Consider, for example, the molecule butane. When you initially draw the molecule, you will undoubtedly not draw a perfectly correct structure in terms of bond lengths, bond angles, and dihedral angles. The computer will calculate the energy of the structure you draw and then make a slight change in the coordinates of the atoms and recalculate the energy of the molecule. If the newly calculated energy is lower than the previously calculated one, it will continue to change the coordinates until a minimum energy is reached. This iterative process is continued until it is found that no change in any bond angle, bond length, torsion angle, or van der Waals interaction can be made without raising the energy of the molecule. At this point the calculation stops.

But there is one possible pitfall in this process: The computer program usually has no way of knowing whether it has calculated the structure with the lowest possible energy for the molecule (the *global minimum*) or just a *local minimum*. A geographical analogy may make this statement more clear. It is as if you were instructing the computer to roll a ball to the lowest place in California (Death Valley). The computer might find the lowest place in the Sacramento Valley, but the algorithm on which it operates has no way of knowing whether that is the global minimum or if there is any way to climb the mountains that separate the local minimum from the global minimum.

Now consider once more the molecule butane. If the initial structure put into the computer has a conformation with a torsion angle (dihedral angle) of between 0° and just a bit less than 120° (margin figure a), the molecular mechanics program will calculate the *gauche* conformation (margin figure b) with a 60° torsion angle as the lowest-energy form. In fact, the lowest-energy form (margin figure c) is *anti* and has a torsion angle of 180°. The global minimum is the *anti*conformer, but the calculation can easily fall into one of the two local minima, the *gauche* conformations, which are not the lowest-energy conformations. The full potential energy diagram for all rotational conformers of butane generated by molecular mechanics, shown in Figure 15.4, illustrates this concept clearly.

There are several approaches to the solution to this problem, but no one approach is completely satisfactory. If one has enough experience and intuition, it is often possible to spot the fact that the minimum-energy conformation calculated by the computer is not the global minimum. A better solution, however, is to calculate energies for an arbitrary set of dihedral angles about one of the bonds. In butane, one could ask for the energies to be calculated at dihedral angles of 0°, 30°, 60°, 90°, 120°, 150°, and 180° using a part of the program called a *dihedral driver*. In this way you would be sure to catch the lowest-energy form.

(a)

(b)

(c)

■ **FIG. 15.4**
The potential energy of butane
as a function of the dihedral
angle between the two central
carbons.

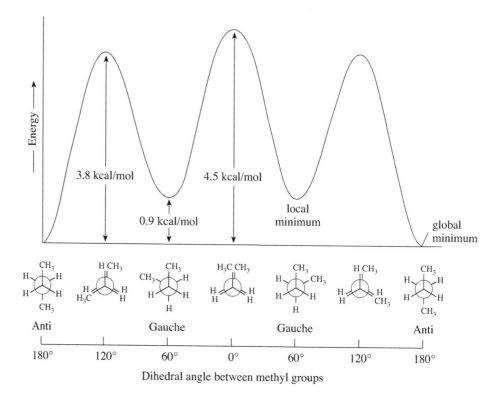

Dihedral angle between methyl groups

Now consider a larger molecule. If one were to search for the best conformation at angles of 0°, 30°, 60°, 90°, 120°, 150°, and 180° around each carbon-carbon bond, it would be necessary to make 7^n calculations, where n is the number of carbon-carbon bonds. In 2-bromononane, for example, this would be $7^8 = 5,764,801$ calculations. The problem gets out of hand very rapidly.

It is not possible to use a dihedral driver within a ring, so the lowest-energy form is found by other means (ring flip and flap, molecular dynamics, and Monte Carlo methods).

An elaboration of this computational method, called *molecular dynamics*, is often used to study the motions of large molecules such as proteins and other large polymers. Molecular dynamics is basically a series of molecular mechanics calculations, with each conformation being a frame in a movie animation of the movements of a molecule in the process of folding into a preferred shape.

Quantum Chemical or ab initio Models

In the 1920s, a more sophisticated theory of atomic and molecular structure was developed; it improved on Lewis structures and yielded a more accurate representation of molecular properties. This new theory relied on quantum mechanics and the Schrödinger equation for predicting the behavior of an electron in the presence of a positively charged nucleus. However, the mathematics for the exact solution of the equations involved was so complex that a full calculation could only be performed on the one-electron hydrogen atom. However, by assuming that the nuclei don't move (Born-Oppenheimer approximation) and that electrons move independently of one another [Hartree-Fock (HF) approximation; ab initio models are often called HF models], the Schrödinger equation could be simplified so that it was practical to perform useful calculations on multiatom systems. Many scientists have worked on further refinements of the equations and computer programs that have helped to gradually improve the quality of the results. Because of the intensive nature of the calculations, the availability of fast desktop computers and efficient computational programs has also been very important. Today, calculations involving more than 25 nonhydrogen atoms are feasible. Two commonly used HF model programs are HF/3-21G and HF/6-31G*.[1] If further accuracy is required, one can enhance the HF models by using perturbation theory to compensate for electron-electron repulsion within a molecular orbital. Approximations that add perturbation theory were developed by Möller and Plesset, and a commonly used MP-level model is MP2/6-31G*. Note that different approximations often lead to different results. The mathematical complexity of quantum calculations puts a full explanation of the principles beyond the scope of this text.

Unlike molecular mechanics, ab initio models use no empirical data. The calculation is truly from first principles, using approximate Schrödinger equations to calculate the locations of the atomic nuclei and *all* electrons: valence, unpaired, and inner shell. The equilibrium (minimum-energy) geometries are usually improved relative to those from molecular mechanics, better reflecting the true structure. Many properties that are defined by electron distribution, such as dipole moments, ionization potentials, and the shape and sign of molecular orbitals, are calculated. It is also possible to calculate vibrational energies to produce an approximate infrared (IR) spectrum. More importantly, the energies of molecules, transition states, and bond-making or bond-breaking reactions can be computed and compared. This information cannot be derived from molecular mechanics, but it can be calculated using the semiempirical models to be discussed in the following section.

These capabilities are best demonstrated by the following example. In the presence of a free radical initiator (AIBN, azoisobutylnitrile; *see* Chapter 18),

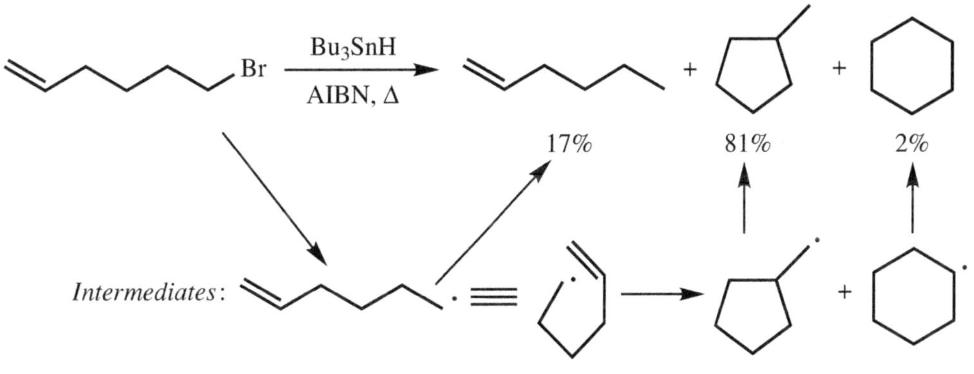

---

1. The asterisk means that the program has been improved.

■ **FIG. 15.5**

The four possible energy–
reaction coordinate diagrams
for the intramolecular cyclization
of the hexenyl radical to either
the cyclopentylmethyl radical
or the cyclohexyl radical.

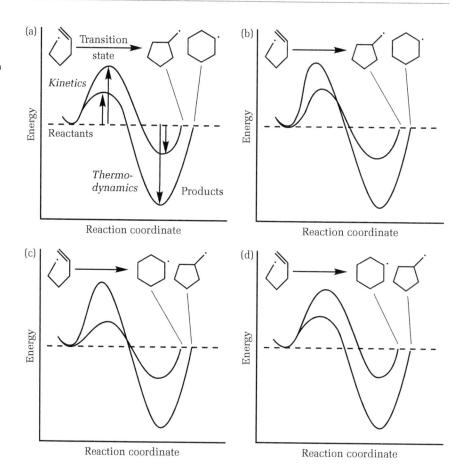

tributyltin hydride cleaves carbon-bromine bonds to give free radicals. In the case
of 6-bromo-1-hexene, the initially formed radical can either abstract a hydrogen
from surrounding molecules to form 1-hexene or undergo intramolecular radical
cyclization to form either a cyclopentylmethyl or cyclohexyl radical, either of
which also picks up hydrogen to give the products shown.

Organic chemistry fundamentals would predict that cyclohexane should be
the major cyclic product because the formation of six-membered rings and secondary radicals in the intermediate cyclohexyl radical would be favored. Contrary
to this prediction, the major product (81%) is derived from the five-membered cyclopentylmethyl radical, with the less stable primary radical. Let us see whether an
analysis of the energetics of this reaction using ab initio methods provides insight
into this apparent contradiction.

As in the energy analysis of the conformers of butane (Fig. 15.4), it is useful
to use energy–reaction coordinate diagrams to understand the possible reaction
pathways. Figure 15.5 shows four diagrams depicting the energy needed to cause
the radical and π bond to react (the transition state) and the energy released as a

new bond is formed. The former energy determines the kinetics or rate of the reaction, and the latter determines the thermodynamic stability of the products relative to the reactant and to each other. There are four different energy pathways possible, as shown in the four diagrams.

Diagrams (a) and (b) in Figure 15.5 show the cyclohexyl radical to be more thermodynamically stable than the cyclopentylmethyl radical, consistent with the chemical fundamentals discussed previously. Diagrams (c) and (d) make the cyclopentylmethyl radical the more stable one, which would be inconsistent. When the energy minima of these two radicals are calculated using a 6-31G* ab initio computation, the cyclohexyl radical is roughly 5 kcal/mol more stable than the cyclopentylmethyl radical, consistent with our chemical knowledge. Energy diagrams (c) and (d) are therefore incorrect depictions of the reaction.

If the cyclohexyl radical is the more stable, as calculated and depicted in diagrams (a) and (b), there must be another explanation of why the cyclopentylmethyl product is formed 40 times more readily. A possible explanation is that the transition state energy necessary to get the initial hexenyl radical to react at carbon 2, forming a five-membered ring, is less than the energy needed to react at carbon 6. Because ab initio methods also allow calculation of the energy maxima of transition states, this question can also be examined computationally. The calculated difference in transition state energies is 2.7 kcal/mol lower in favor of ring closure to the cyclopentylmethyl radical, making diagram (b) the only correct representation of the energetics. Computational chemistry shows that the final product ratio is determined by kinetics, the rate of cyclopentylmethyl closure being faster than the rate of cyclohexyl closure due to the lower transition state energy.

Note that the addition of the initial hexenyl radical to the double bond is for this case *irreversible* because the reverse transition state energy is quite high. If the energy difference were low enough, a dynamic equilibrium between product and reactant radicals might eventually lead to the formation of the more thermodynamically stable product (cyclohexyl radical), given sufficient time and energy. Most *reversible* reaction product ratios are determined by thermodynamics, not kinetics.

Product ratios and reactant:product ratios, whether determined by thermodynamics or transition state energy differences, are roughly proportional to the energy differences, as shown in Table 15.1. Notice that even small energy differences lead to large differences in these ratios.

TABLE 15.1 • The Relationship of Product Ratio to Energy Differences		
Energy Difference (kcal/mol)	Energy Difference (kJ/mol)	Product Ratio, Major:Minor
0.5	2	80:20
1	4	90:10
2	8	95:5
3	12	99:1

Semiempirical Models

Most molecular computations done by organic chemists, especially those examining minimum energy geometries, are done using a semiempirical model because such a model provides the best compromise between speed and accuracy. This kind of model can be thought of as a hybrid of molecular mechanics models based on experimentally measured *empirical* data and pure quantum chemical theory or ab initio models, thus the name *semiempirical*. Semiempirical models use Schrödinger equation approximations like those described previously but, in order to make the calculations less time consuming, *only* the locations of *valence electrons* are calculated, *not all electrons*. For the inner shell electrons, empirical data from typical organic molecules is used to estimate their locations. This and other approximations allow semiempirical models to be used for molecules containing up to 100 or more nonhydrogen atoms, much larger than those that can be modeled by the more rigorous ab initio models. However, semiempirical models are often less accurate, especially for calculations on novel molecules with unusual bonding or reactivity. Two commonly used programs are AM1 and PM3.

Most computational chemistry texts not only provide explanations of the different mathematical model equations but also do a large number of comparisons of calculated properties (bond lengths, angles, dipoles, various energies, etc.) from the different mathematical models versus actual properties measured experimentally in the lab. The data can be quite sobering, with differences between calculated and experimental values averaging 7%–10% and with many examples having much larger errors. As seen from Table 15.1, these types of errors in energy calculations would lead to even larger discrepancies in calculated and experimental product ratios. As stated earlier, every computational result has to be examined critically to make sure it fits with both chemical logic and experimental results. Nevertheless, being able to model transient intermediates such as free radicals or transition states, species that are virtually impossible to isolate and study in the lab, provides some exciting ways to extend our understanding of chemistry.

Types of Calculations

The model used to perform the calculation of any particular property or graphical representation determines the amount and accuracy of information generated in the output. The first two types of calculations, geometry optimization and single point calculations, can be performed with any of the three computational models described previously. The last three types of calculations, involving electron locations and energies, are not solvable by molecular mechanics.

1. Geometry Optimization

Geometry optimization is a standard computational chemistry calculation to find the lowest-energy, or most relaxed, conformation for a molecule. The approach is the same for all levels of calculation, involving an iterative jiggling process like that described for molecular mechanics. At each step, the molecular geometry is modified slightly, and the energy of the molecule is compared with the last cycle.

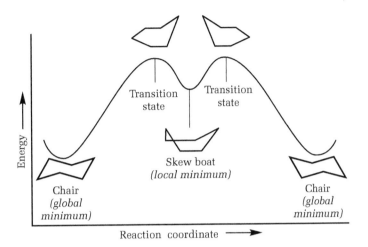

The computer moves the atoms in the molecule a little, calculates the energy, moves them a little more, and keeps going until it finds the lowest energy. This is the energy minimum of the molecule and is obtained at the optimized geometry. Recall that the energies from molecular mechanics can only be used in a relative sense, while those from quantum electronic structure models can be compared in an absolute sense, like heats of formation.

Be wary of structures that may be stuck in a local, not global, energy minimum. As you may know, the chair conformation of cyclohexane is lower in energy than the boat conformation (Fig. 15.6). If you entered the boat structure, geometry optimization should find the chair as the structure with the lowest energy. Remember that sometimes, however, the computer gets "stuck" trying to find a minimum due to the presence of a local minimum. For example, if you start with a boat structure of cyclohexane, the computer may get to a "skew-boat" local energy minimum but may not flip the molecule completely into the chair conformation that corresponds to the *global minimum*. This can happen when the energy jump from skew boat to chair is too large. A knowledge of fundamental organic chemistry as presented in introductory textbooks helps to recognize these types of problems.

2. Single Point Calculations

Single point calculations are often used in combination with geometry optimization to investigate steric hindrance. In this case the method only performs one computational cycle to calculate the energy of a particular fixed geometry. In a thermodynamically controlled reaction, the energetic difference between two conformations is often due to steric hindrance. If the product molecule optimizes in one conformation, you can use single point calculations to determine how much more energy is needed to form the nonpreferred conformation. The structure drawing and manipulation part of the software will allow you to move only that part of the molecule that changes in the higher energy form, leaving the rest of the molecule optimized. The single point calculation performed on this modified molecule will give an energy that you can directly compare with the optimized energy to find the energy difference between the lower and higher energy conformers. For example, the energetic difference between having a constituent in the axial or equatorial position on a cyclohexane ring can be determined.

3. Transition State Calculations

Transition state calculations can be thought of as the reverse of geometry optimizations. In this case, the method searches for a structure of *maximum* energy, a transient intermediate that cannot be isolated experimentally. For example, this type of calculation allows one to examine transition state energies and geometries of intermediates involved in carbocation rearrangements. The literature contains standard models that should used as the starting point for these calculations. It takes an appreciable amount of effort and experience to properly analyze transition state structures and energies.

4. Vibrational Frequency Calculations

Vibrational frequency calculations allow the calculation of IR stretching and bending absorption frequencies, and it is a lot of fun to view animations of these types of motions in molecules. The vibrational frequency of a two-atom system is proportional to the square root of the force constant (the second derivative of the energy with respect to the interatomic distance) divided by the reduced mass of the system (which depends on the masses of the two atoms). In fact, the analysis of transition states (as well as other essential quantities such as entropy) really starts by exploring slight movements of atoms that can be considered as vibrations.

The frequencies calculated using semiempirical models are usually higher than the frequencies measured on an IR instrument by about 10%, and the accuracy depends on the computational model used. The best way to check the calculated values is see whether the vibration animations make sense; for example, if an animation shows a $C{=}O$ stretching vibration, the calculated frequency can be matched with the measured frequency of that type, which is usually easy to assign in the acquired IR spectrum.

5. Electronic Density and Spin Calculations, Graphical Models, and Property Maps

Electronic density and spin calculations, graphical models, and property maps allow the visualization of electronic properties such as electron densities, electrostatic potentials, spin densities, and the shapes and signs of molecular orbitals. The values for a particular property at each point in the three-dimensional space around a molecule are displayed on a two-dimensional computer screen as a surface of constant numerical value, often called an *isosurface*, which can be rotated in any direction to study it. Alternately, numerical variations of a given property (such as electron density) at a defined distance from the molecule can be displayed as a property map using color as a key. Carrying out surface calculations and viewing their graphical representations are major activities in computational chemistry and can provide useful insight into the mechanisms of organic reactions. A few examples of isosurfaces, orbitals, and property maps are shown in Figure 15.1 (on page 308). More examples and explanations, including color images, are available on the website for this book.

Online Study Center

General Resources
Molecular Property Maps

Examples of Calculations

Try to reduce computational time by understanding the various computational methods and answering the question "What kind of information do I need?" If only geometry or ring strain information is needed, using the more intensive ab initio calculation to produce high-precision molecular orbitals is not necessary. Even if you need the high accuracy of an ab initio calculation, do not start there. First, approximate the molecular geometry using a molecular mechanics calculation. Then use this structure to run an AM1 semiempirical calculation for a good approximation of the valence electron properties. Finally, if even more accuracy is required, use the AM1 output for the 3-21G input to see how the inner shell electrons affect the electronic properties.

An instructive exercise is to calculate the energy of butane at dihedral angles of 0°, 60°, 120°, and 180°, using single point energy calculations for each. You should be able to reproduce the *form* of the curve shown in Figure 15.4 (on page 314), but you usually will not obtain the experimentally derived energies shown.

Remember that the calculated steric energy has no physical significance. Differences in steric energies can be useful, however, if the molecules being compared have the same bonding pattern. The molecule with the lower energy will be the more stable isomer (geometric, conformational, or stereo). Other types of comparisons cannot be made. A molecular mechanics calculation cannot be used to compare the stabilities of unrelated molecules.

If the program you are using breaks down the steric energy into the six terms E_{steric}, E_{str}, E_{bnd}, E_{tor}, E_{oop}, and E_{vdW}, do not give the individual terms much credence. Differing parameter sets in other programs may weight these differently. You should pay the most attention to the total steric energy.

Molecular mechanics calculations, like IR and NMR spectroscopy, are a tool available to the chemist. As such, they are presented in this text as adjuncts to experimental work in the laboratory, not as an end in themselves. In most of the computational problems presented in the following chapters, you are asked to carry out computations of the steric energies or the heats of formations of isomers.

One must be careful when using molecular mechanics computations to draw conclusions. The computation can be made on any molecule, but the resulting steric energy cannot always be compared to the steric energy of another molecule, even though they are isomers. For example, *cis*- and *trans*-butene and other similar *cis*- and *trans*-isomers can be compared to each other because they are both 1,2-disubstituted alkenes, but a valid comparison cannot be made between 1,2- and 1,1-disubstituted alkenes.

The increase in time needed for a calculation on a molecule having n atoms:

molecular mechanics, n^2

semiempirical, n^3

ab initio, $>n^4$

Although not intending to mislead you deliberately, we have included computations on reactions and products that for one reason or another are not expected to give valid comparisons or answers. It is for you to decide whether the computations are in agreement with your experiment.

Higher levels of computation, the semiempirical molecular orbital methods and the even higher ab initio methods, will give heats of formation that can be compared to one another. The problem with these higher levels of computation is that they require more time and more computer power. But often an AM1 semiempirical molecular orbital calculation on a relatively small molecule can be done in a few seconds. The cost (in computer time) of ab initio calculations goes up more than the fourth power of the number of atoms, whereas molecular mechanics calculations go up as the square of the number of atoms in the molecule.

The radical chlorination of 1-chlorobutane gives four isomeric dichlorobutanes. In Chapter 18 these chlorobutanes are synthesized, and their relative amounts are measured by gas chromatography with the object of determining the relative reactivity of the various hydrogens in the starting material. Using a semiempirical program, you can calculate the energies of the four products to see whether or not these energies correlate with the product distributions. A correlation or lack thereof may give information regarding the mechanism of the reaction.

Similarly, the dehydration of 2-butanol (*see* Chapter 10) gives three isomeric butenes that are separated and analyzed by gas chromatography. Calculation of their heats of formation or steric energies may or may not correlate with the percentage of each isomer formed in the reaction. Again, a correlation or lack thereof may give information about the mechanism of the reaction or about the applicability of computational chemistry to problems of this type.

Treatment of *cis*-norbomene-5,6-endo-dicarboxylic acid with concentrated sulfuric acid gives the isomeric compound X (Chapter 48). The energies of any proposed structures for X can be calculated and compared with that of the starting material. If the reaction conditions (hot concentrated sulfuric acid) are regarded as favoring an equilibrium between product and starting material, then X would be expected to have the lower energy if the correct structure is proposed.

In Chapter 50 you are asked to calculate the most stable conformation of the product *p*-terphenyl. Hexaphenylethane is a molecule that does not exist, but this does not preclude calculation of its steric energy and bond lengths (Chapter 31). One can calculate the energy of the isomeric dimer that does form. This calculation gives a clear picture of why the triphenylmethyl radical does not simply dimerize, even though the steric energy may not be correct.

The oxidation of citronellol can give four possible isomeric isopulegols (*see* Chapter 23). Molecular mechanics calculations can indicate which of these is the most stable. Similarly, oxidation of these alcohols can give two possible isopulegones that theoretically can equilibrate. Molecular mechanics calculations can again indicate which is the more stable. These rearrange to pulegone. Again, calculation may disclose whether this product is more stable than the starting material, throwing light on the mechanism of the isomerization reaction.

The Wittig reaction usually gives a mixture of *cis-* and *trans-*isomers. In Chapter 39 two Wittig reactions are carried out, in which one isomer predominates in each reaction. A molecular mechanics calculation discloses not only what these isomers look like but also their energies.

A very easy reaction to carry out is the aldol condensation of benzaldehyde with acetone to give in high yield dibenzalacetone (mp 110°C–111°C). It is interesting to explore the products of this reaction with molecular mechanics because three geometric isomers can be formed, and seven single-bond *cis-* or *trans-*isomers of the geometric isomers lie at the energy minima. The question is, which of these 10 isomers is formed in the reaction? Computational aspects of this question are dealt with at length in Chapter 37.

In all of these calculations, not only will you calculate the energies of the molecules, you will also be able to see what the lowest-energy conformation of the molecule looks like. In this regard it is interesting to look at the calculated low-energy conformation of benzophenone (Chapter 38), tetraphenylcyclopentadienone (Chapter 51), and pseudopellitierene (Chapter 66). *Z-* and *E-*stilbene are synthesized in Chapters 58 and 60, and in Chapter 59 the Perkin reaction is used to make a mixture of *Z-* and *E-*phenylcinnamic acids. Energy calculations, as well as a picture of the molecular conformations, help rationalize the relative amounts of the isomers formed in these synthetic reactions as well as their ultraviolet (UV) spectra.

Other examples of synthetic experiments with computational chemistry supplements are available on this book's website.

Online Study Center

General Resources
Additional Experiments

REFERENCES

Burkert, Ulrich, and Norman L. Allinger, *Molecular Mechanics* (ACS Monograph 177). Washington, DC: American Chemical Society, 1982.

Clark, Tim, *A Handbook of Computational Chemistry: A Practical Guide to Chemical Structure and Energy Calculations.* New York: John Wiley & Sons, 1985.

Earl, Boyd L., David W. Emerson, Brian J. Johnson, and Richard L. Titus, "Teaching Practical Computer Skills to Chemistry Majors," *J. Chem. Educ.* 71 (1994): 1065.

Freeman, Fillmore, Zufan M.Tsegai, K. Marc Lasner, Warren J. Hehre, "A Comparison of the ab Initio Calculated and Experimental Conformational Energies of Alkylcyclohexanes," *J. Chem. Educ.* 77 (2000): 661.

Grant, Guy H., and W. Graham Richards, *Computational Chemistry.* Oxford, UK: Oxford University Press, 1995.

Hehre, Warren J., *A Guide to Molecular Mechanics and Quantum Chemical Calculations.* Irvine, CA: Wavefunction Press, 2003.

Hehre, Warren J., *A Laboratory Book of Computational Organic Chemistry.* Irvine, CA: Wavefunction Press, 1998.

Hehre, Warren J., Lonnie D. Burke, Alan J. Shusterman, and William J. Pietro, *Experiments in Computational Organic Chemistry.* Irvine, CA: Wavefunction Press, 1993.

Hehre, Warren J., *The Molecular Modeling Workbook for Organic Chemistry*, 2nd ed. Irvine, CA: Wavefunction Press, 2005.

Jarret, Ronald M., and Ny Sin, "Molecular Mechanics as an Organic Chemistry Laboratory Exercise," *J. Chem. Educ.* 67 (1990): 153–155.

Jensen, Frank, *Introduction to Computational Chemistry*. New York: John Wiley & Sons, 1999.

Lipkowitz, Kenny B., Raima Larter, and Tom Cundari, eds., *Reviews in Computational Chemistry*, Vol. 20. New York: John Wiley & Sons, 2004.

Lipkowitz, Kenny B., and Daniel Robertson, "Conformer Hunting: An Open-Ended Computational Chemistry Exercise That Expresses Real-World Complexity and Student Forethought," *J. Chem. Educ.* 77 (2000): 206.

The *Journal of Chemical Education*. Since 1989, many articles on molecular mechanics calculations and computational chemistry in general have been published. Use electronic journal searches to find these articles.

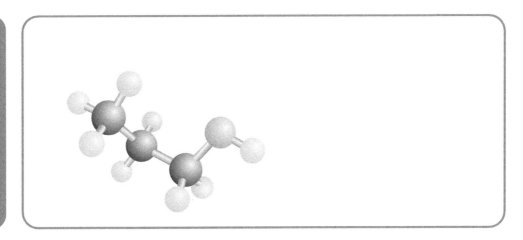

CHAPTER
17

Nucleophilic Substitution
Reactions of Alkyl Halides

Online Study Center

This icon will direct you to
techniques, equipment setups,
and online resources at
http://college.hmco.com/PIC/
williamsonMMOE5e.

PRELAB EXERCISE: Predict the outcomes of the two sets of
experiments to be carried out with the 11 halides used in this
chapter.

The alkyl halides, R—X (where X = Cl, Br, I, and sometimes F), play a
central role in organic synthesis. These can easily be prepared from, among
others, alcohols, alkenes, and, industrially, alkanes. In turn, they are the starting
materials for the synthesis of a large number of new functional groups. The syntheses are often carried out by nucleophilic substitution reactions in which the
halide is replaced by another group such as cyano, hydroxyl, ether, ester, alkyl—
the list is long. As a consequence of the importance of this substitution reaction, it
has been studied carefully by employing reactions such as the two used in this
chapter. Some of the questions that can be asked are as follows:

- How does the structure of the alkyl part of the alkyl halide affect the
  reaction?
- What is the effect of changing the . . .

  nature of the halide?
  nature of the solvent?
  relative concentrations of the reactants?
  temperature of the reaction?
  nature of the nucleophile?

In this chapter we explore the answers to some of these questions.

In free radical reactions the covalent bond undergoes homolysis when it
breaks:

$$R:\ddot{C}l: \longrightarrow R\cdot + \cdot\ddot{C}l:$$

333

A carbocation is planar and has an empty *p*-orbital.

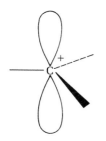

whereas in ionic reactions it undergoes heterolysis:

$$R:\ddot{C}l: \longrightarrow R^+ + :\ddot{C}l:^-$$

A carbocation is often formed as a reactive intermediate in these reactions. This carbocation is sp^2 hybridized and trigonal-planar in structure with a vacant *p*-orbital. Much experimental evidence of the type obtained in this chapter indicates that the order of stability of carbocations is

Order of carbocation stability

$$
\begin{array}{cccc}
R & R & H & H \\
| & | & | & | \\
R-C^+ > & R-C^+ > & R-C^+ > & H-C^+ \\
| & | & | & | \\
R & H & H & H
\end{array}
$$

The alkyl (R) groups stabilize the positive charge of the carbocation by displacing or releasing electrons toward the positive charge. As the number of alkyl groups attached to the carbocation increases, the stabilization increases.

Nucleophilic substitution

Many organic reactions occur when a nucleophile (a species with an unshared pair of electrons) reacts with an alkyl halide to replace the halogen with the nucleophile.

$$Nu:^- + R:\ddot{X}: \longrightarrow Nu:R + :\ddot{X}:^-$$

This substitution reaction can occur in one smooth step:

One step

$$Nu:^- + R:\ddot{X}: \longrightarrow \left[\overset{\delta-}{Nu} \cdots\cdots R \cdots\cdots \overset{\delta-}{\ddot{X}:} \right] \longrightarrow Nu:R + :\ddot{X}:^-$$

or it can occur in two discrete steps:

Two steps

$$R:\ddot{X}: \longrightarrow R^+ + :\ddot{X}:^-$$
$$Nu:^- + R^+ \longrightarrow Nu:R$$

depending primarily on the structure of the R group. The nucleophile ($Nu:^-$) can be a substance with a full negative charge, such as $:\ddot{I}:^-$ or $H:\ddot{O}:^-$, or an uncharged molecule with an unshared pair of electrons such as exists on the oxygen atom in water, $H-\ddot{O}-H$. Not all of the halides, $:\ddot{X}:^-$, depart with equal ease in nucleophilic substitution reactions. In this chapter we investigate the ease with which the different halogens leave in one of the substitution reactions.

Reaction kinetics

To distinguish between the reaction that occurs as one smooth step and the reaction that occurs as two discrete steps, it is necessary to study the kinetics of the reaction. If the reaction were carried out with several different concentrations of $R:\ddot{X}:^-$ and $Nu:^-$, we could determine if the reaction is bimolecular or unimolecular.

In the case of the smooth, one-step reaction, the nucleophile must collide with the alkyl halide. The kinetics of the reaction

$$Nu:^- + R:\ddot{X}: \longrightarrow \left[\overset{\delta-}{Nu} \cdots R \cdots \overset{\delta-}{\ddot{X}}: \right] \longrightarrow Nu:R + :\ddot{X}:^-$$

$$Rate = k\left[Nu:^- \right]\left[R:\ddot{X}: \right]$$

are found to depend on the concentration of both the nucleophile and the halide. Such a reaction is said to be a bimolecular nucleophilic substitution reaction, S_N2. If the reaction occurs as a two-step process,

$$R:\ddot{X}: \xrightarrow{\text{slow}} R^+ + :\ddot{X}:^-$$

$$Nu:^- + R^+ \xrightarrow{\text{fast}} Nu:R$$

$$Rate = k\left[R:\ddot{X}: \right]$$

the rate of the first step, the slow step, depends only on the concentration of the halide, and it is said to be a unimolecular nucleophilic substitution reaction, S_N1.

The S_N1 reaction proceeds through a planar carbocation. Even if the starting material were chiral, the product would be a 50:50 mixture of enantiomers because the planar intermediate can form a bond with the nuceophile on either face.

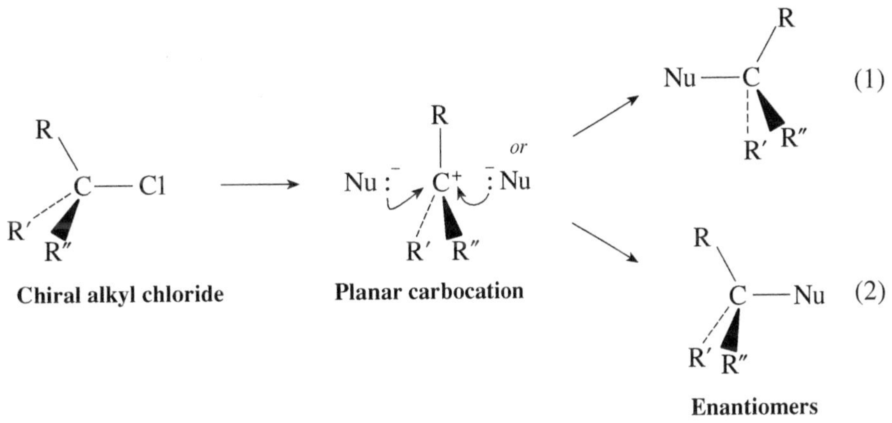

Racemization: S_N1

Chiral alkyl chloride   **Planar carbocation**

(1)

(2)

Enantiomers

Inversion: S_N2

The S_N2 reaction occurs with inversion of configuration to give a product of the opposite chirality from the starting material.

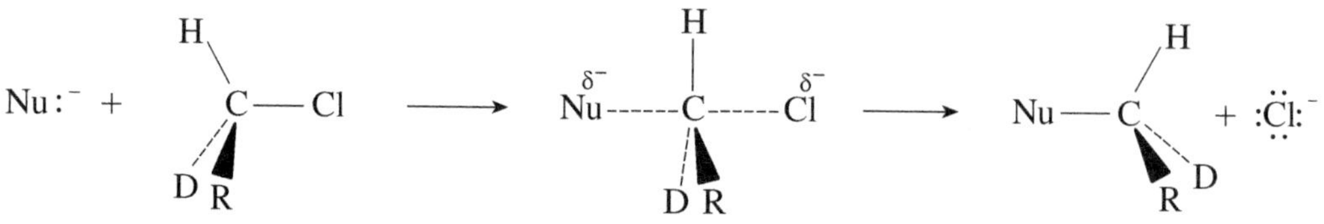

$$\text{Nu}:^- \ + \quad \overset{H}{\underset{D\ \ R}{C}}{-}\text{Cl} \quad \longrightarrow \quad \text{Nu}\text{-----}\overset{\overset{H}{|}}{\underset{\underset{D\ \ R}{|}}{C}}\text{-----}\overset{\delta^-}{\text{Cl}} \quad \longrightarrow \quad \text{Nu}{-}\overset{H}{\underset{\underset{R}{D}}{C}} \ + \ :\ddot{\text{Cl}}:^-$$

The order of reactivity for *simple* alkyl halides in the S_N2 reaction is

Order of S_N2 reactivity

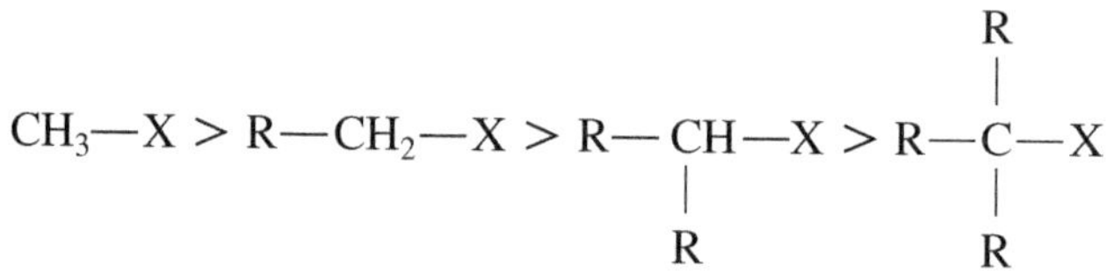

$$CH_3{-}X > R{-}CH_2{-}X > R{-}\underset{\underset{R}{|}}{\overset{\overset{}{|}}{CH}}{-}X > R{-}\underset{\underset{R}{|}}{\overset{\overset{R}{|}}{C}}{-}X$$

The tertiary halide is in parentheses because it usually does not react by an S_N2 mechanism. The primary factor in this order of reactivity is steric hindrance, that is, the ease with which the nucleophile can come within bonding distance of the alkyl halide; 2,2-dimethyl-1-bromopropane, even though it is a primary halide, reacts 100,000 times slower than ethyl bromide (CH_3CH_2Br) because of steric hindrance to attack on the bromine atom in the dimethyl compound.

Steric hindrance toward S_N2

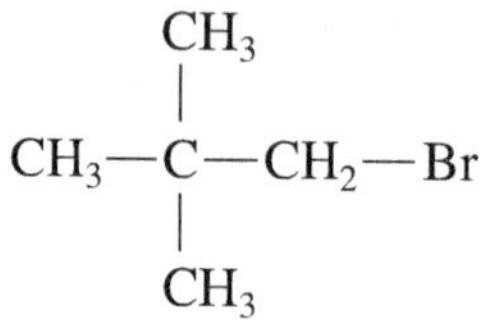

$$CH_3{-}\underset{\underset{CH_3}{|}}{\overset{\overset{CH_3}{|}}{C}}{-}CH_2{-}Br \qquad\qquad CH_3CH_2Br$$

Rate = 1   Rate = 100,000

The primary factor in S_N1 reactivity is the relative stability of the carbocation that is formed. For simple alkyl halides, this means that only tertiary halides react by this mechanism. The tertiary halide must be able to form a planar carbocation. Only slightly less reactive are the allyl carbocations, which derive their great stability from the delocalization of the charge on the carbon by resonance.

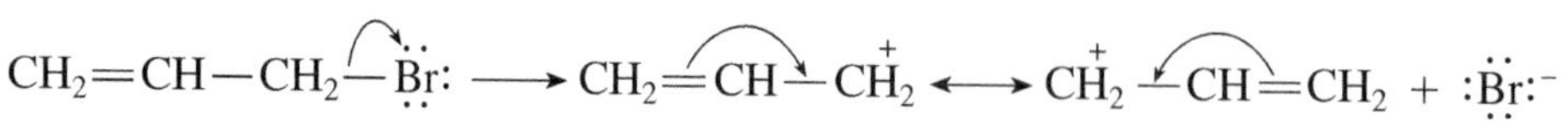

The allyl carbocation

$$CH_2{=}CH{-}CH_2{-}\ddot{\text{Br}}: \ \longrightarrow \ CH_2{=}CH{-}\overset{+}{C}H_2 \ \longleftrightarrow \ \overset{+}{C}H_2{-}CH{=}CH_2 \ + \ :\ddot{\text{Br}}:^-$$

Solvent effects

The nature of the solvent has a large effect on the rates of S_N2 reactions. In a solvent with a hydrogen atom attached to an electronegative atom such as oxygen, the protic solvent forms hydrogen bonds to the nucleophile. These solvent molecules get in the way during an S_N2 reaction.

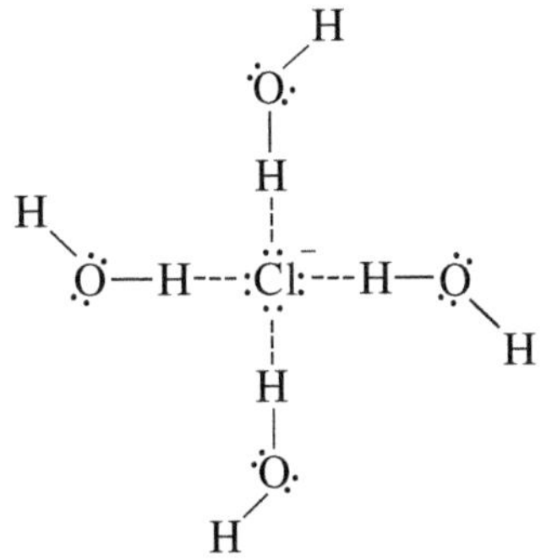

Aprotic: no ionizable protons

If the solvent is polar and aprotic, solvation of the nucleophile cannot occur, and the S_N2 reaction can occur up to a million times faster. Some common polar, aprotic solvents are N,N-dimethylformamide and dimethylsulfoxide:

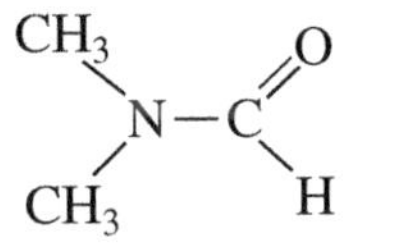

N,N-Dimethylformamide 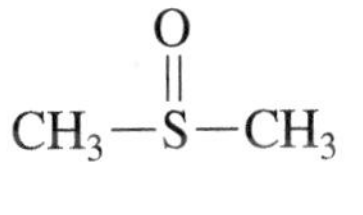   **Dimethylsulfoxide**

In the S_N1 reaction a polar protic solvent such as water stabilizes the transition state more than it does the reactants, lowering the energy of activation for the reaction and thus increasing the rate, relative to the rate in a nonpolar solvent. Acetic acid, ethanol, and acetone are relatively nonpolar solvents and have lower dielectric constants than the polar solvents water, dimethylsulfoxide, and N,N-dimethylformamide.

The leaving group

The rate of S_N1 and S_N2 reactions depends on the nature of the leaving group, the best leaving groups being the ones that form stable ions. Among the halogens we find that the iodide ion (I^-) is the best leaving group as well as the best nucleophile in the S_N2 reaction.

Vinylic and aryl halides

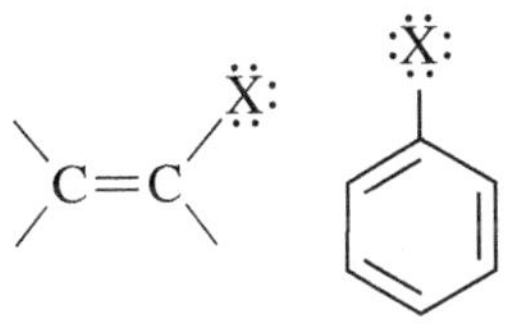

do not normally react by S_N1 reactions because the resulting carbocations

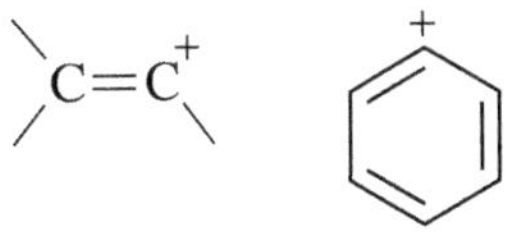

TABLE 17.1 • Summary of S$_N$1 and S$_N$2 Reactions		
	Unimolecular Nucleophilic Substitution (S$_N$1)	**Second-Order Nucleophilic Substitution (S$_N$2)**
Kinetics	First order	Second order
Mechanism	Two steps; unimolecular in the rate-determining step via a carbocation	One-step, bimolecular
Stereochemistry	Racemization predominates	Inversion of configuration
Reactivity of structure	3° > 2° > 1° > CH$_3$ > vinyl	3° < 2° < 1° < CH$_3$ < vinyl
Rearrangements	May occur	No rearrangements because no carbocation intermediate
Effect of leaving group	—I > —Br > —Cl >> —F	—I > —Br > —Cl >> —F
Effect of nucleophile	Not important because it is not in the rate-determining step	I$^-$ > Br$^-$ > Cl$^-$ > F$^-$
Concentration of nucleophile	S$_N$1 is favored by low concentration	S$_N$2 is favored by high concentration
Solvent polarity	High, favors S$_N$1	Low, favors S$_N$2

are relatively difficult to form. For S$_N$2 reactions, electrons in the nearby double bonds repel the nucleophile, which is either an ion or a polarized neutral species.

Temperature dependence

The rates of both S$_N$1 and S$_N$2 reactions depend on the temperature of the reaction. As the temperature increases, the kinetic energy of the molecules increases, leading to a greater rate of reaction. The rate of many organic reactions will approximately double when the temperature increases about 10°C. This information is summarized in Table 17.1.

EXPERIMENTS

In the experiments that follow, 11 representative alkyl halides are treated with sodium iodide in acetone and with an ethanolic solution of silver nitrate.

1. Sodium Iodide in Acetone

Acetone, with a dielectric constant of 21, is a relatively nonpolar solvent that will readily dissolve sodium iodide. The iodide ion is an excellent nucleophile, and the nonpolar solvent (acetone) favors the S$_N$2 reaction; it does not favor ionization of the alkyl halide. The extent of reaction can be observed because sodium bromide and sodium chloride are not soluble in acetone and precipitate from solution if a reaction occurs.

Organic halides that can react by an S$_N$2 mechanism give a precipitate of NaX with sodium iodide in acetone.

$$Na^+I^- + R\!-\!Cl \longrightarrow R\!-\!I + NaCl\!\downarrow$$
$$Na^+I^- + R\!-\!Br \longrightarrow R\!-\!I + NaBr\!\downarrow$$

2. Ethanolic Silver Nitrate Solution

When an alkyl halide is treated with an ethanolic solution of silver nitrate, the silver ion coordinates with an electron pair of the halogen. This weakens the carbon-halogen bond because a molecule of insoluble silver halide is formed, thus promoting an S_N1 reaction of the alkyl halide. The solvent, ethanol, favors ionization of the halide, and the nitrate ion is a very poor nucleophile, so alkyl nitrates do not form by an S_N2 reaction.

Organic halides that can react by an S_N1 mechanism give a precipitate of AgX with ethanolic silver nitrate solution.

$$R-\overset{..}{\underset{..}{X}}: \xrightleftharpoons{Ag^+} \overset{\delta^+}{R}\overset{..}{\underset{..}{X}}\overset{\delta^+}{Ag} \longrightarrow R^+ + AgX \downarrow$$

On the basis of the foregoing discussion, tertiary halides would be expected to react with silver nitrate most rapidly and primary halides least rapidly. The R^+ ion can react with the solvent to give either an alkene or an ether.

Microscale Procedure

Label 11 small containers (reaction tubes, 3-mL centrifuge tubes, 10×75-mm test tubes, or 1-mL vials), and place 0.1 mL or 100 mg of each of the following halides in the tubes.

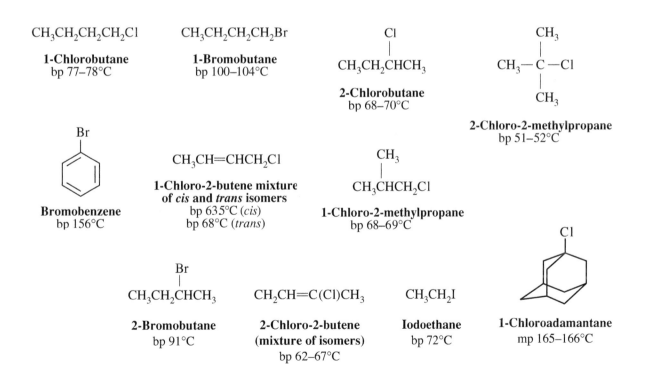

$CH_3CH_2CH_2CH_2Cl$

1-Chlorobutane
bp 77–78°C

$CH_3CH_2CH_2CH_2Br$

1-Bromobutane
bp 100–104°C

$CH_3CH_2CHClCH_3$

2-Chlorobutane
bp 68–70°C

$(CH_3)_3C-Cl$

2-Chloro-2-methylpropane
bp 51–52°C

Bromobenzene
bp 156°C

$CH_3CH=CHCH_2Cl$

1-Chloro-2-butene mixture of *cis* and *trans* isomers
bp 63.5°C (*cis*)
bp 68°C (*trans*)

$CH_3CHCH_3CH_2Cl$

1-Chloro-2-methylpropane
bp 68–69°C

$CH_3CH_2CHBrCH_3$

2-Bromobutane
bp 91°C

$CH_2CH=C(Cl)CH_3$

2-Chloro-2-butene (mixture of isomers)
bp 62–67°C

CH_3CH_2I

Iodoethane
bp 72°C

1-Chloroadamantane
mp 165–166°C

To each tube then rapidly add 1 mL of an 18% solution of sodium iodide in acetone, stopper each tube, mix the contents thoroughly, and note the time. Note the time of first appearance of any precipitate. If no reaction occurs within about 5 min, place those tubes in a 50°C water bath and watch for any reaction over the next 5 or 6 min.

Empty the tubes, rinse them with ethanol, place the same amount of each of the alkyl halides in each tube as in the first part of the experiment, add 1 mL of 1% ethanolic silver nitrate solution to each tube, mix the contents well, and note the time of addition as well as the time of appearance of the first traces of any precipitate. If a precipitate does not appear in 5 min, heat the tubes containing these unreactive halides in a 50°C water bath for 5 to 6 min and watch for any reaction.

To test the effect of solvent on the rate of S_N1 reactivity, compare the time needed for a precipitate to appear when 2-chlorobutane is treated with 1% ethanolic silver nitrate solution and when treated with 1% silver nitrate in a mixture of 50% ethanol and 50% water.

In your analysis of the results from these experiments, consider the following for both S_N1 and S_N2 conditions:

- the nature of the leaving group (Cl vs. Br) in the 1-halobutanes
- the effect of structure, that is, compare simple primary, secondary, and tertiary halides, unhindered primary vs. hindered primary halides, a simple tertiary halide vs. a complex tertiary halide, and an allylic halide vs. a tertiary halide
- the effect of solvent polarity on the S_N1 reaction
- the effect of temperature on the reaction

Online Study Center

General Resources
Additional Experiments

Cleaning Up. Because all of the test solutions contain halogenated material, all test solutions and washes as well as unused starting materials should be placed in the halogenated organic waste container.

QUESTIONS

1. What would be the effect of carrying out the sodium iodide in acetone reaction with the alkyl halides using an iodide solution half as concentrated?
2. The addition of sodium or potassium iodide catalyzes many S_N2 reactions of alkyl chlorides or bromides. Explain.
3. In S_N1 reactions, the intermediate carbocations can eliminate a proton to yield alkenes or react with the solvent to yield ethers. Draw the structures of the byproducts of this type that would be derived from the reaction of the carbocation derived from 2-bromo-2-methylbutane in ethanol.

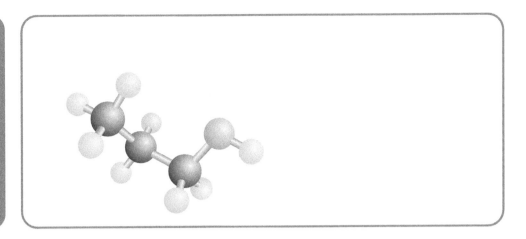

CHAPTER 26

Sodium Borohydride Reduction of 2-Methylcyclohexanone: A Problem in Conformational Analysis

Online Study Center

This icon will direct you to techniques, equipment setups, and online resources at http://college.hmco.com/PIC/williamsonMMOE5e.

PRELAB EXERCISE: Using carefully prepared drawings, try to predict whether *cis-* or *trans*-2-methylcyclohexanol will predominate in this reduction. Then use molecular models to check your conclusion. Finally, if you have access to the appropriate software, see if semi-empirical molecular orbital calculations will change your conclusions.

The objective of this experiment is to determine the structures and relative percentage yields of the products formed when 2-methylcyclohexanone is reduced with sodium borohydride. We also compare these results to predictions made using molecular mechanics and semiempirical molecular orbital calculations.

In the reduction of 2-methylcyclohexanone, both the *cis* and *trans* isomers of 2-methylcyclohexanol can be formed:

2-Methylcyclohexanone   2-Methylcyclohexanol

An examination of the models reveals that the two cyclohexanols can, in principle, exist in four chair conformations:

trans, diequatorial   *trans*, diaxial

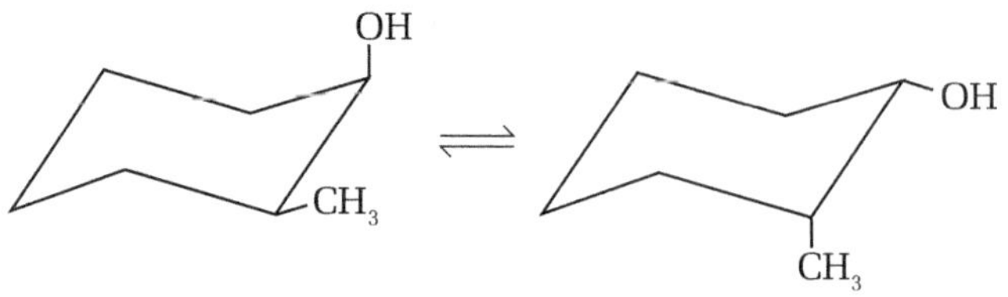

OH

CH₃   ⇌   OH

CH₃

cis, equatorial methyl,
axial hydroxyl

cis, axial methyl,
equatorial hydroxyl

A molecular mechanics program can be used to calculate the relative energies of these four isomers and help you to predict the most stable *trans* and *cis* conformations. Even without calculation, you should be able to predict which of the two *trans* conformations is the more stable.

By carrying out semiempirical molecular orbital calculations on the starting material (*see* Chapter 15), you may be able to decide, based on the shape and location of the lowest unoccupied molecular orbital (LUMO), whether the borohydride anion will attack the carbonyl group from the top, to give predominantly the *trans* isomer, or from the bottom, to give mostly the *cis* isomer. You may also be able to make this prediction by studying a molecular model or even a drawing of 2-methylcyclohexanone:

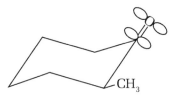

You can determine the structures and relative percentage yields of the products in this reaction using nuclear magnetic resonance (NMR) spectroscopy. The 250-MHz ^{1}H NMR spectrum of a 50:50 mixture of methylcyclohexanols is given in Figure 26.1. The two lowest field groups of peaks are from the hydrogen atom on the hydroxyl-bearing carbon atom.

The reduction of 2-methylcyclohexanone with sodium borohydride will give a mixture of products but not necessarily a 50:50 mixture. Integration of the two low-field groups of peaks will allow determination of the actual percentages of products in this reaction. First, however, the peaks must be assigned to the *cis* and *trans* isomers. The groups of peaks have been expanded and numbered on the spectrum in Figure 26.1, and the frequencies of the 11 peaks are listed in the caption.

The NMR coupling constant of the proton on the hydroxyl-bearing carbon is a function of the dihedral angle between that proton and an adjacent vicinal proton. This is seen most easily by examining molecular models and Newman projections. For the *trans* isomer of 2-methylcyclohexanol, the predominant conformer is the one in which both the methyl group and hydroxyl group are equatorial, and H_A and H_B are diaxial with a dihedral angle of 180° between them. The coupling constant for dihedral hydrogens is normally in the 9–12 Hz range. For the less favorable conformer, H_A and H_B are both equatorial, and the dihedral angle between them is 60°. This results in a coupling constant in the 2–5 Hz range.

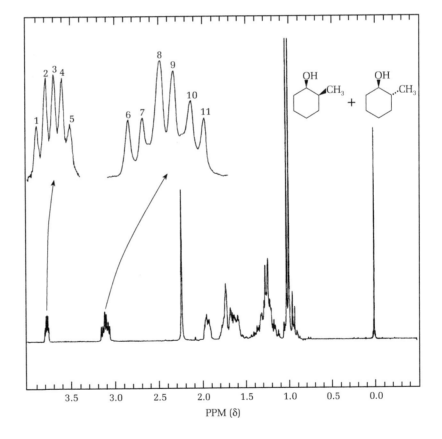

■ **FIG. 26.1**

The 250-MHz NMR spectrum of a mixture of *cis*- and *trans*-2-methylcyclohexanol. The frequencies of peaks 1 to 5 are 950.3, 947.5, 944.8, 942.2, and 939.5 Hz, respectively. The frequencies of peaks 6 to 11 are 791.2, 786.9, 781.6, 777.1, 771.6, and 767.4 Hz, respectively.

For *cis*-2-methylcyclohexanol, the two possible conformers are about equal in energy. In either conformer, H_A and H_B are in a 60° axial-equatorial stereochemical relationship, which results in a coupling constant in the 4–7 Hz range.

From this information, it should be possible to assign the groups of peaks at 3.1 and 3.8 ppm to either *cis*- or *trans*-2-methylcyclohexanol. From the areas of the two sets of peaks, the relative percentages of the isomers can be determined.

EXPERIMENT

Borohydride Reduction of 2-Methylcyclohexanone

This reaction can be run on a scale four times larger in a 25-mL Erlenmeyer flask.

IN THIS EXPERIMENT a liquid ketone in methanol is reduced with solid sodium borohydride. Base is added, and the product is extracted into dichloromethane. In the usual way this organic layer is dried and evaporated to leave the liquid alcohol. The same procedure could be used to reduce most ketones to their corresponding alcohols.

To a reaction tube add 1.25 mL of methanol and 300 mg of 2-methyl-cyclohexanone. Cool this solution in an ice bath contained in a small beaker. While the reaction tube is in the ice bath, carefully add 50 mg of sodium borohydride to the solution. After the vigorous reaction has ceased, remove the tube from the ice bath and allow it to stand at room temperature for 10 min, at which time the reaction should appear to be finished. To decompose the borate ester, add 1.25 mL of 3 *M* sodium hydroxide solution. To the resulting cloudy

■ FIG. 26.2
The IR spectrum of a
mixture of *cis-* and *trans-*
2-methylcyclohexanol (thin film).

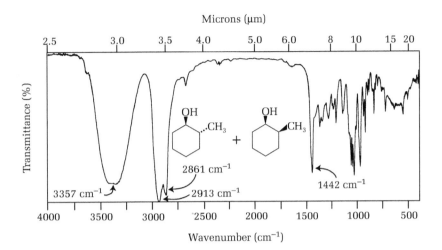

■ FIG. 26.2
The IR spectrum of a
mixture of *cis-* and *trans-*
2-methylcyclohexanol (thin film).

solution, add 1 mL of water. The product will separate as a small, clear upper layer. Remove as much of this as possible, place it in a reaction tube, and then extract the remainder of the product from the reaction mixture with two 0.5-mL portions of dichloromethane. Add these dichloromethane extracts to the small product layer and dry the combined extracts over anhydrous sodium sulfate (not calcium chloride). After a few minutes, transfer the solution to a dry reaction tube containing a boiling chip. In the hood, boil off the dichloromethane (and any accompanying methanol) and use the residue to run an NMR spectrum in the usual way.

Integrate the two low-field sets of peaks at 3.1 and 3.8 ppm, analyze the coupling constant patterns to assign the two sets of peaks, and report the percentage distribution of *cis-* and *trans-*2-methylcyclohexanol formed in this reaction. The relative amounts of the products can, of course, also be determined by gas chromatography. How do these results compare to your predictions and calculations? Which *cis* and which *trans* conformation is the more stable?

Run an infrared (IR) spectrum of the product as a thin film to determine whether the reduction of the starting material has been completed (Fig. 26.2).

Cleaning Up. The reaction mixture is neutralized with acetic acid (to react with sodium borohydride) and flushed down the drain with water.

Computational Chemistry

Online Study Center

General Resources
Additional Experiments

Using a molecular mechanics program, calculate the steric energies of *cis-* and *trans-*2-methylcyclohexanol. Each of these isomers has two principal conformations. Does reduction of 2-methylcyclohexanone give the more stable isomer? Explain.

QUESTIONS

1. Draw the NMR peak expected for H_A when H_A couples with H_B in the following structure.

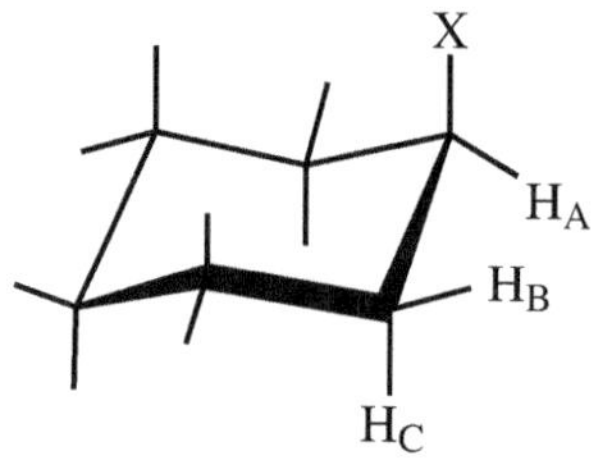

2. Draw the NMR peak expected when H_A couples with both H_B and H_C. Remember that the coupling constants are not equal.

3. What is the approximate frequency of the most important peak in the starting material that should be absent in the IR spectrum of the product?

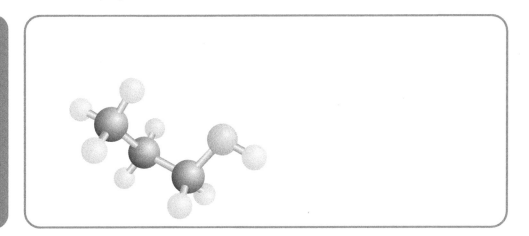

Friedel–Crafts Acylation of Ferrocene: Acetylferrocene

Online Study Center

This icon will direct you to techniques, equipment setups, and online resources at http://college.hmco.com/PIC/ williamsonMMOE5e.

PRELAB EXERCISE: How many possible isomers could exist for diacetylferrocene? Explain. Calculate the volume of 3 *M* aqueous sodium hydroxide needed to neutralize the acetic and phosphoric acids in the synthesis of acetylferrocene.

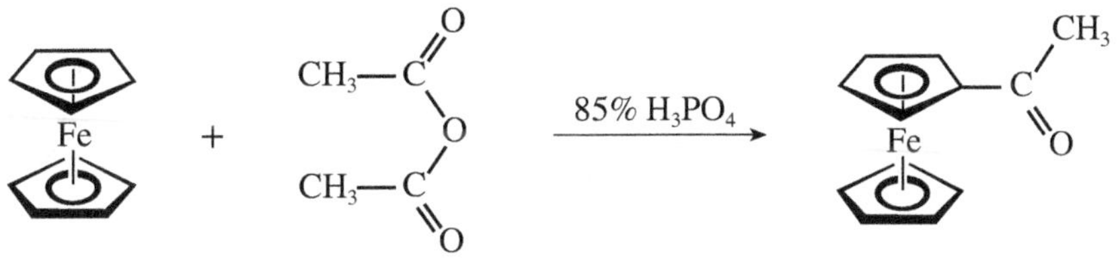

Ferrocene
[Bis(cyclopentadienyl)iron]
MW 186.04

Acetic anhydride
bp 139.5°C, den. 1.08
MW 102.09

85% H_3PO_4

Acetylferrocene
(Acetylcyclopentadienyl)cyclopentadienyliron
mp 85–86°C
MW 228.08

The Friedel–Crafts acylation of benzene requires aluminum chloride as the catalyst. However, ferrocene, with a very high π-electron density that has been referred to as a superaromatic compound, can be acylated under much milder conditions with phosphoric acid as catalyst. Because the acetyl group is a deactivating substituent, the addition of a second acetyl group, which requires more vigorous conditions, will occur in the nonacetylated cyclopentadienyl ring to give 1,1'-diacetylferrocene. Because ferrocene gives just one monoacetyl derivative and just one diacetyl derivative, it was assigned an unusual sandwich structure.

Acetylferrocene and ferrocene (both highly colored) are easily separated by column chromatography (*see* Chapter 9).

EXPERIMENT

Acetylferrocene
Microscale Procedure

> **IN THIS EXPERIMENT** ferrocene reacts with acetic anhydride in an acid-catalyzed Friedel–Crafts reaction. First water is added, which reacts with the anhydride; then base is added to react with the acetic acid. The crude product is isolated by filtration and then chromatographed on alumina. This is an excellent system to observe separation on a chromatography column because the unreacted ferrocene is yellow, and the acetyl product is orange.

Handle acetic anhydride and phosphoric acid with care. Wipe up spills immediately.

Online Study Center

Photo: Capping a Reaction Tube with a Septum; Video: Microscale Filtration on the Hirsch Funnel

Online Study Center

Photo: Column Chromatography; Video: Column Chromatography

Hexanes are extremely flammable. No flames!

To 93 mg of dry sublimed ferrocene (*see* Chapter 49) in a 10 × 100 mm reaction tube, add 0.35 mL (0.38 g) of acetic anhydride followed by 0.1 mL (170 mg) of 85% phosphoric acid. Cap the tube with a septum bearing an empty syringe needle and warm it on a steam bath or in a beaker of boiling water while agitating the mixture to dissolve the ferrocene. Heat the mixture for an additional 10 min and then cool the tube thoroughly in ice. Carefully add to the solution 0.5 mL of ice-water dropwise with thorough mixing, followed by the dropwise addition of 3 *M* aqueous sodium hydroxide solution until the mixture is neutral (test with indicator paper and avoid an excess of base). Collect the product on a Hirsch funnel, wash it thoroughly with water, and press it as dry as possible between sheets of filter paper. Save a sample of this material for melting-point and thin-layer chromatographic (TLC) analyses and purify the remainder by column chromatography.

Column Chromatography

Follow the procedure as given in Chapter 9 for packing a microscale chromatography column and adding the sample (about 90 mg of a ferrocene/acetylferrocene mixture). Use Brockman activity III alumina as the adsorbent.

Elution

Carefully add hexanes to the column and begin to elute the product from the column. Unreacted ferrocene will move down the column as an orange-yellow band. Collect this in a 10-mL flask. Wash any crystalline material that collects on the tip of the valve into the flask with a few drops of ether. Then elute the column with a 50:50 mixture of hexanes and ether. This will move down the acetylferrocene as an orange-red band. Collect this in a separate 10-mL Erlenmeyer flask. Ferrocene is highly symmetrical and nonpolar, while the acetyl compound is the opposite. These properties govern the order of elution from the chromatography column. Spot a TLC plate with these two solutions as well as the crude acetylferrocene and the ferrocene starting material, and analyze as described in the next section. Any diacetylferrocene will be seen as a dark band at the top of the column. Evaporate

■ FIG. 32.1
The evaporation of a solvent under reduced pressure. Heat and swirling motion is supplied by one hand; the vacuum is controlled by the thumb of the other hand.

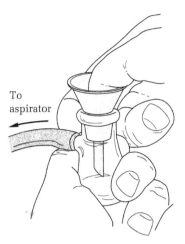

To aspirator

the solvents under reduced pressure and determine the weights of the residues (Fig. 32.1).

Flash Chromatography

Equilibrium between the solutes adsorbed on the alumina and the eluting solvent is established rapidly, so it does no harm to increase the flow rate of the solvent through the column. This can be done by applying pressure with a rubber bulb above the solvent. There is, however, a possibility that air will be sucked into the bottom of the column when the bulb is removed, which will destroy the uniform packing of the column. Practice this technique on a column that has been eluted in the usual way using gravity flow of the solvent. On a macroscale, special fittings are available for applying air pressure to larger columns.

Online Study Center

Photos: Crystallization, Filtration Using a Pasteur Pipette, Flash Chromatography; Videos: Recrystallization; Filtration of Crystals Using the Pasteur Pipette

Recrystallize the acetylferrocene from hot hexanes. Dissolve the residue in a filter flask in the minimum quantity of boiling solvent (about 1 mL) and transfer this hot solution with a Pasteur pipette to a reaction tube. Allow the tube to cool slowly to room temperature and then cool it in ice for at least 10 min. Acetylferrocene crystallizes as dark-red rosettes of needles. Remove the solvent with a Pasteur pipette and then, under vacuum, scrape out the product onto filter paper; after it is dry, determine the weight, calculate the percent yield, and determine the melting point.

Perform a TLC analysis on your product. Dissolve very small samples of pure ferrocene, the crude reaction mixture, and recrystallized acetylferrocene, each in a few drops of toluene; spot the three solutions with microcapillaries on silica gel plates; and develop the chromatogram with 30:1 toluene and absolute ethanol mixture. Visualize the spots under an ultraviolet (UV) lamp if the silica gel has a fluorescent indicator or by adsorption of iodine vapor. Do you detect unreacted ferrocene in the reaction mixture and/or a spot that might be attributed to diacetylferrocene?

Cleaning Up. The reaction mixture filtrate can be flushed down the drain. Unused chromatography, recrystallization, and TLC solvents should be placed in the organic solvents container. The alumina from the column should be placed in the hood to allow the hexanes and ether to evaporate from it. Once free of organic solvents, the alumina can be placed in the nonhazardous solid waste container. If local regulations do not allow for the evaporation of solvents, dispose the wet alumina in a special container.

 Macroscale Procedure

> **IN THIS EXPERIMENT** ferrocene reacts with acetic anhydride in an acid-catalyzed Friedel–Crafts reaction. First water is added, which reacts with the anhydride, then base is added to react with the acetic acid. The crude product is isolated by filtration and then purified by crystallization from hexane. This is an excellent system to observe separation by chromatography because the unreacted ferrocene is yellow, and the acetyl product is orange.

Both acetic anhydride and phosphoric acid are corrosive to tissue. Handle both with care and wipe up any spills immediately.

In a 25-mL round-bottomed flask place 3.0 g of ferrocene, 10.0 mL of acetic anhydride, and 2.0 mL of 85% phosphoric acid. Equip the flask with a reflux condenser and a calcium chloride drying tube. Warm the flask gently on a hot water or steam bath while swirling to dissolve the ferrocene; then heat strongly for an additional 10 min. Pour the reaction mixture onto 50 g of crushed ice in a 400-mL beaker and rinse the flask with 10 mL of ice water. Stir the mixture for a few minutes with a glass rod, add 75 mL of 3 M sodium hydroxide solution (the solution should still be acidic), and then add solid sodium bicarbonate (be careful of foaming) until the remaining acid has been neutralized. Stir and crush all lumps, allow the mixture to stand for 20 min, and then collect the product by suction filtration. Press the crude material as dry as possible between sheets of filter paper, save a few crystals for TLC analysis, transfer the remainder to an Erlenmeyer flask, and add 40 mL of hexanes to the flask. Boil the solvent on a hot water or steam bath for a few minutes and then decant the dark-orange solution into another Erlenmeyer flask, leaving a gummy residue of polymeric material. Treat the solution with decolorizing charcoal and filter it through a fluted filter paper placed in a warm stemless funnel into an appropriately sized Erlenmeyer flask. Evaporate the solvent (use an aspirator tube; *see* Fig. 8.11 on page 186) until the volume is about 20 mL. Set the flask aside to cool slowly to room temperature. Beautiful rosettes of dark orange-red needles of acetylferrocene will form. After the product has been cooled in ice, collect it on a Büchner funnel and wash the crystals with a small quantity of cold solvent. Pure acetylferrocene has a melting point of 84°C–85°C. Your yield should be about 1.8 g.

Perform a TLC analysis on your product. Dissolve very small samples of pure ferrocene, the crude reaction mixture, and recrystallized acetylferrocene, each in a few drops of toluene; spot the three solutions with microcapillaries on silica gel plates; and develop the chromatogram with 3:1 toluene and absolute ethanol mixture. Visualize the spots under a UV lamp, if the silica gel has a fluorescent indicator, or by adsorption of iodine vapor. Do you detect unreacted ferrocene in the reaction mixture and/or a spot that might be attributed to diacetylferrocene?

Hexanes are extremely flammable. No flames!

Online Study Center

Video: Macroscale Crystallization

Online Study Center

General Resources
Additional Experiments

Cleaning Up. The reaction mixture filtrate can be flushed down the drain. Unused recrystallization and TLC solvents should be placed in the organic solvents container. The decolorizing charcoal, once free of solvent, can be placed in the non-hazardous solid waste container. If local regulations do not allow for the evaporation of solvents, dispose of the wet alumina in a special container.

QUESTIONS

1. What is the structure of the intermediate species that attacks ferrocene to form acetylferrocene? What other organic molecule is formed?
2. Why does the second acetyl group enter the unoccupied ring to form diacetylferrocene?

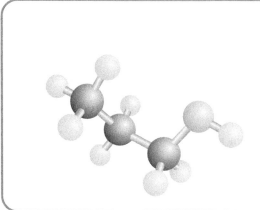

Aldehydes and Ketones

Online Study Center

This icon will direct you to techniques, equipment setups, and online resources at http://college.hmco.com/PIC/williamsonMMOE5e.

PRELAB EXERCISE: Outline a logical series of experiments designed to identify an unknown aldehyde or ketone with the least effort. Consider the time required to complete each identification reaction.

The carbonyl group occupies a central place in organic chemistry. Aldehydes and ketones—compounds such as formaldehyde, acetaldehyde, acetone, and 2-butanone—are very important industrial chemicals used by themselves and as starting materials for a host of other substances. For example, more than 10 billion pounds (4.5 billion kilograms) of formaldehyde-containing plastics are produced in the United States each year.

The carbonyl carbon is sp^2 hybridized, the bond angles between adjacent groups are 120°, and the four atoms R, R′, C, and O lie in one plane:

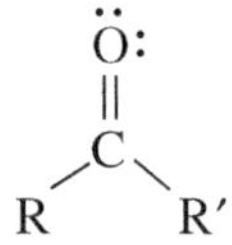

$$\ddot{O}:$$
$$\|$$
$$C$$
$$R \qquad R'$$

The electronegative oxygen polarizes the carbon-oxygen bond, rendering the carbon electron deficient and hence subject to nucleophilic substitution.

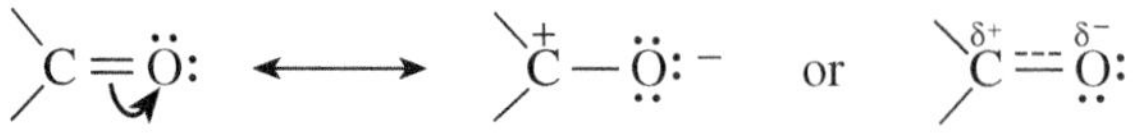

$$\begin{array}{ccccc} \diagdown & & & & \\ C = \ddot{O}: & \longleftrightarrow & \overset{+}{C} - \ddot{O}:^{-} & \text{or} & \overset{\delta+}{C} = \overset{\delta-}{O}: \\ \diagup & & \diagup & & \diagup \end{array}$$

Geometry of the carbonyl group

Attack on the sp^2 hybridized carbon occurs via the π-electron cloud above and below the plane of the carbonyl group:

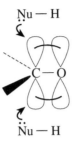

Reactions of the Carbonyl Group

Many reactions of carbonyl groups are acid catalyzed. The acid attacks the electronegative oxygen, which bears a partial negative charge, to create a carbocation that subsequently reacts with the nucleophile:

$$R_2C=\ddot{O}: + H^+ \rightleftharpoons R_2C=\overset{+}{\ddot{O}}-H \longleftrightarrow R_2\overset{+}{C}-\ddot{O}-H$$

$$R_2C=\overset{+}{\ddot{O}}-H + \ddot{N}u-H \rightleftharpoons \underset{R'}{\overset{R}{C}}-\ddot{O}-H \rightleftharpoons \underset{R'}{\overset{R}{C}}-\ddot{O}-H + H^+$$

The strength of the nucleophile and the structure of the carbonyl compound determine whether the equilibrium lies on the side of the carbonyl compound or the tetrahedral adduct. Water, a weak nucleophile, does not usually add to the carbonyl group to form a stable compound:

$$\underset{R'}{\overset{R}{C}}=\ddot{O}: + H_2O \rightleftharpoons \underset{R'}{\overset{:\ddot{O}-H}{C}}\ddot{O}-H$$

In the special case of trichloroacetaldehyde, however, the electron-withdrawing trichloromethyl group allows a stable hydrate to form:

$$\underset{H}{\overset{Cl_3C}{C}}=\ddot{O}: + H_2O \rightleftharpoons \underset{Cl_3C,\ H}{\overset{:\ddot{O}-H}{C}}\ddot{O}-H$$

A stable hydrate: chloral hydrate

The compound so formed, chloral hydrate, was discovered by Justus von Liebig in 1832 and was introduced as one of the first sedatives and hypnotics (sleep-inducing substances) in 1869. It is now most commonly encountered in detective fiction as a "Mickey Finn" or "knockout drops."

Reaction with an alcohol; hemiacetals

In an analogous manner, an aldehyde or ketone can react with an alcohol. The product, a hemiacetal or hemiketal, is usually not stable, but in the case of certain cyclic hemiacetals the product can be isolated. Glucose is an example of a stable hemiacetal.

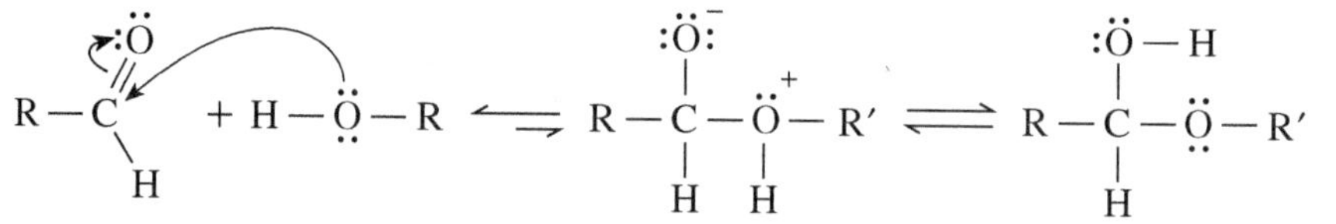

Hemiacetal
usually not stable

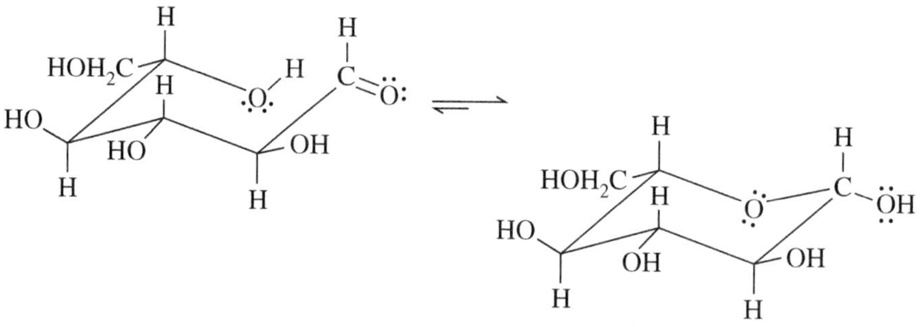

Glucose
a stable cyclic hemiacetal

Bisulfite Addition

The bisulfite ion is a strong nucleophile but a weak acid. It will attack the unhindered carbonyl group of an aldehyde or methyl ketone to form an addition product:

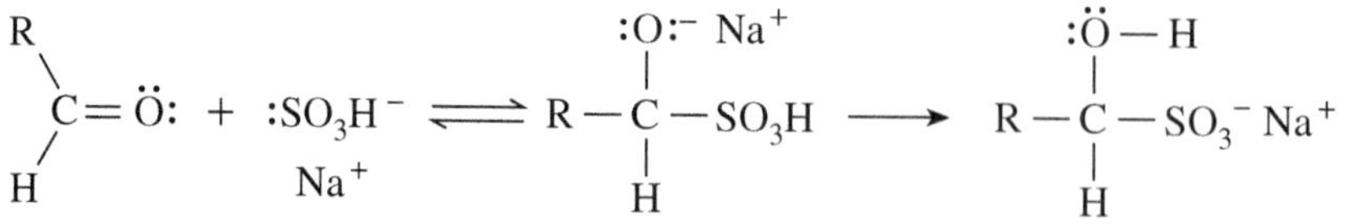

Because these bisulfite addition compounds are ionic water-soluble compounds and can be formed in up to 90% yield, they serve as a useful means of separating aldehydes and methyl ketones from mixtures of organic compounds. At high sodium bisulfite concentrations these adducts crystallize and can be isolated by filtration. The aldehyde or ketone can be regenerated by adding either a strong acid or base:

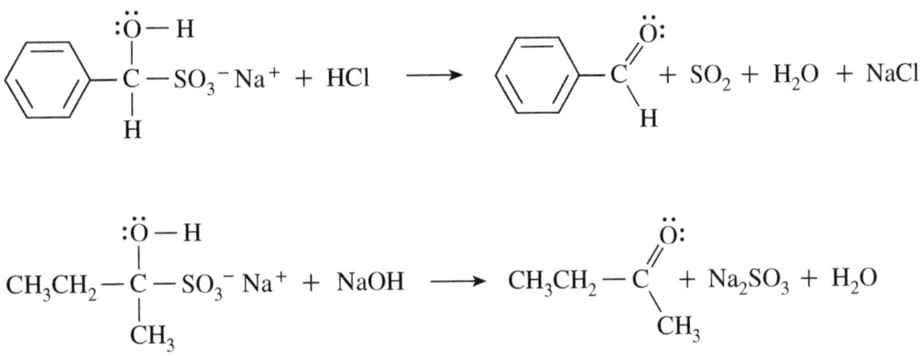

Cyanide Addition

A similar reaction occurs between aldehydes and ketones and hydrogen cyanide, which, like bisulfite, is a weak acid but a strong nucleophile. The reaction is hazardous to carry out because of the toxicity of cyanide, but the cyanohydrins are useful synthetic intermediates:

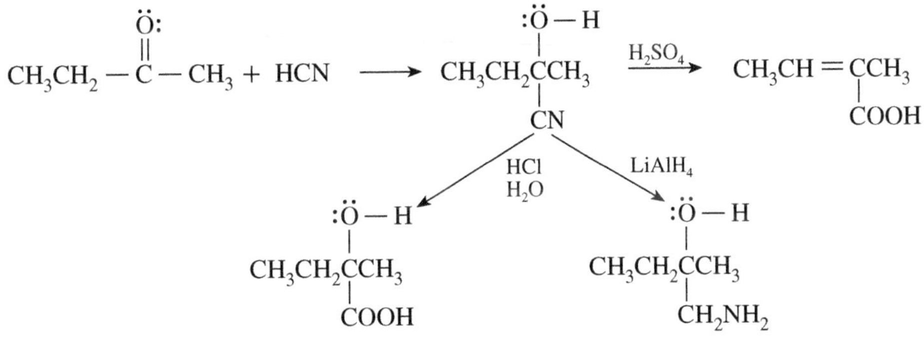

Cyanohydrin formation and reactions

Amines are good nucleophiles and readily add to the carbonyl group:

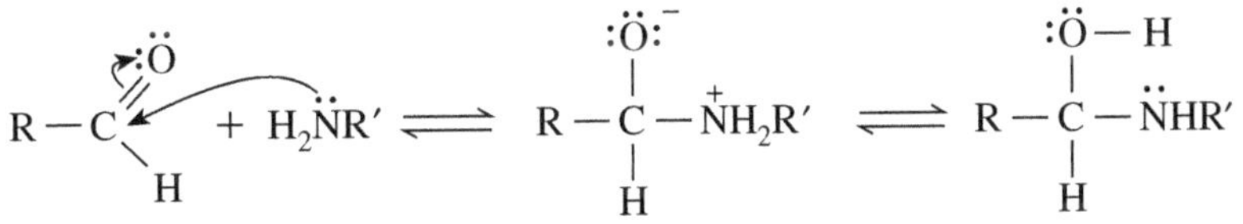

The reaction is strongly dependent on the pH. In acid the amine is protonated (RN^+H_3) and is no longer a nucleophile. In strong base there are no protons available to catalyze the reaction. But in weak acid solution (pH 4–6) the equilibrium between acid and base (**a**) is such that protons are available to protonate the carbonyl (**b**), and yet there is free amine present to react with the protonated carbonyl (**c**):

(**a**) $CH_3\ddot{N}H_2 + HCl \rightleftharpoons CH_3\overset{+}{N}H_3 + Cl^-$

(**b**) $CH_3-\overset{\ddot{O}}{\underset{H}{C}} + HCl \rightleftharpoons \left[CH_3-\overset{\overset{+}{\ddot{O}}-H}{\underset{H}{C}} \longleftrightarrow CH_3-\overset{\ddot{O}-H}{\underset{H}{C^+}} \right] + Cl^-$

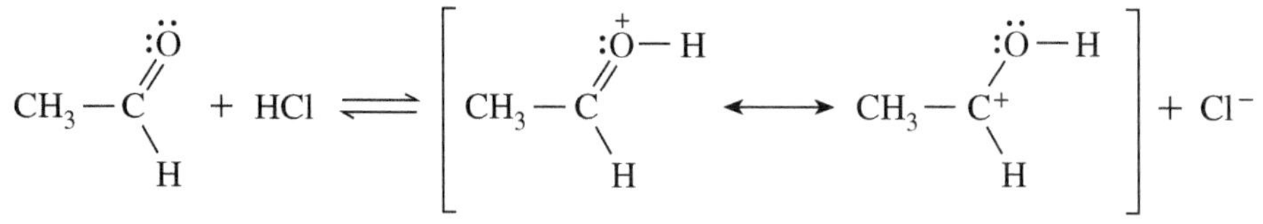

(**c**) $CH_3-\overset{\overset{+}{\ddot{O}}-H}{\underset{H}{C}} + CH_3\ddot{N}H_2 \rightleftharpoons CH_3-\overset{\ddot{O}-H}{\underset{H}{C}}-\overset{+}{N}H_2CH_3 \underset{}{\overset{-H^+}{\rightleftharpoons}} CH_3-\overset{\ddot{O}-H}{\underset{H}{C}}-\overset{}{\underset{H}{\ddot{N}}}-CH_3$

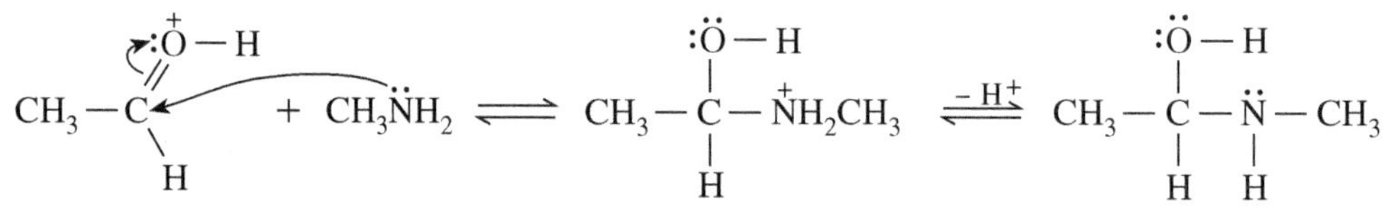

Schiff Bases

The intermediate hydroxyamino form of the adduct is not stable and spontaneously dehydrates under the mildly acidic conditions of the reaction to give an imine, commonly referred to as a *Schiff base*:

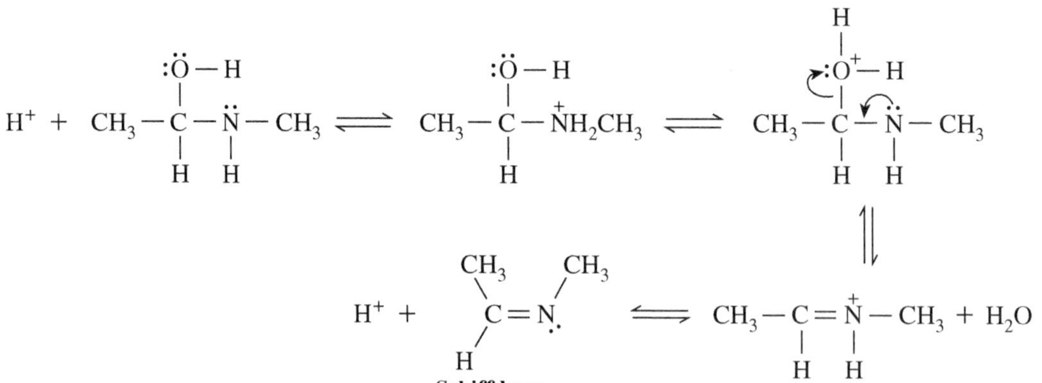

$H^+ + CH_3-\overset{\ddot{O}-H}{\underset{H\ \ H}{C}}-\overset{}{N}-CH_3 \rightleftharpoons CH_3-\overset{\ddot{O}-H}{\underset{H}{C}}-\overset{+}{N}H_2CH_3 \rightleftharpoons CH_3-\overset{\overset{H}{\underset{}{\overset{+}{O}-H}}}{\underset{H\ \ H}{C}}-\overset{}{\ddot{N}}-CH_3$

$H^+ + \overset{CH_3}{\underset{H}{}}\diagdown C=N\diagup^{CH_3} \rightleftharpoons CH_3-\overset{}{\underset{H\ \ H}{C}}=\overset{+}{N}-CH_3 + H_2O$

Schiff base

Imine or Schiff base formation

The biosynthesis of most amino acids proceeds through Schiff base intermediates.

Oximes, Semicarbazones, and 2,4-Dinitrophenylhydrazones

Three rather special amines form useful stable imines:

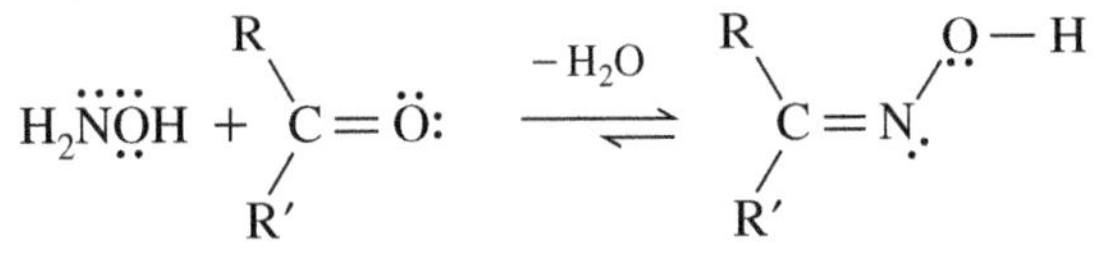

$$\text{H}_2\ddot{\text{N}}\ddot{\text{O}}\text{H} + \underset{\text{R}'}{\overset{\text{R}}{\diagdown}}\text{C}=\ddot{\text{O}}: \quad \xrightarrow{-\text{H}_2\text{O}} \quad \underset{\text{R}'}{\overset{\text{R}}{\diagdown}}\text{C}=\ddot{\text{N}}\diagup\overset{\diagup\text{O}-\text{H}}{}$$

Hydroxylamine   Oxime

$$\text{H}_2\text{NNHCNH}_2 + \underset{\text{R}'}{\overset{\text{R}}{\diagdown}}\text{C}=\text{O} \quad \xrightarrow{-\text{H}_2\text{O}} \quad \underset{\text{R}'}{\overset{\text{R}}{\diagdown}}\text{C}=\text{N}-\underset{\overset{|}{\text{H}}}{\text{N}}-\overset{\overset{\text{O}}{\|}}{\text{C}}-\text{NH}_2$$

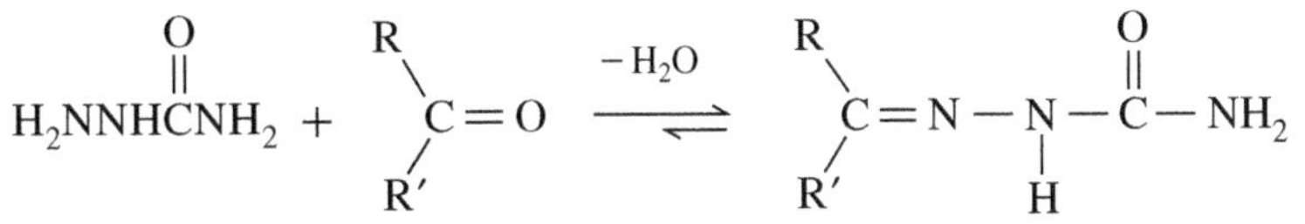

Semicarbazide   Semicarbazone

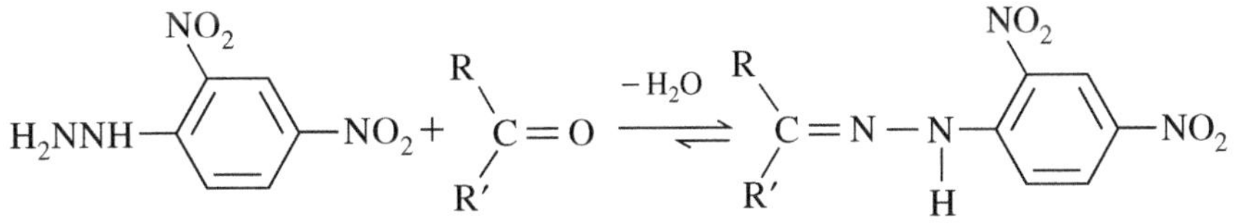

2,4-Dinitrophenylhydrazine   2,4-Dinitrophenylhydrazone

These imines are solids and are useful for the characterization of aldehydes and ketones. For example, IR (infrared) and NMR (nuclear magnetic resonance) spectroscopies may indicate that a certain unknown is acetaldehyde. It is difficult to determine the boiling point of a few milligrams of a liquid, but if it can be converted to a solid derivative, the melting point *can* be determined with that amount. The 2,4-dinitrophenylhydrazones are usually the derivatives of choice because they are crystalline compounds with well-defined melting or decomposition points, and they increase the molecular weight by 180. Ten milligrams of acetaldehyde will give 51 mg of the 2,4-dinitrophenylhydrazone.

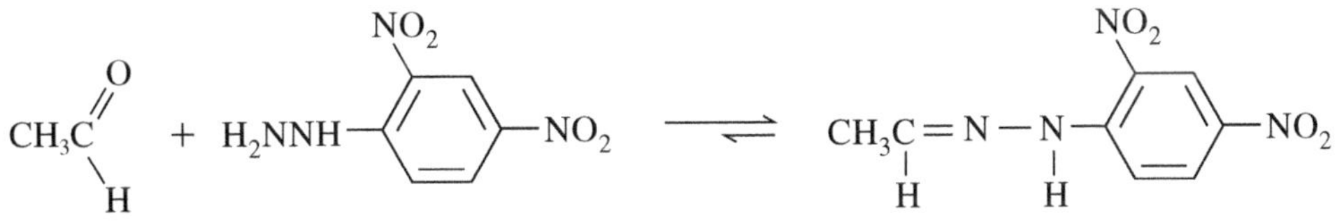

Acetaldehyde
MW 44.05
bp 20.8°C

2,4-Dinitrophenylhydrazine
MW 198.14
mp 196°C

Acetaldehyde 2,4-dinitrophenylhydrazone
MW 224.19
mp 168.5°C

Tollens' Reagent

Before the advent of NMR and IR spectroscopy and mass spectrometry, the chemist was often called on to identify aldehydes and ketones by purely chemical means. Aldehydes can be distinguished chemically from ketones by their ease of oxidation to carboxylic acids. The oxidizing agent, an ammoniacal solution of silver nitrate, Tollens' reagent, is reduced to metallic silver, which is deposited on the inside of a test tube as a silver mirror.

$$2\,Ag(NH_3)_2OH + R-C\overset{O}{\underset{H}{\diagup\!\!\!\diagdown}} \longrightarrow 2\,Ag + R-C\overset{O}{\underset{O^-}{\diagup\!\!\!\diagdown}} \ NH_4^+ + H_2O + 3\,NH_3$$

Schiff's Reagent

Another way to distinguish aldehydes from ketones is to use Schiff's reagent. This is a solution of the red dye Basic Fuchsin, which is rendered colorless on treatment with sulfur dioxide. In the presence of an aldehyde, the colorless solution turns magenta.

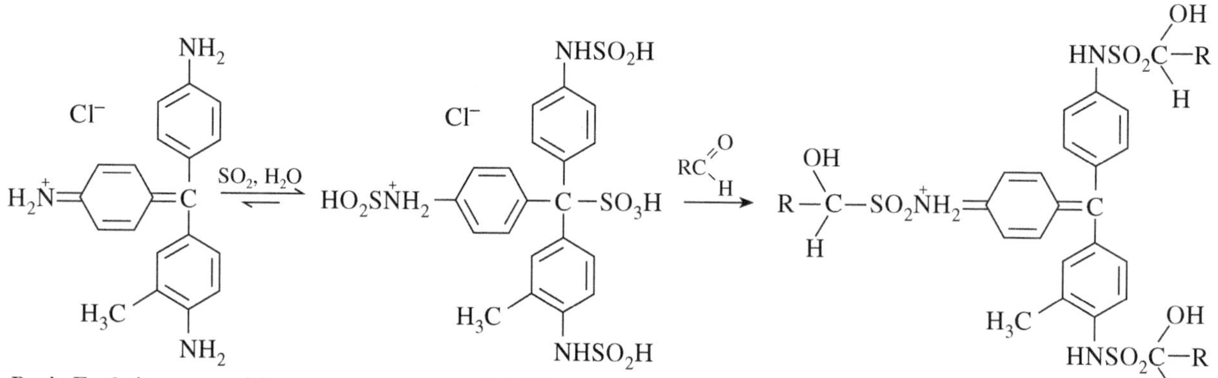

Basic Fuchsin, *p*-rosaniline hydrochloride   **Schiff's reagent, colorless**   **Magenta in color**

Iodoform Test

A test for methyl ketones

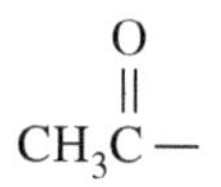

$$\underset{\textbf{Methyl ketone}}{CH_3\overset{O}{\overset{\|}{C}}-}$$

Methyl ketones can be distinguished from other ketones by the iodoform test. The methyl ketone is treated with iodine in a basic solution. Introduction of the first iodine atom increases the acidity of the remaining methyl protons, so halogenation stops only when the triiodo compound has been produced. The base then allows the relatively stable triiodomethyl carbanion to leave, and a subsequent proton transfer gives iodoform, a yellow crystalline solid with a melting point of 119°C–123°C. The test is also positive for fragments or compounds easily oxidized to methyl ketones, such as the fragment $CH_3\overset{|}{C}HOH$ or the compound ethanol. Acetaldehyde also gives a positive test because it is both a methyl ketone and an aldehyde.

$$R - \overset{\overset{\overset{\displaystyle \cdot \cdot}{\displaystyle O} :}{\|}}{C} - CH_3 + OH^- \rightleftharpoons R - \overset{\overset{\overset{\displaystyle \cdot \cdot}{\displaystyle O} :}{\|}}{C} - \overset{\overset{\displaystyle - }{}}{\overset{\displaystyle \cdot \cdot}{C}H_2} + H_2O$$

$$R - \overset{\overset{\overset{\displaystyle \cdot \cdot}{\displaystyle O} :}{\|}}{C} - \overset{\overset{\displaystyle \cdot \cdot}{}}{\overset{\displaystyle - }{C}H_2} + I_2 \longrightarrow R - \overset{\overset{\displaystyle O}{\|}}{C} - CH_2I + I^- \xrightarrow{OH^-} \xrightarrow{I_2} \xrightarrow{OH^-} \xrightarrow{I_2} R - \overset{\overset{\overset{\displaystyle \cdot \cdot}{\displaystyle O} :}{\|}}{C} - CI_3$$

$$R - \overset{\overset{\overset{\displaystyle \cdot \cdot}{\displaystyle O} :}{\|}}{C} - CI_3 + OH^- \longrightarrow R - \overset{\overset{\displaystyle :\overset{\cdot \cdot}{O}:^-}{|}}{\underset{\displaystyle :\underset{\cdot \cdot}{O}H}{C}} - CI_3 \longrightarrow R - \overset{\displaystyle :\overset{\cdot \cdot}{O}}{\underset{\displaystyle :\underset{\cdot \cdot}{O} - H}{C}} + :\overset{\displaystyle -}{C}I_3 \longrightarrow R - \overset{\displaystyle :\overset{\cdot \cdot}{O}}{\underset{\displaystyle \underset{\cdot \cdot}{O}:^-}{C}} + CHI_3$$

Iodoform
mp 123°C

EXPERIMENTS

1. Unknowns

You will be given an unknown that may be any of the aldehydes or ketones listed in Table 36.1. At least one derivative of the unknown is to be submitted to your instructor; but if you first do the bisulfite and iodoform characterizing tests, the results may suggest derivatives whose melting points will be particularly revealing.

You can further characterize the unknown by determining its boiling point, which is best done with a digital thermometer and a reaction tube (*see* Chapter 5). The boiling points of the unknowns are given on this book's website.

Carry out three tests:

Known positive
Known negative
Unknown

In conducting the following tests, you should perform three tests simultaneously: on a compound known to give a positive test, on a compound known to give a negative test, and on the unknown. In this way you will be able to determine whether the reagents are working as they should as well as interpret a positive or a negative test.

2. 2,4-Dinitrophenylhydrazones

$$\underset{R'}{\overset{R}{\diagdown}}C = \overset{\cdot \cdot}{O}: + H_2\overset{\cdot \cdot}{N} - \overset{\overset{\displaystyle }{|}}{\underset{H}{N}} - \underset{NO_2}{\overset{NO_2}{\bigcirc}} - NO_2 \xrightarrow{H^+} \underset{R'}{\overset{R}{\diagdown}}C = \overset{\cdot \cdot}{N} \overset{\cdot \cdot}{\underset{H}{N}} - \underset{NO_2}{\overset{NO_2}{\bigcirc}} - NO_2$$

An easily prepared derivative of aldehydes and ketones

To 5 mL of the stock solution[1] of 2,4-dinitrophenylhydrazine in phosphoric acid add about 0.05 g of the compound to be tested. Five milliliters of the 0.1 *M* solution contains 0.5 mmol (0.0005 mol) of the reagent. If the compound to be tested

---

1. Dissolve 2.0 g of 2,4-dinitrophenylhydrazine in 50 mL of 85% phosphoric acid by heating, cool, add 50 mL of 95% ethanol, cool again, and clarify by suction filtration from a trace of residue.

TABLE 36.1 • Melting Points of Derivatives of Some Aldehydes and Ketones

Compound[a]	Formula	n_D^{20}	MW	Water Solubility	Phenyl-hydrazone	2,4-dinitro-phenyl-hydrazone	Semi-carbazone
Acetone	CH_3COCH_3	1.3590	58.08		42	126	187
n-Butanal	$CH_3CH_2CH_2CHO$	1.3790	72.10	4 g/100 g	Oil	123	95 (106)[b]
3-Pentanone (diethyl ketone)	$CH_3CH_2COCH_2CH_3$	1.3920	86.13	4.7 g/100 g	Oil	156	138
2-Furaldehyde (furfural)	$C_4H_3O \cdot CHO$	1.5260	96.08	9 g/100 g	97	212 (230)[b]	202
Benzaldehyde	C_6H_5CHO	1.5450	106.12	Insol.	158	237	222
Hexane-2,5-dione	$CH_3COCH_2CH_2COCH_3$	1.4260	114.14	∞	120[c]	257[c]	224[c]
2-Heptanone	$CH_3(CH_2)_4COCH_3$	1.4080	114.18	Insol.	Oil	89	123
3-Heptanone	$CH_3(CH_2)_3COCH_2CH_3$	1.4080	114.18	Insol.	Oil	81	101
n-Heptanal	$n\text{-}C_6H_{13}CHO$	1.4125	114.18	Insol.	Oil	108	109
Acetophenone	$C_6H_5COCH_3$	1.5325	120.66	Insol.	105	238	198
2-Octanone	$CH_3(CH_2)_5COCH_3$	1.4150	128.21	Insol.	Oil	58	122
Cinnamaldehyde	$C_6H_5CH{=}CHCHO$	1.6220	132.15	Insol.	168	255	215
Propiophenone	$C_6H_5COCH_2CH_3$	1.5260	134.17	Insol.	About 48	191	182

[a]Visit this book's website for data on additional aldehydes and ketones.

[b]Both melting points have been found, depending on crystalline form of derivative.

[c]Monoderivative or diderivative.

Online Study Center

Video: Microscale
Crystallization

has a molecular weight of 100, then 0.05 g is 0.5 mmol. Warm the reaction mixture for a few minutes in a water bath and then let crystallization proceed. Collect the product by suction filtration (Fig. 36.1), wash the crystals with a large amount of water to remove all of the phosphoric acid, press a piece of moist litmus paper onto the crystals, and wash them with more water if they are acidic. Press the product between sheets of filter paper until it is as dry as possible and recrystallize from ethanol. Occasionally a high molecular weight derivative will not dissolve in a reasonable quantity (10 mL) of ethanol. In that case cool the hot suspension and isolate the crystals by suction filtration. The boiling ethanol treatment removes impurities so that an accurate melting point can be obtained on the isolated material.

An alternative procedure is applicable when the 2,4-dinitrophenylhydrazone is known to be sparingly soluble in ethanol. Measure 0.5 mmol (0.1 g) of crystalline 2,4-dinitrophenylhydrazine into a 50-mL Erlenmeyer flask, add 15 mL of 95% ethanol, digest on a steam bath until all the solid particles are dissolved, and then add 0.5 mmol of the compound to be tested and continue warming. If there is no immediate change, add, from a Pasteur pipette, 3–4 drops of concentrated hydrochloric acid as a catalyst and note the result. Warm for a few minutes, then

■ FIG. 36.1
A Hirsch funnel filtration apparatus.

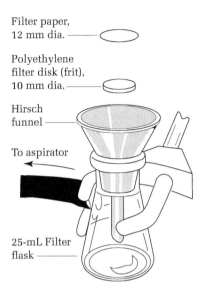

Filter paper,
12 mm dia. ——

Polyethylene
filter disk (frit),
10 mm dia. ——

Hirsch
funnel ——

To aspirator

25-mL Filter
flask ——

cool and collect the product. This procedure would be used for an aldehyde like cinnamaldehyde (C_6H_5CH=CHCHO).

The alternative procedure strikingly demonstrates the catalytic effect of hydrochloric acid, but it is not applicable to a substance like diethyl ketone, whose 2,4-dinitrophenylhydrazone is much too soluble to crystallize from the large volume of ethanol. The first procedure is obviously the one to use for an unknown.

Cleaning Up. The filtrate from the preparation of the 2,4-dinitrophenylhydrazone should have very little 2,4-dinitrophenylhydrazine in it, so after dilution with water and neutralization with sodium carbonate, it can be flushed down the drain. Similarly, the mother liquor from crystallization of the phenylhydrazone should have very little product in it and so should be diluted and flushed down the drain. If solid material is detected, it should be collected by suction filtration, the filtrate flushed down the drain, and the filter paper placed in the solid hazardous waste container because hydrazines are toxic.

3. Semicarbazones

Semicarbazide hydrochloride   **Pyridine**   **Semicarbazone**   **Pyridine hydrochloride**

Semicarbazide (mp 96°C) is not very stable in the free form and is used as the crystalline hydrochloride (mp 173°C). Because this salt is insoluble in methanol or ethanol and does not react readily with typical carbonyl compounds in alcohol-water mixtures, pyridine, a basic reagent, is added to liberate free semicarbazide.

To 0.5 mL of the stock solution[2] of semicarbazide hydrochloride, which contains 1 mmol of the reagent, add 1 mmol of the compound to be tested and enough methanol (1 mL) to produce a clear solution; then add 10 drops of pyridine (a twofold excess) and warm the solution gently on the steam bath for a few minutes. Cool the solution slowly to room temperature. It may be necessary to scratch the inside of the test tube in order to induce crystallization. Cool the tube in ice, collect the product by suction filtration, and wash it with water followed by a small amount of cold methanol. Recrystallize the product from methanol, ethanol, or ethanol/water. The product can easily be collected on a Wilfilter.

Online Study Center

Photo: Use of the Wilfilter

Cleaning Up. Combine the filtrate from the reaction and the mother liquor from the crystallization, dilute with water, make very slightly acidic with dilute hydrochloric acid, and flush the mixture down the drain.

4. Tollens' Test

Test for aldehydes

$$R-C\overset{\displaystyle O}{\underset{\displaystyle H}{\Big\backslash}} + 2\ Ag(NH_3)_2OH \longrightarrow 2\ Ag + RCOO^-NH_4^+ + H_2O + 3\ NH_3$$

Clean four or five test tubes by adding a few milliliters of 3 M sodium hydroxide solution to each and heating them in a water bath while preparing the Tollens' reagent.

To 2.0 mL of 0.03 M silver nitrate solution, add 1.0 mL of 3 M sodium hydroxide in a test tube. To the gray precipitate of silver oxide, Ag_2O, add 0.5 mL of a 2.8% aqueous ammonia solution (10 mL of concentrated ammonium hydroxide diluted to 100 mL). Stopper the tube and shake it. Repeat the process until *almost* all of the precipitate dissolves (3.0 mL of ammonia at most); then dilute the solution with water to 10 mL. Empty the test tubes of sodium hydroxide solution, rinse them, and add 1 mL of Tollens' reagent to each. Add 1 drop (no more) of the substance to be tested by allowing it to run down the inside of the inclined test tube. Set the tubes aside for a few minutes without agitating the contents. If no reaction occurs, warm the mixture briefly on a water bath. As a known aldehyde, try 1 drop of a 0.1 M solution of glucose. A more typical aldehyde to test is benzaldehyde.

At the end of the reaction, promptly destroy any excess Tollens' reagent with nitric acid: It can form an explosive fulminate on standing. Nitric acid can also be used to remove silver mirrors from test tubes.

---

2. Prepare a stock solution by dissolving 1.11 g of semicarbazide hydrochloride in 5 mL of water; 0.5 mL of this solution contains 1 mmol of reagent.

Cleaning Up. Place all solutions used in this experiment in a beaker (unused ammonium hydroxide, sodium hydroxide solution used to clean out the tubes, Tollens' reagent from all tubes). Remove any silver mirrors from reaction tubes with a few drops of nitric acid, which is added to the beaker. Make the mixture acidic with nitric acid to destroy unreacted Tollens' reagent and then neutralize the solution with sodium carbonate and add some sodium chloride solution to precipitate silver chloride (about 40 mg). The whole mixture can be flushed down the drain, or the silver chloride can be collected by suction filtration, and the filtrate flushed down the drain. The silver chloride would go in the nonhazardous solid waste container.

5. Schiff's Test

Very sensitive test for aldehydes

Add 3 drops of the unknown to 2 mL of Schiff's reagent.[3] A magenta color will appear within 10 min with aldehydes. As in all of these tests, compare the colors produced by a known aldehyde, a known ketone, and the unknown compound.

Cleaning Up. Neutralize the solution with sodium carbonate and flush it down the drain. The amount of *p*-rosaniline in this mixture is negligible (1 mg).

6. Iodoform Test

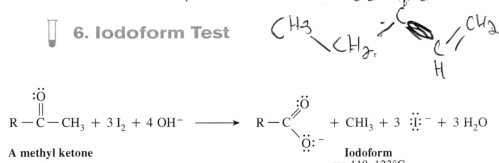

$$R - \overset{\overset{\ddot{O}}{\|}}{C} - CH_3 + 3\,I_2 + 4\,OH^- \longrightarrow R - C\overset{\ddot{O}}{\underset{\ddot{O}:^-}{\diagup}} + CHI_3 + 3\,:\!\ddot{I}:^- + 3\,H_2O$$

A methyl ketone   **Iodoform**
                                                          mp 119–123°C

Test for methyl ketones

The reagent contains iodine in potassium iodide solution[4] at a concentration such that 2 mL of solution, on reaction with excess methyl ketone, will yield 174 mg of iodoform. If the substance to be tested is water soluble, dissolve 4 drops of a liquid or an estimated 50 mg of a solid in 2 mL of water in a 20 × 150-mm test tube; add 2 mL of 3 *M* sodium hydroxide, and then slowly add 3 mL of the iodine solution. In a positive test the brown color of the reagent disappears, and yellow iodoform separates. If the substance to be tested is insoluble in water, dissolve it in 2 mL of 1,2-dimethoxyethane, proceed as above, and at the end dilute with 10 mL of water.

Suggested test substances are hexane-2,5-dione (water soluble), *n*-butyraldehyde (water soluble), and acetophenone (water insoluble).

---

3. Schiff's reagent is prepared by dissolving 0.1 g Basic Fuchsin (*p*-rosaniline hydrochloride) in 100 mL of water and then adding 4 mL of a saturated aqueous solution of sodium bisulfite. After 1 h, add 2 mL of concentrated hydrochloric acid.

4. Dissolve 25 g of iodine in a solution of 50 g of potassium iodide in 200 mL of water.

Iodoform can be recognized by its odor and yellow color and, more securely, from its melting point (119°C–123°C). The substance can be isolated by suction filtration of the test suspension or by adding 0.5 mL of dichloromethane, shaking the stoppered test tube to extract the iodoform into the small lower layer, withdrawing the clear part of this layer with a Pasteur pipette, and evaporating it in a small tube on a steam bath. The crude solid is crystallized from a methanol-water mixture (*see* Chapter 4). It can be collected on a Wilfilter.

Cleaning Up. Combine all reaction mixtures in a beaker, add a few drops of acetone to destroy any unreacted iodine in potassium iodide reagent, remove the iodoform by suction filtration, and place the iodoform in the halogenated organic waste container. The filtrate can be flushed down the drain after neutralization (if necessary).

7. Bisulfite Test

Forms with unhindered carbonyls

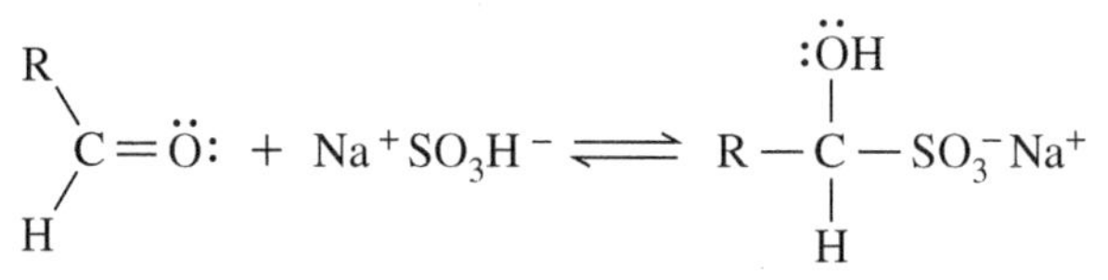

$$\underset{H}{\overset{R}{>}}C=\ddot{O}: \ + \ Na^+SO_3H^- \ \rightleftharpoons \ R-\underset{H}{\overset{:\ddot{O}H}{\underset{|}{\overset{|}{C}}}}-SO_3^-Na^+$$

Put 1 mL of the stock solution[5] into a 13 × 100-mm test tube and add 5 drops of the substance to be tested. Shake each tube during the next 10 min and note the results. A positive test will result from aldehydes, unhindered cyclic ketones such as cyclohexanone, and unhindered methyl ketones.

If the bisulfite test is applied to a liquid or solid that is very sparingly soluble in water, formation of the addition product is facilitated by adding a small amount of methanol before the addition to the bisulfite solution.

Cleaning Up. Dilute the bisulfite solution or any bisulfite addition products (they will dissociate) with a large volume of water and flush the mixture down the drain. The amount of organic material being discarded is negligible.

8. IR and NMR Spectroscopy

IR spectroscopy is extremely useful in analyzing all carbonyl-containing compounds, including aldehydes and ketones (Fig. 36.4 and Fig. 36.11). Refer to the extensive discussion in Chapter 11. In the modern laboratory, spectroscopy has almost completely supplanted the qualitative tests described in this chapter. Figures 36.4 through 36.11 present IR and NMR spectra of typical aldehydes and ketones. Compare these spectra with those of your unknown. Also compare the IR and ^{1}H NMR spectra of the hydrocarbon fluorene (Fig. 36.2 and Fig. 36.3) with those of the ketone derivative, fluorenone (Fig. 36.4 and Fig. 36.5).

5. Prepare a stock solution from 50 g sodium bisulfite dissolved in 200 mL of water with brief swirling.

■ FIG. 36.2
The IR spectrum of fluorene in CS_2.

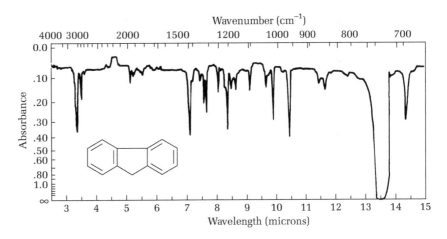

■ FIG. 36.3
The ^1H NMR spectrum of fluorene (250 MHz).

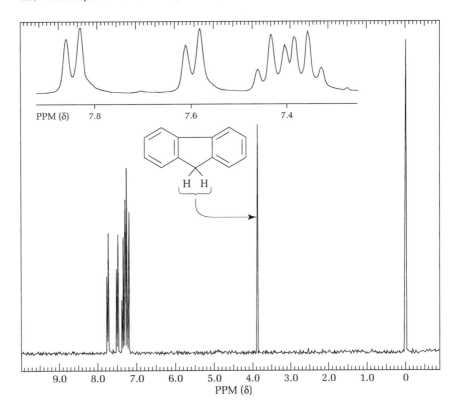

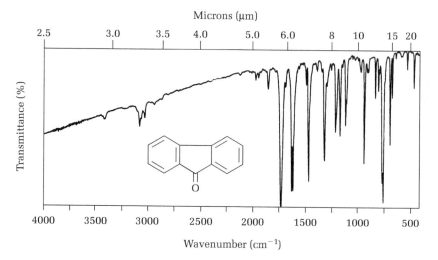

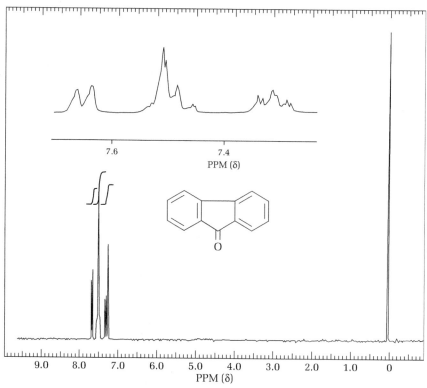

A peak at 9.6–10 ppm in the ¹H NMR spectrum is highly characteristic of aldehydes because almost no other peaks appear in this region (Fig. 36.7 and Fig. 36.10). Similarly, a sharp singlet at 2.2 ppm is very characteristic of methyl ketones; beware of contamination of the sample by acetone, which is often used to clean glassware.

■ FIG. 36.6
The ^1H NMR spectrum of
2-butanone, $CH_3COCH_2CH_3$
(90 MHz).

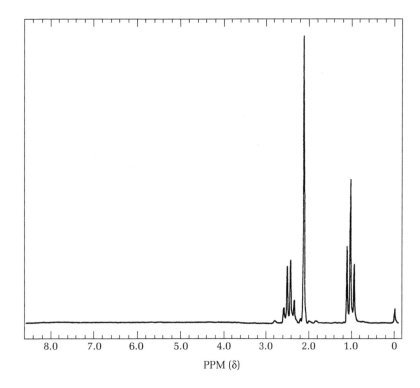

PPM (δ)

■ FIG. 36.7
The ^1H NMR spectrum of
crotonaldehyde, $CH_3CH=CHCHO$
(90 MHz).

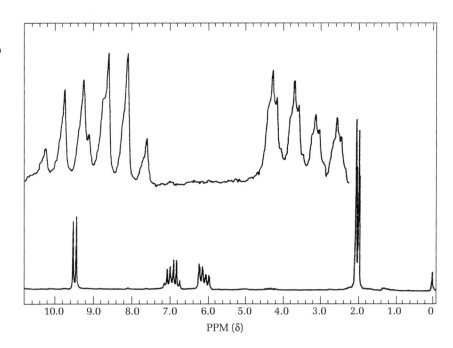

PPM (δ)

FIG. 36.8
The ^{13}C NMR spectrum of
2-butanone, $CH_3COCH_2CH_3$
(22.6 MHz).

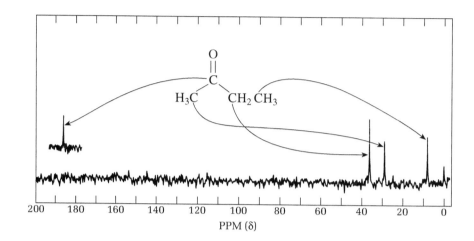

FIG. 36.9
The ^{13}C NMR spectrum of
crotonaldehyde (22.6 MHz).

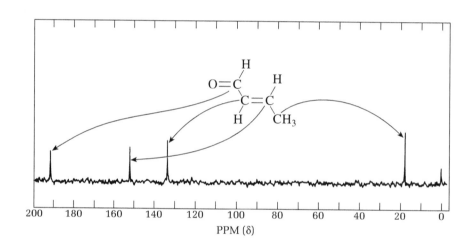

QUESTIONS

1. What is the purpose of making derivatives of unknowns?
2. Why are 2,4-dinitrophenylhydrazones better derivatives than phenylhydrazones?
3. Using chemical tests, how would you distinguish among 2-pentanone, 3-pentanone, and pentanal?
4. Draw the structure of a compound with the empirical formula C_5H_8O that gives a positive iodoform test and does not decolorize permanganate.
5. Draw the structure of a compound with the empirical formula C_5H_8O that gives a positive Tollens' test and does not react with bromine in dichloromethane.

■ **FIG. 36.10**
The ¹H NMR spectrum of
benzaldehyde (250 MHz).

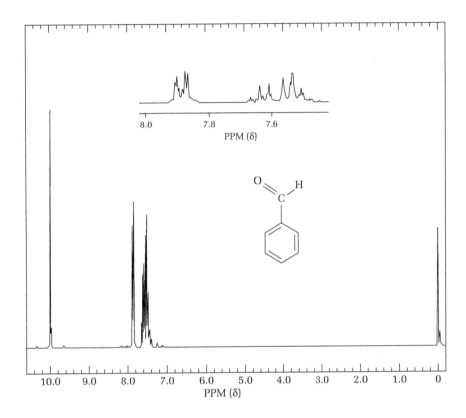

■ **FIG. 36.11**
The IR spectrum of benzaldehyde
(thin film).

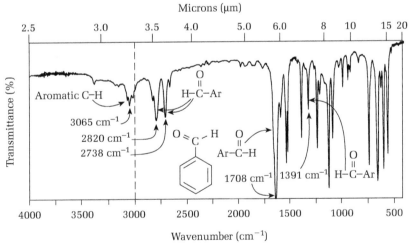

6. Draw the structure of a compound with the empirical formula C_5H_8O that reacts with phenylhydrazine, decolorizes bromine in dichloromethane, and does not give a positive iodoform test.

7. Draw the structure of two geometric isomers with the empirical formula C_5H_8O that give a positive iodoform test.

8. What vibrations cause the peaks at about 3.6 μm (2940 cm^{-1}) in the IR spectrum of fluorene (Fig. 36.2)?

9. Locate the carbonyl peak in Figure 36.4.

10. Assign the various peaks in the ^{1}H NMR spectrum of 2-butanone to specific protons in the molecule (Fig. 36.6).

11. Assign the various peaks in the ^{1}H NMR spectrum of crotonaldehyde to specific protons in the molecule (Fig. 36.7).

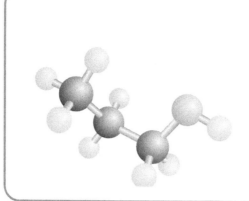

CHAPTER

38

Grignard Synthesis of Triphenylmethanol and Benzoic Acid

Online Study Center

This icon will direct you to techniques, equipment setups, and online resources at http://college.hmco.com/PIC/ williamsonMMOE5e.

PRELAB EXERCISE: Prepare a flow sheet for the preparation of triphenylmethanol. Using your knowledge of the physical properties of the solvents, reactants, and products, show how the products can be purified. Indicate which layer should contain the product in the liquid/liquid extraction steps.

In 1912 Victor Grignard received the Nobel Prize in Chemistry for his work on the reaction that bears his name, a carbon-carbon bond-forming reaction by which almost any alcohol may be formed from appropriate alkyl halides and carbonyl compounds. The Grignard reagent is easily formed by reacting an alkyl halide, in particular a bromide, with magnesium metal in anhydrous diethyl ether. The reaction can be written and thought of as simply

$$R—Br + Mg \longrightarrow R—Mg—Br$$

However, the structure of the material in solution is rather more complex. There is evidence that dialkylmagnesium is present:

$$2\ R—Mg—Br \rightleftharpoons R—Mg—R + MgBr_2$$

Structure of the Grignard reagent

The magnesium atoms, which have the capacity to accept two electron pairs from donor molecules to achieve a four-coordinated state, are solvated by the unshared pairs of electrons on diethyl ether:

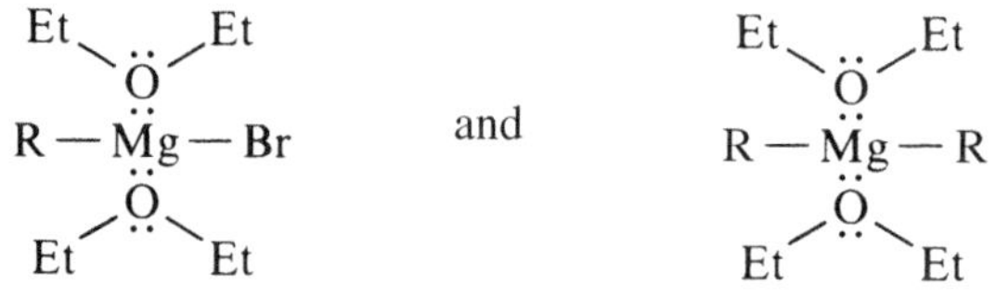

$$
\begin{array}{ccc}
\text{Et}\diagdown\ddot{\text{O}}\diagup\text{Et} & & \text{Et}\diagdown\ddot{\text{O}}\diagup\text{Et} \\
R—\overset{|}{\underset{|}{\ddot{\text{M}}\text{g}}}—Br & \text{and} & R—\overset{|}{\underset{|}{\ddot{\text{M}}\text{g}}}—R \\
\text{Et}\diagup\ddot{\text{O}}\diagdown\text{Et} & & \text{Et}\diagup\ddot{\text{O}}\diagdown\text{Et}
\end{array}
$$

528

A strong base and strong nucleophile

The Grignard reagent is both a strong base and a strong nucleophile. As a base it will react with all protons that are more acidic than those found on alkenes and alkanes. Thus, Grignard reagents react readily with water, alcohols, amines, thiols, and so on to regenerate an alkane:

$$R{-}Mg{-}Br + H_2O \longrightarrow R{-}H + MgBrOH$$

$$R{-}Mg{-}Br + R'OH \longrightarrow R{-}H + MgBrOR'$$

$$R{-}Mg{-}Br + R'NH_2 \longrightarrow R{-}H + MgBrNHR'$$

$$R{-}Mg{-}Br + R'C \equiv C{-}H \longrightarrow R{-}H + R'C \equiv CMgBr$$
(an acetylenic Grignard reagent)

The starting material for preparing the Grignard reagent cannot contain any acidic protons. The reactants and apparatus must be completely and absolutely dry; otherwise the reaction will not start. If proper precautions are taken, however, the reaction proceeds smoothly.

Magnesium metal, in the form of a coarse powder, has a coat of oxide on the outside. A fresh surface can be exposed by crushing the powder under absolutely dry ether in the presence of an organic halide. The reaction will begin at exposed surfaces, as evidenced by a slight turbidity in the solution and evolution of bubbles. Once the exothermic reaction starts, it proceeds easily, the magnesium dissolves, and a solution of the Grignard reagent is formed. The solution is often turbid and gray due to impurities in the magnesium. The reagent is not isolated but reacted immediately with, most often, an appropriate carbonyl compound

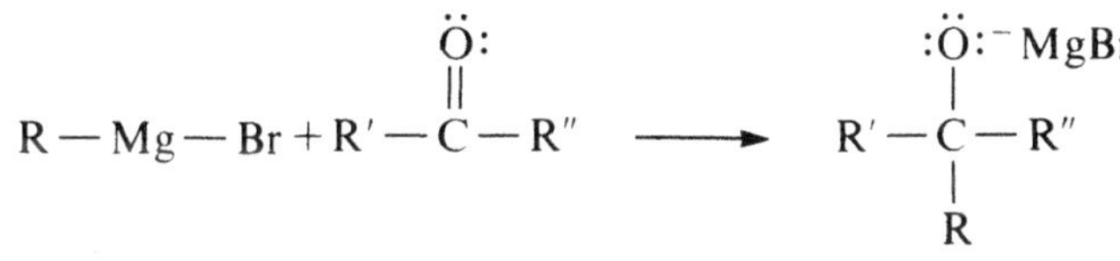

$$R-Mg-Br + R'-\overset{\overset{\ddot{O}:}{\|}}{C}-R'' \longrightarrow R'-\overset{\overset{:\ddot{O}:^-MgBr^+}{|}}{\underset{\underset{R}{|}}{C}}-R''$$

to give, in another exothermic reaction, magnesium alkoxide, a salt insoluble in ether. In a simple acid-base reaction, this alkoxide is reacted with acidified ice water to give the covalent, ether-soluble alcohol and the ionic water-soluble magnesium salt:

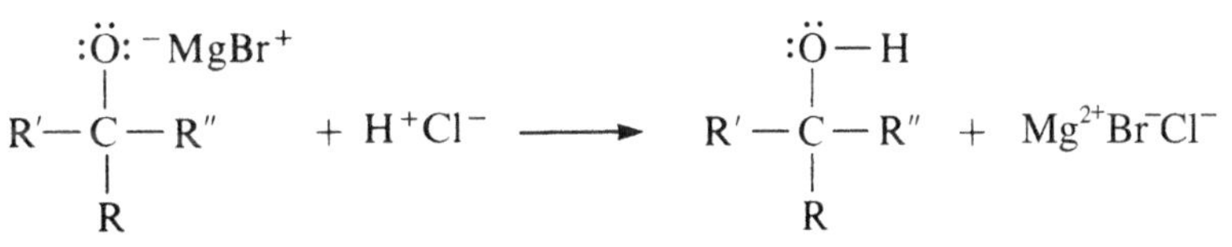

$$R'-\overset{\overset{:\ddot{O}:^-MgBr^+}{|}}{\underset{\underset{R}{|}}{C}}-R'' + H^+Cl^- \longrightarrow R'-\overset{\overset{:\ddot{O}-H}{|}}{\underset{\underset{R}{|}}{C}}-R'' + Mg^{2+}Br^-Cl^-$$

A versatile reagent

The great versatility of this reaction lies in the wide range of reactants that undergo it. Thirteen representative reactions are shown in Figure 38.1. In every case except reaction 1, the intermediate alkoxide must be hydrolyzed to give the product. The reaction with oxygen (reaction 2) is usually not a problem because the ether vapor over the reagent protects it from attack by oxygen, but this reaction is one reason why the reagent cannot usually be stored without special precautions. The reaction with solid carbon dioxide (dry ice) occurs readily to produce a carboxylic acid (reaction 3). Reactions 5, 6, and 7 with aldehydes and ketones giving, respectively, primary, tertiary, and secondary alcohols are among the most common. Ring-opening of an epoxide via Grignard addition (reaction 4) yields an alcohol as well. Reactions 8–13 are not nearly so common.

■ FIG. 38.1
The versatility of the Grignard reaction is illustrated by the wide range of reactants that undergo it.

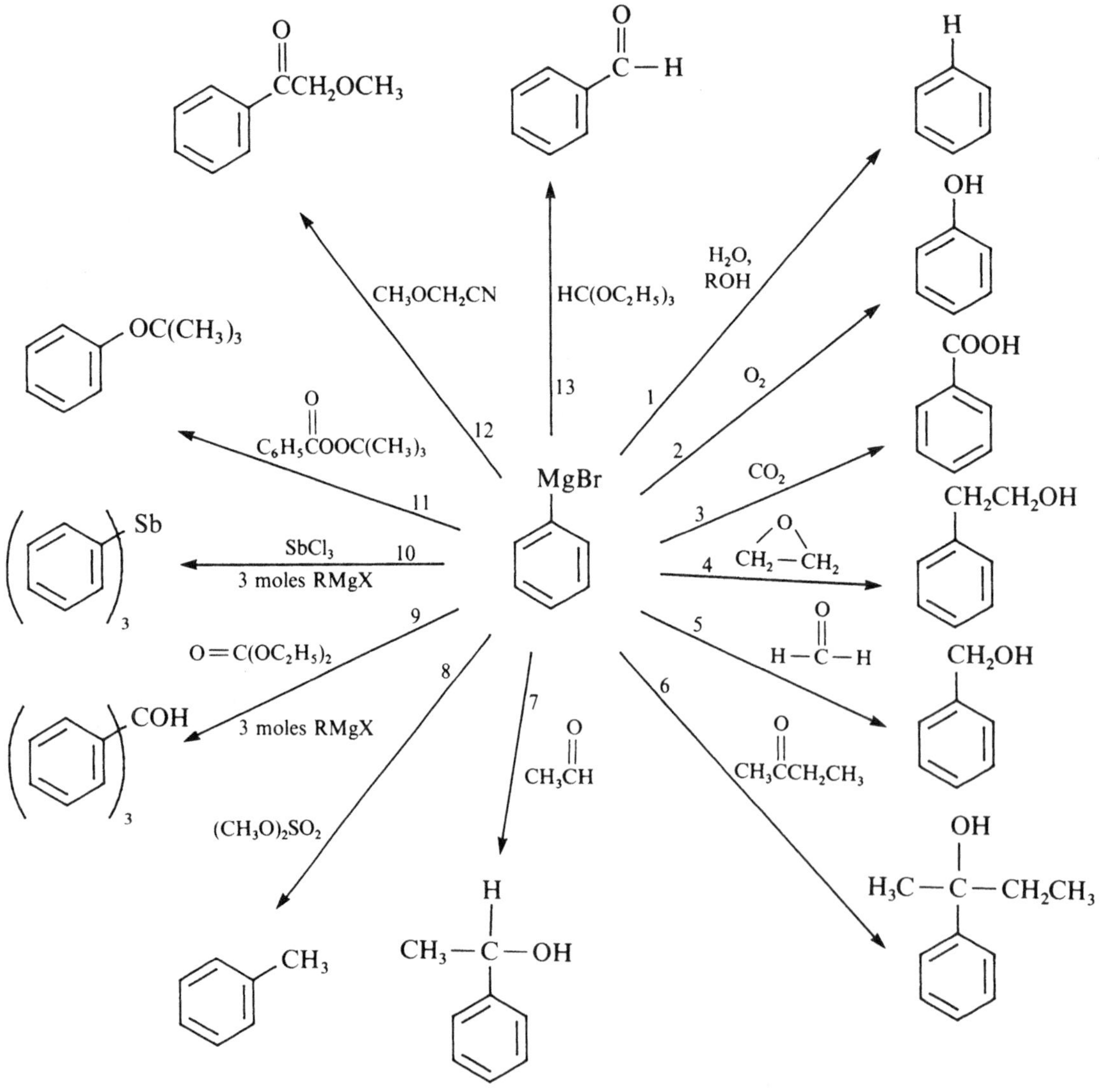

In one of the experiments in this chapter, we will carry out another common type of Grignard reaction, the formation of a tertiary alcohol from 2 moles of the reagent and 1 mole of an ester. The ester employed is the methyl benzoate, which can be synthesized in Chapter 40. The initially formed product is unstable and decomposes to a ketone, which, being more reactive than an ester, immediately reacts with more Grignard reagent:

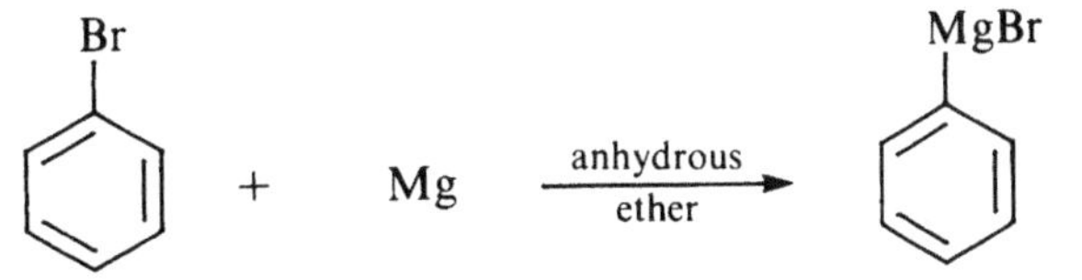

Bromobenzene[1]
MW 157.02
bp 156.4°C
den. 1.491

Magnesium
At Wt 24.1

**Phenylmagnesium
bromide**

Methyl benzoate
MW 136.16
bp 199.6°C
den. 1.09

Triphenylmethanol
MW 260.34
mp 164.2°C

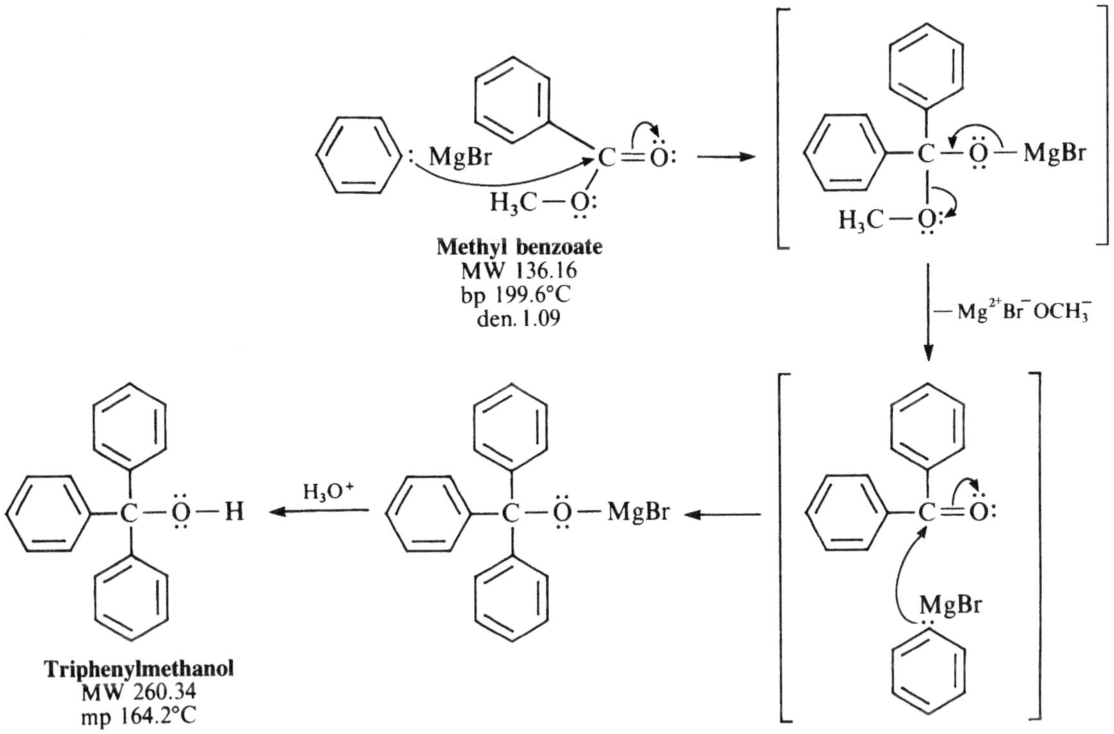

The primary impurity in these experiments is biphenyl, formed by reacting phenylmagnesium bromide with unreacted bromobenzene. (Figure 38.2 shows the ^{13}C NMR spectrum of bromobenzene.) The most effective way to lessen this side reaction is to add the bromobenzene slowly to the reaction mixture so it will

■ FIG. 38.2
The ¹³C NMR spectrum of bromobenzene (22.6 MHz).

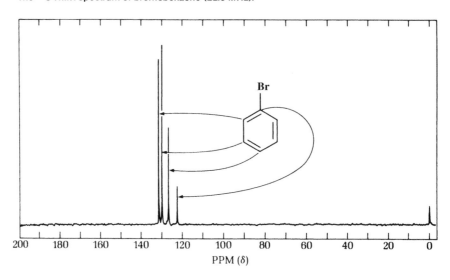

react with the magnesium and not be present in high concentration to react with previously formed Grignard reagent. The impurity is easily eliminated because it is much more soluble in hydrocarbon solvents than triphenylmethanol.

Biphenyl has a characteristic odor. Triphenylmethanol is odorless.

Biphenyl
mp 72°C

Triphenylmethanol can also be prepared from benzophenone.

Phenylmagnesium bromide

Benzophenone
MW 182.22
mp 48°C

Triphenylmethanol
MW 260.34
mp 164.2°C

EXPERIMENTS

1. Phenylmagnesium Bromide (Phenyl Grignard Reagent)

Bromobenzene	**Magnesium**	**Phenylmagnesium bromide**
MW 157.02	At Wt	not isolated, used in situ
bp 156°C	24.31	
den. 1.491		

Br + Mg → (anhydrous ether) → MgBr

Advance Preparation

It is imperative that all equipment and reagents be absolutely dry. The magnesium and the glassware to be used—two reaction tubes, two 1-dram vials (1 dram = 1.78 mL), and a stirring rod—can be dried in a 110°C oven for at least 30 min. Alternatively, if the glassware, syringe, septa, and magnesium appear to be perfectly dry, they can be used without special drying. The plastic and rubber components should be rinsed with acetone if either appears to be dirty or wet with water; then place these components in a desiccator for at least 12 h. Do not place plastic or rubber components in the oven. New, factory-sealed packages of syringes can be used without prior drying. The ether used throughout this reaction must be absolutely dry (absolute ether).

To prepare the Grignard reagent, absolute diethyl ether must be used; elsewhere, ordinary ether (diethyl or *t*-butyl methyl) can be used; for example, ether extractions of aqueous solutions do not need to be carried out with dry ether. It is strongly recommended that *t*-butyl methyl ether be used in all cases except in the preparation of the Grignard reagent itself.

A very convenient container for absolute diethyl ether is a 50-mL septum-capped bottle. This method of dispensing the solvent has three advantages: The ether is kept anhydrous, the exposure to oxygen is minimized, and there is little possibility of its catching fire. Ether is extremely flammable; do not work with this solvent near flames.

To remove ether from a septum-capped bottle, inject a volume of air into the bottle equal to the amount of ether being removed. Pull more ether than needed into the syringe and then push the excess back into the bottle before removing the syringe. In this way there will be no air bubbles in the syringe, and it will not dribble (Fig. 38.3).

Procedure

CAUTION: Ether is extremely flammable. Extinguish all flames before using ether.

Online Study Center

Video: The Grignard Reaction: Removing a Liquid from a Septum-Capped Bottle with Syringe; Photos: Removing a Reagent from a Septum-Capped Bottle, Polypropylene Syringe Containing Ether

IN THIS EXPERIMENT magnesium and absolutely dry diethyl ether in a dry, septum-capped tube are reacted with an aromatic halide to give phenylmagnesium bromide. The exothermic reaction is started by crushing the magnesium. If the reaction does not start in 2–3 min, it will be necessary to start again, using equipment that was not used in the first attempt.

Remove a reaction tube from the oven and immediately cap it with a septum. In the operations that follow, keep the tube capped except when it is necessary to open it. After the tube cools to room temperature, add about 2 mmol (about 50 mg) of magnesium powder. Record the weight of magnesium used to the nearest milligram. The magnesium will become the limiting reagent by using a 5% molar excess of bromobenzene (about 2.1 mmol).

Using a dry syringe, add to the magnesium by injection through the septum 0.5 mL of anhydrous diethyl ether. Your laboratory instructor will demonstrate the transfer from the storage container used in your laboratory.

Into an oven-dried vial weigh about 2.1 mmol (about 330 mg) of dry (stored over molecular sieves) bromobenzene. Using a syringe, add to this vial 0.7 mL anhydrous diethyl ether and *immediately*, with the same syringe, remove all the solution from the vial. This can be done virtually quantitatively so you do not need to rinse the vial. Immediately cap the empty vial to keep it dry for later use. Inject

Diethyl ether can be made and kept anhydrous by storing over Linde 5A molecular sieves. Discard diethyl ether within 90 days because of peroxide formation. *t*-Butyl methyl ether does not have this problem.

FIG. 38.3
A polypropylene syringe (1 mL, with 0.01-mL graduations). The needle is blunt. When there are no air bubbles in the syringe, it will not dribble ether.

FIG. 38.4
Once the reaction has started, bromobenzene in ether is added slowly from the syringe. The empty needle is for pressure relief, but if condensation is complete (aided by the damp pipe cleaner), it will not be needed. Once the reaction slows down, stir it with a magnetic stirrer and stirring bar.

FIG. 38.5
To start the Grignard reaction, remove the septum and apply pressure to the stirring rod while rotating the reaction tube on a hard surface that will not scratch the tube, such as a book.

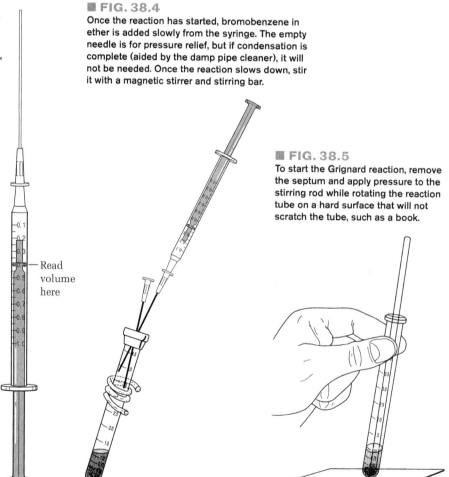

Read volume here

about 0.1 mL of the bromobenzene-ether mixture into the reaction tube and mix the contents by flicking the tube. Pierce the septum with another syringe needle for pressure relief (Fig. 38.4).

The reaction will not ordinarily start at this point, so remove the septum, syringe, and empty syringe needle and crush the magnesium with a dry stirring rod. You can do this easily in the confines of the 10-mm-diameter reaction tube while it is positioned on a hard surface. There is little danger of poking the stirring rod through the bottom of the tube (Fig. 38.5). Immediately replace the septum, syringe, and empty syringe needle (for pressure relief). The reaction should start within seconds. The formerly clear solution becomes cloudy and soon begins to boil as the magnesium metal reacts with the bromobenzene to form the Grignard reagent—phenylmagnesium bromide.

If the reaction does not start within 1 min, begin again with completely different, dry equipment (syringe, syringe needle, reaction tube, etc.). Once the Grignard reaction begins, it will continue. To prevent the ether from boiling away, wrap a pipe cleaner around the top part of the reaction tube. Dampen the pipe cleaner with water or, if the room temperature is very hot, with alcohol.

To the refluxing mixture add slowly and dropwise over a period of several minutes the remainder of the bromobenzene-ether solution at a rate such that the reaction remains under control at all times. After all the bromobenzene solution is added, spontaneous boiling of the diluted mixture may be slow or become slow. At this point, add a magnetic stirring bar to the reaction tube and stir the reaction mixture with a magnetic stirrer. If the rate of reaction is too fast, slow down the stirrer. The reaction is complete when none or a very small quantity of the metal remains. Check to see that the volume of ether has not decreased. If it has, add more anhydrous diethyl ether. Because the solution of the Grignard reagent deteriorates on standing, Experiment 2 should be started at once. The phenylmagnesium bromide can be converted to triphenylmethanol or to benzoic acid.

It takes much force to crush the magnesium. Place the tube on a hard surface and bear down with the stirring rod while twisting the reaction tube. Do not pound the magnesium.

Online Study Center

Video: The Grignard Reaction: Starting the Reaction

Online Study Center

Video: The Grignard Reaction: Addition of Bromobenzene and Refluxing of Reaction Mixture

2. Triphenylmethanol

Phenylmagnesium   **Benzophenone**   **Triphenylmethanol**
bromide   MW 182.22   MW 260.34
                           mp 48°C   mp 164.2°C

> **IN THIS EXPERIMENT** the Grignard reagent prepared in Experiment 1 is added to a dry ether solution of benzophenone. Very thorough mixing is required. When the red color disappears, the salt is hydrolyzed by adding hydrochloric acid. More ether is added, the layers separated, and the ether dried and evaporated to give crude product. An impurity, biphenyl, is removed by dissolving it in hexanes, and the product is recrystallized from 2-propanol. The product recrystallizes very slowly, so do not collect the product immediately.

Make *t*-butyl methyl ether anhydrous by storing it over molecular sieves.

Online Study Center

Videos: The Grignard Reaction: Dissolution of Benzophenone in Ether, Addition to Grignard Reagent; The Grignard Reaction: Mixing Reaction Mixture by Flicking Reaction Tube

Mixing the reaction mixture is very important.

Online Study Center

Videos: The Grignard Reaction: Final Addition of Benzophenone Solution; The Grignard Reaction: Color Change at Completion of Reaction

Online Study Center

Videos: The Grignard Reaction: Addition of Hydrochloric Acid; The Grignard Reaction: Solution Treated with Saturated Sodium Chloride Solution

In an oven-dried vial dissolve 2.0 mmol (0.364 g) of benzophenone in 1.0 mL of anhydrous ether by capping the vial and mixing the contents thoroughly. With a dry syringe, remove all the solution from the vial and add it dropwise with *thorough* mixing (magnetic stirring, flicking of the tube, or stirring with a stirring rod near the end of the addition) after each drop to the solution of the Grignard reagent. Add the benzophenone at a rate so as to maintain the ether at a gentle reflux. Rinse the vial with a few drops of anhydrous ether after all the first solution has been added; then add this rinse to the reaction tube.

After all the benzophenone has been added, the mixture should be homogeneous. If not, mix it again, using a stirring rod if necessary. The syringe can be removed but leave the pressure-relief needle in place. Allow the reaction mixture to stand at room temperature. The reaction apparently is complete when the red color disappears.

At the end of the reaction period, cool the tube in ice and add to it dropwise with stirring (use a glass rod or a spatula) 2 mL of 3 *M* hydrochloric acid. A creamy-white precipitate of triphenylmethanol will separate between the layers. Add more ether (it need not be anhydrous) to the reaction tube and shake the contents to dissolve all the triphenylmethanol. The result should be two perfectly clear layers. Remove a drop of the ether layer for thin-layer chromatographic (TLC) analysis. Any bubbling seen at the interface or in the lower layer is leftover magnesium reacting with the hydrochloric acid. Remove the aqueous layer and shake the ether layer with an equal volume of saturated aqueous sodium chloride solution to remove water and any remaining acid. Carefully remove the entire aqueous layer; then dry the ether layer by adding anhydrous calcium chloride pellets to the reaction tube until the drying agent no longer clumps together. Cork the tube and shake it from time to time over 5–10 min to complete the drying.

Using a Pasteur pipette, remove the ether from the drying agent and place it in another tared, dry reaction tube or a centrifuge tube. Use more ether to wash off the drying agent and combine these ether extracts. Evaporate the ether in a hood by blowing nitrogen or air onto the surface of the solution while warming the tube in a beaker of water or in the hand.

After all the solvent has been removed, determine the weight of the crude product. Note the odor of the biphenyl, the product of the side reaction that takes place between bromobenzene and phenylmagnesium bromide during the first reaction.

Trituration (grinding) of the crude product with petroleum ether will remove the biphenyl. Stir the crystals with 0.5 mL of petroleum ether in an ice bath, remove

■ **FIG. 38.6**

An apparatus for drying crystals in a reaction tube under vacuum.

■ **FIG. 38.7**

A semimicroscale, research-type apparatus for the Grignard reaction, with provision for a motor-driven stirrer and an inlet and outlet for dry nitrogen.

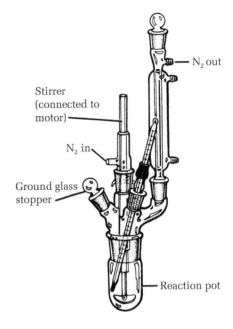

Stirrer (connected to motor)

N₂ out

N₂ in

Ground glass stopper

Reaction pot

Online Study Center

Videos: The Grignard Reaction: Crude Product Triturated and Recrystallized—Pure Triphenylcarbinol Isolated; Microscale Filtration on the Hirsch Funnel

Online Study Center

Photo: Filtration Using a Pasteur Pipette; Video: Filtration of Crystals Using the Pasteur Pipette

the solvent as thoroughly as possible, add a boiling stick, and recrystallize the residue from boiling 2-propanol (no more than 2 mL). Allow the solution to cool slowly to room temperature and then cool it thoroughly in ice. Triphenylmethanol crystallizes slowly, so allow the mixture to remain in the ice as long as possible. Stir the ice-cold mixture well and collect the product by vacuum filtration on a Hirsch funnel. Save the filtrate. Concentration may give a second crop of crystals.

An alternative method for purifying the triphenylmethanol utilizes a mixed solvent. Dissolve the crystals in the smallest possible quantity of warm ether and add 1.5 mL of hexanes to the solution. Add a boiling stick to the solution and boil off some of the ether until the solution becomes slightly cloudy, indicating that it is saturated. Allow the solution to cool slowly to room temperature. Triphenylmethanol is deposited slowly as large, thick prisms. Cool the solution in ice; after allowing time for complete crystallization to occur, remove the ether with a Pasteur pipette and wash the crystals once with a few drops of a cold 1:4 ether-hexanes mixture. Dry the crystals in the tube under a vacuum (Fig. 38.6).

Determine the weight, melting point, and percent yield of the triphenylmethanol. Analyze the crude and recrystallized product by TLC on silica gel (*see* Chapter 8), developing the plate with a 1:5 mixture of dichloromethane and petroleum ether. An infrared (IR) spectrum can be determined in chloroform solution or by preparing a mull or KBr disk (*see* Chapter 11). Compare the apparatus used in this experiment with the research-type apparatus shown in Figure 38.7.

Cleaning Up. Combine the acidic aqueous layer and saturated sodium chloride layers, dilute with water, neutralize with sodium carbonate, and flush down the drain with excess water. Ether is allowed to evaporate from the drying agent in the hood, and the drying agent is then discarded in the nonhazardous solid waste container. If local regulations do not allow for evaporating solvents in a hood, the wet drying agent should be discarded in a special container. The petroleum ether and 2-propanol or ether-hexanes mother liquor are placed in the organic solvents container.

3. Benzoic Acid

IN THIS EXPERIMENT the Grignard reagent prepared in Experiment 1 is squirted onto a piece of dry ice. The resulting white carboxylate salt is hydrolyzed with hydrochloric acid, releasing benzoic acid, which is extracted into the ether layer. The ether solution could be dried and evaporated to give the product, but a better product is obtained by adding base to make the benzoate salt and then adding acid to this basic solution to cause benzoic acid to crystallize. It can be recrystallized from hot water.

Phenylmagnesium bromide

Carbon dioxide
MW 44.01
mp $-78.5°C$ (sublimes)

Benzoic acid
MW 122.12
mp 123°C

CAUTION: Handle dry ice with a towel or gloves. Contact with the skin can cause frostbite because dry ice sublimes at $-78.5°C$.

Prepare 2 mmol of phenylmagnesium bromide exactly as described in Experiment 1. Wipe off the surface of a small piece of dry ice (solid carbon dioxide) with a dry towel to remove frost and place it in a dry 30-mL beaker. Remove the pressure-relief needle from the reaction tube; then insert a syringe through the septum, turn the tube upside down, and draw into the syringe as much of the reagent solution as possible. Squirt this solution onto the piece of dry ice; then, using a clean needle, rinse out the reaction tube with 1 mL of anhydrous diethyl ether and squirt this onto the dry ice. Allow excess dry ice to sublime; then hydrolyze the salt by adding 2 mL of 3 M hydrochloric acid.

Transfer the mixture from the beaker to a reaction tube and shake it thoroughly. Two homogeneous layers should result. Add 1–2 mL of acid or ordinary (not anhydrous) ether if necessary. Remove the aqueous layer and shake the ether layer with 1 mL of water, which is removed and discarded. Then extract the benzoic acid by adding to the ether layer 0.7 mL of 3 M sodium hydroxide solution, shaking the mixture thoroughly, and withdrawing the aqueous layer, which is placed in a very small beaker or vial. The extraction is repeated with another 0.5-mL portion of base and finally 0.5 mL of water. Now that the extraction is complete, the ether, which can be discarded, contains primarily biphenyl, the byproduct formed during the preparation of phenylmagnesium bromide.

The combined aqueous extracts are heated briefly to about 50°C to drive off dissolved ether from the aqueous solution and then made acidic by adding

Online Study Center
Video: Extraction with Ether

concentrated hydrochloric acid (test with indicator paper). Cool the mixture thoroughly in an ice bath. Collect the benzoic acid on a Hirsch funnel and wash it with about 1 mL of ice water while on the funnel. A few crystals of this crude material are saved for a melting-point determination; the remainder of the product is recrystallized from boiling water.

The solubility of benzoic acid in water is 68 g/L at 95°C and 1.7 g/L at 0°C. Dissolve the acid in very hot water. Let the solution cool slowly to room temperature; then cool it in ice for several minutes before collecting the product by vacuum filtration on a Hirsch funnel. Use the ice-cold filtrate in the filter flask to complete the transfer of benzoic acid from the reaction tube. Turn the product out onto a piece of filter paper, squeeze out excess water, and allow it to dry thoroughly. Once dry, weigh it, calculate the percent yield, and determine the melting point along with the melting point of the crude material. The IR spectrum may be determined as a solution in chloroform (1 g of benzoic acid dissolves in 4.5 mL of chloroform) or as a mull or KBr disk (*see* Chapter 11).

Cleaning Up. Combine all aqueous layers, dilute with a large quantity of water, and flush the slightly acidic solution down the drain.

4. Phenylmagnesium Bromide (Phenyl Grignard Reagent)

Bromobenzene	Magnesium	Phenylmagnesium bromide
MW 157.02	At Wt	not isolated, used in situ
bp 156°C	24.31	
den. 1.491		

Br + Mg → (anhydrous ether) → MgBr

Diethyl ether can be kept anhydrous by storing it over Linde 5A molecular sieves. Discard the ether after 90 days because of peroxide formation.

CAUTION: Ether is extremely flammable. Extinguish all flames before using ether.

All equipment and reagents must be *absolutely dry*. The Grignard reagent is prepared in a dry, 100-mL round-bottomed flask fitted with a long reflux condenser. A calcium chloride drying tube inserted in a cork that will fit either the flask or the top of the condenser is also made ready (Fig. 38.8a). The flask, condenser, and magnesium (2.00 g = 0.082 mol of magnesium turnings) should be as dry as possible to begin with and then should be further dried in a 110°C oven for at least 35 min. Alternatively, the magnesium is placed in the flask, the calcium chloride tube is attached directly, and the flask is heated gently but thoroughly with a cool luminous flame.[1] Do not overheat the magnesium. It will become deactivated through oxidation or, if strongly overheated, can burn. The flask on cooling pulls dry air through the calcium chloride. Cool to room temperature before proceeding! Extinguish all flames! Ether vapor is denser than air and can travel along bench tops and into sinks. Use care.

---

1. Alternatively, if nitrogen gas is available, the reaction can be run under nitrogen using a bubbler.

■ **FIG. 38.8**

(a) A calcium chloride drying tube fitted with a rubber stopper. Store for future use with a cork in the top and a pipette bulb on the bottom. (b) An apparatus for refluxing the Grignard reaction.

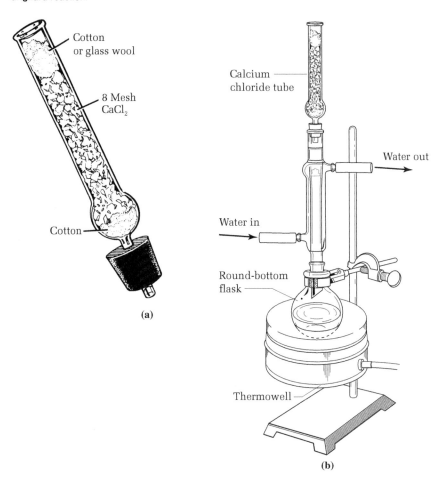

(a)

(b)

Specially dried ether is required.

Prepare an ice bath in case control of the reaction becomes necessary, although this is usually not the case. Remove the drying tube and fit it to the top of the condenser. Then pour into the flask through the condenser 15 mL of *absolute* ether (absolutely dry, anhydrous) and 9 mL (13.5 g = 0.086 mol) of bromobenzene. Be sure the graduated cylinders used to measure the ether and bromobenzene are absolutely dry. (More ether is to be added as soon as the reaction starts, but the concentration of bromobenzene is kept high at the outset to promote easy starting.) If there is no immediate sign of reaction, insert a *dry* stirring rod with a flattened end and crush a piece of magnesium firmly against the bottom of the flask under the surface of the liquid, giving a twisting motion to the rod. When this is done properly, the liquid becomes slightly cloudy, and ebullition commences at the surface of the compressed metal. Be careful not to punch a hole in the bottom of the flask. Attach the condenser at once and swirl the flask to provide fresh surfaces for contact. As soon as you are sure that the reaction has started, add an additional 25 mL of absolute ether through the top of the condenser before spontaneous boiling becomes too vigorous (replace the drying tube). Note the volume of ether in the flask. Cool in ice if necessary to slow the reaction but do not overcool the mixture; the reaction can be stopped by too much cooling. Any difficulty in initiating the reaction can be dealt with by trying the following prompts in succession.

Starting the Grignard reaction

1. Warm the flask with your hands or in a beaker of warm water. Then see if boiling continues when the flask (condenser attached) is removed from the warmth.
2. Try further mashing of the metal with a dry stirring rod.
3. Add a tiny crystal of iodine as a starter (in this case the ethereal solution of the final reaction product should be washed with sodium bisulfite solution to remove the yellow color).
4. Add a few drops of a solution of phenylmagnesium bromide or methylmagnesium iodide (which can be made in a test tube).
5. Start afresh, taking greater care with the dryness of apparatus, measuring tools, and reagents, and sublime a crystal or two of iodine on the surface of the magnesium to generate Gattermann's activated magnesium before beginning the reaction again.

Use caution when heating to avoid condensation on the outside of the condenser.

Once the reaction begins, spontaneous boiling in the diluted mixture may be slow or become slow. If so, mount the flask and condenser in a heating mantle or Thermowell (one clamp supporting the condenser suffices; Fig. 38.8b) and reflux gently until the magnesium has disintegrated and the solution has acquired a cloudy or brownish appearance. The reaction is complete when only a few remnants of metal (or metal contaminants) remain. Check to see that the volume of ether has not decreased. If it has, add more anhydrous ether. Because the solution of Grignard reagent deteriorates on standing, Experiment 5 should be started at once.

5. Triphenylmethanol from Methyl Benzoate

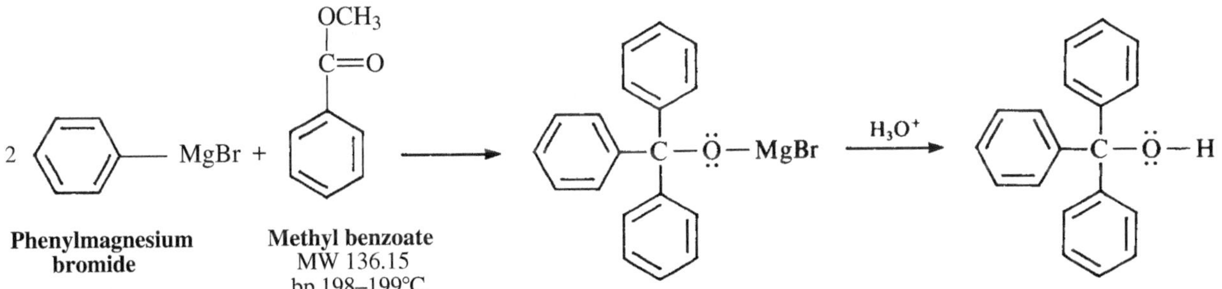

2 [C₆H₅]—MgBr +   Methyl benzoate (OCH₃, C=O) ⟶ [intermediate] C—Ö—MgBr —H₃O⁺→ Triphenylmethanol C—Ö—H

Phenylmagnesium bromide

Methyl benzoate
MW 136.15
bp 198–199°C

Triphenylmethanol
MW 260.34
mp 164.2°C

Mix 5 g (0.037 mol) of methyl benzoate and 15 mL of absolute ether in a separatory funnel, cool the flask containing the phenylmagnesium bromide solution briefly in an ice bath, remove the drying tube, and insert the stem of a separatory funnel into the top of the condenser. Run in the methyl benzoate solution *slowly* with only such cooling as is required to control the mildly exothermic reaction, which affords an intermediate salt that separates as a white solid. Replace the calcium chloride tube; swirl the flask until it is at room temperature and the reaction has subsided. Go to Experiment 7.

6. Triphenylmethanol from Benzophenone

| Phenylmagnesium bromide | Benzophenone MW 182.22 mp 48°C | | | Triphenylmethanol MW 260.34 mp 164.2°C |

Dissolve 6.75 g (0.037 mol) of benzophenone in 25 mL of absolute ether in a separatory funnel and cool the flask containing *half* the phenylmagnesium bromide solution (0.041 mol) briefly in an ice bath. (The other half can be used to make benzoic acid.) Remove the drying tube and insert the stem of the separatory funnel into the top of the condenser. Add the benzophenone solution *slowly* with swirling and only such cooling as is required to control the mildly exothermic reaction, which gives a bright-red solution and then precipitates a white salt. Replace the calcium chloride tube; swirl the flask until it is at room temperature and the reaction has subsided. Go to Experiment 7.

7. Completion of Grignard Reaction

This is a suitable stopping point.

In this part of the experiment, ordinary (not anhydrous) diethyl ether or *t*-butyl methyl ether may be used.

The reaction is then completed by either refluxing the mixture for 30 min or stoppering the flask with the calcium chloride tube and letting the mixture stand overnight (subsequent refluxing is then unnecessary).[2]

Pour the reaction mixture into a 250-mL Erlenmeyer flask containing 50 mL of 10% sulfuric acid and about 25 g of ice and use both ordinary ether and 10% sulfuric acid to rinse the flask. Swirl well to promote hydrolysis of the addition compound; basic magnesium salts are converted into water-soluble neutral salts;

---

2. A rule of thumb for organic reactions: A 10°C rise in temperature will double the rate of the reaction.

triphenylmethanol is distributed into the ether layer. An additional amount of ordinary ether may be required. Pour the mixture into a separatory funnel (rinse the flask with ether), shake, and draw off the aqueous layer. Shake the ether solution with 10% sulfuric acid to further remove magnesium salts and wash with saturated sodium chloride solution to remove water that has dissolved in the ether. The amounts of liquid used in these washing operations are not critical. In general, an amount of wash liquid equal to one-third of the ether volume is adequate.

Saturated aqueous sodium chloride solution removes water from ether.

To effect final drying of the ether solution, pour the ether layer out of the neck of the separatory funnel into an Erlenmeyer flask, add about 5 g of calcium chloride pellets, swirl the flask intermittently, and after 5 min remove the drying agent by gravity filtration (using a filter paper and funnel) into a tared Erlenmeyer flask. Rinse the drying agent with a small amount of ether. Add 25 mL of 66°C–77°C hexanes and concentrate the ether-hexanes solutions (steam bath or hot plate) in an Erlenmeyer flask under an aspirator tube (*see* Fig. 9.6 on page 206). Evaporate slowly until crystals of triphenylmethanol just begin to separate; then let crystallization proceed, first at room temperature and then at 0°C. The product should be colorless and should melt at no lower than 160°C. Concentration of the mother liquor may yield a second crop of crystals. A typical student yield is 5.0 g. Evaporate the mother liquors to dryness and save the residue for later isolation of the components by chromatography.

Perform a TLC analysis of the first crop of triphenylmethanol and the residue from the evaporation of the mother liquors. Dissolve equal quantities of the two solids (a few crystals) and also biphenyl in equal quantities of dichloromethane (1 or 2 drops). Using a microcapillary, spot equal quantities of material on silica gel TLC plates and develop the plates in an appropriate solvent system. Try a 1:3 dichloromethane–petroleum ether mixture first and adjust the relative quantities of solvent as needed. The spots can be seen by examining the TLC plate under a fluorescent lamp or by treating the TLC plate with iodine vapor. From this analysis decide how pure each of the solids is and whether it would be worthwhile to attempt to isolate more triphenylmethanol from the mother liquors.

Turn in the product in a vial labeled with your name, the name of the compound, its melting point, and the overall percent yield from benzoic acid.

Cleaning Up. Combine the acidic aqueous layer and saturated sodium chloride layers, dilute with water, neutralize with sodium carbonate, and flush down the drain with excess water. Ether is allowed to evaporate from the drying agent in the hood, and the drying agent is then discarded in the nonhazardous solid waste container. If local regulations do not allow for evaporating solvents in a hood, the wet drying agent should be discarded in a special container. The ether-hexanes mother liquor is placed in the organic solvents container.

Dispose of recovered and waste solvents in the appropriate containers.

8. Benzoic Acid

Wipe the frost from a piece of dry ice, transfer the ice to a cloth towel, and crush it with a hammer. Without delay (so moisture will not condense on the cold solid), transfer about 10 g of dry ice to a 250-mL beaker. Cautiously pour one-half of the

Online Study Center

Video: Macroscale
Crystallization

solution of phenylmagnesium bromide prepared in Experiment 1 onto the dry ice. A vigorous reaction will ensue. Allow the mixture to warm up and stir it until the dry ice has evaporated. To the beaker add 20 mL of 3 M hydrochloric acid; then heat the mixture over a steam bath in the hood to boil off the ether. Cool the beaker thoroughly in an ice bath and collect the solid product by vacuum filtration on a Büchner funnel.

Transfer the solid back to the beaker and dissolve it in a minimum quantity of saturated sodium bicarbonate solution (2.8 M). Note that a small quantity of a byproduct remains suspended and floating on the surface of the solution. Note the odor of the mixture. Transfer it to a separatory funnel and shake it briefly with about 15 mL of ether. Discard the ether layer, place the clear aqueous layer in the beaker, and heat it briefly to drive off dissolved ether. Carefully add 3 M hydrochloric acid to the mixture until the solution tests acidic to pH paper. Cool the mixture in ice and collect the product on a Büchner funnel. Recrystallize it from a minimum quantity of hot water and isolate it in the usual manner. Determine the melting point and the weight of the benzoic acid; calculate its yield based on the weight of magnesium used to prepare the Grignard reagent.

Cleaning Up. Combine all aqueous layers, dilute with a large quantity of water, and flush the slightly acidic solution down the drain. The ether-hexanes mother liquor from the recrystallization goes in the organic solvents container. The TLC developer, which contains dichloromethane, is placed in the halogenated organic waste container. Calcium chloride from the drying tube should be dissolved in water and flushed down the drain.

QUESTIONS

1. Triphenylmethanol can also be prepared by reacting ethyl benzoate with phenylmagnesium bromide and by reacting diethylcarbonate with phenylmagnesium bromide. Write stepwise reaction mechanisms for these two reactions.

$$\overset{\displaystyle O}{\underset{\displaystyle C_2H_5OCOC_2H_5}{\|}}$$

2. If the ethyl benzoate used to prepare triphenylmethanol is wet, what byproduct is formed?

3. Exactly what weight of dry ice is needed to react with 2 mmol of phenylmagnesium bromide? Would an excess of dry ice be harmful?

4. In the synthesis of benzoic acid, benzene is often detected as an impurity. How does this come about?

5. In Experiment 3, the benzoic acid could have been extracted from the ether layer using sodium bicarbonate solution. Write equations showing how this might be done and how the benzoic acid would be regenerated. What practical reason makes this extraction method less desirable than sodium hydroxide extraction?

6. What is the weight of frost (ice) on the dry ice that will react with all of the Grignard reagent used in Experiment 8?

7. How many moles of carbon dioxide are contained in 10 g of dry ice?

8. Just after the dry ice has evaporated from the beaker, what is the white solid remaining?

9. Write an equation for the reaction of the white solid with 3 *M* hydrochloric acid.

10. Write an equation for the reaction of the product with sodium bicarbonate.

11. Would you expect sodium benzoate to have an odor? Why or why not?

12. What odor do you detect after the product has dissolved in sodium bicarbonate solution?

13. What is the purpose of the ether extraction?

14. "Isolate the product in the usual way." What is the meaning of this sentence?

Online Study Center

General Resources

Web Links

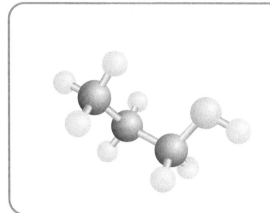

CHAPTER 46

Dyes and Dyeing[1]

Online Study Center

This icon will direct you to techniques, equipment setups, and online resources at http://college.hmco.com/PIC/williamsonMMOE5e.

PRELAB EXERCISE: Operating on the simple hypothesis that the intensity of a dye on a fiber will depend on the number of strongly polar or ionic groups in the fiber molecule, predict the relative intensities produced by Methyl Orange when it is used to dye a variety of different fibers, such as are found in the Multifiber Fabric. Examine the structures of the fiber molecules and try to order them in terms of polarity. Wool is a protein.

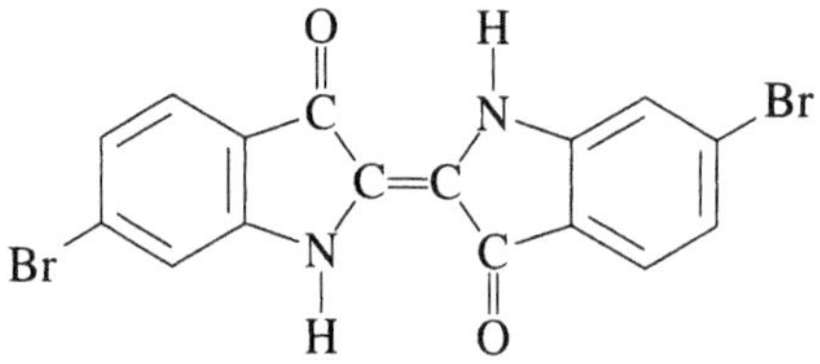

Tyrian purple
(6,6′-Dibromoindigo)

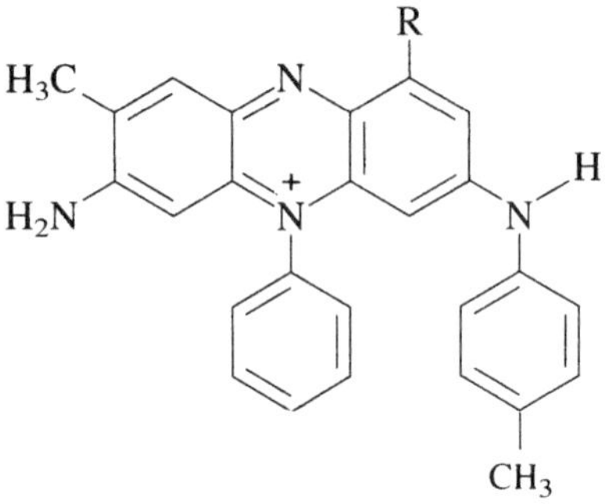

Perkin's Mauve
R = H or CH$_3$

Since prehistoric times humans have been dyeing cloth. The "wearing of the purple" has long been synonymous with royalty, attesting to the cost and rarity of Tyrian purple, a dye derived from the sea snail *Murex brandaris*. The organic chemical industry originated in 1856 with William Henry Perkin's discovery of the first synthetic dye, Perkin's Mauve. The correct structure of this dye has only recently been elucidated.[2]

---

1. For a detailed discussion of the chemistry of dyes and dyeing, see Fieser, L. F.; Fieser, M. *Topics in Organic Chemistry*; Reinhold Publishing Corp.: New York, 1963; 357–417.

2. Meth-Cohn, O.; Smith, M. *J. Chem. Soc., Perkin Trans.* 1, **1994**, 5.

632

A fiber usually absorbs dyes from an aqueous solution. A natural fiber such as cotton has a surface area of about 4.4 hectares per kilogram (5 acres per pound). The dye penetrates the pores in the fiber and is bound to the fiber by electrostatic forces, van der Waals attraction, hydrogen bonding, and covalent bonds (in the case of fiber-reactive dyes). A good dye must be fast to light, heat, and washing; that is, it must not fade, sublime away, or come off during washing.

Among the newest of the dyes are the fiber-reactive compounds, which form a covalent link to the hydroxyl groups of cellulose. This reaction involves an amazing and little-understood nucleophilic displacement of a chloride ion from the triazine part of the molecule by the hydroxyl groups of cellulose, yet the reaction occurs in aqueous solution.

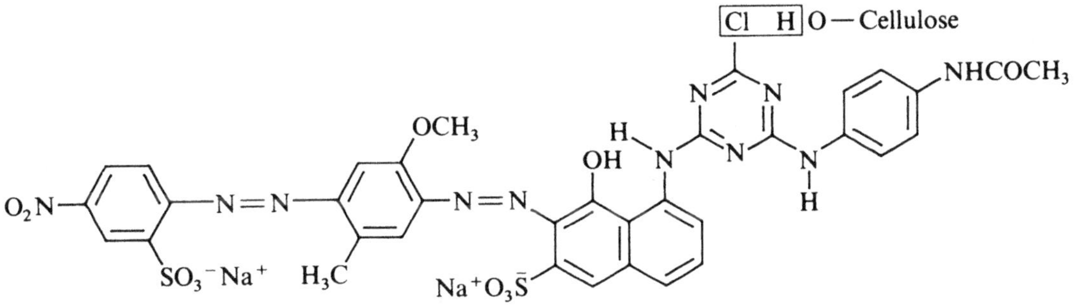

Chlorantin Light Blue 8G

■ FIG. 46.1
Multifiber Fabric.

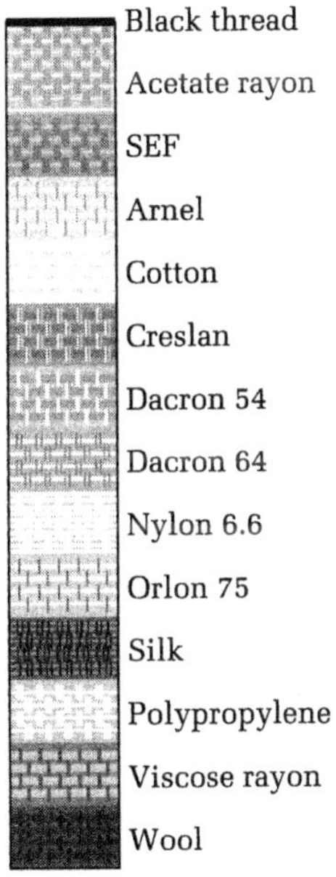

Black thread
Acetate rayon
SEF
Arnel
Cotton
Creslan
Dacron 54
Dacron 64
Nylon 6.6
Orlon 75
Silk
Polypropylene
Viscose rayon
Wool

Several dyes are synthesized in this chapter; these and other dyes are used to dye a representative group of natural and synthetic fibers. You will receive several pieces of Multifiber Fabric (Fig. 46.1), which has 13 strips of different fibers woven into it. Below the black thread at the top, the fibers are acetate rayon (cellulose di- or triacetate), SEF (Solutia's modacrylic Self-Extinguishing Fiber), Arnel (cellulose triacetate), cotton, Creslan (polyacrylonitrile), Dacron 54 and 64 (polyester without and with a brightener), nylon 6.6 (polyamide), Orlon 75 (polyacrylonitrile), silk (polyamide), polypropylene, viscose rayon (regenerated cellulose), and wool (polyamide).

Acetate rayon is cellulose (from any source) in which about two of the hydroxyl groups in each unit have been acetylated. This renders the polymer soluble in acetone, from which it can be spun into fiber. The smaller number of hydroxyl groups in acetate rayon compared to cotton makes direct dyeing of rayon more difficult than cotton.

Cotton is pure cellulose. Nylon is a polyamide made by polymerizing adipic acid and hexamethylenediamine. The nylon polymer chain can be prepared with one acid and one amine group at the termini or with both acids or both amines. Except for these terminal groups, there are no polar centers in nylon; consequently, it is difficult to dye. Similarly, Dacron, a polyester made by polymerizing ethylene glycol and terephthalic acid, has few polar centers within the polymer and is also difficult to dye. Even more difficult to dye is Orlon, a polymer of acrylonitrile. Wool and silk are polypeptides cross-linked with disulfide bridges. The acidic and basic amino acids (e.g., glutamic acid and lysine) provide many polar groups in wool and silk to which a dye can bind, making these fabrics easy to dye. In this experiment note the marked differences in shade produced by the same dye on different fibers.

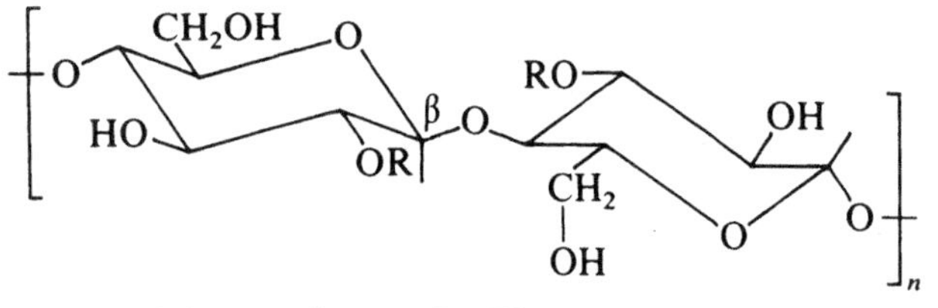

Cellulose (Cotton, R = H)
Acetylated cellulose (Acetate rayon, R = OAc)

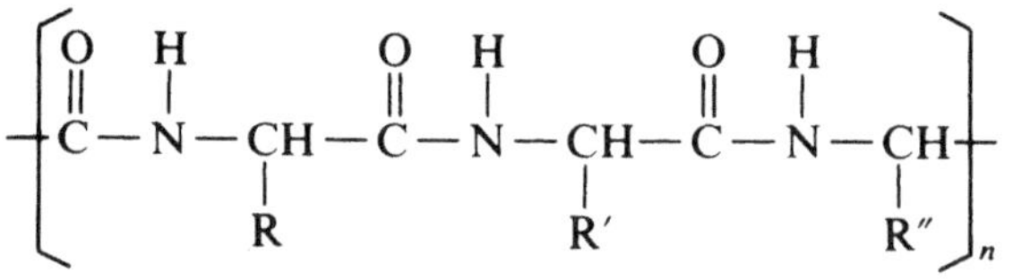

Wool (R = amino acid residue)

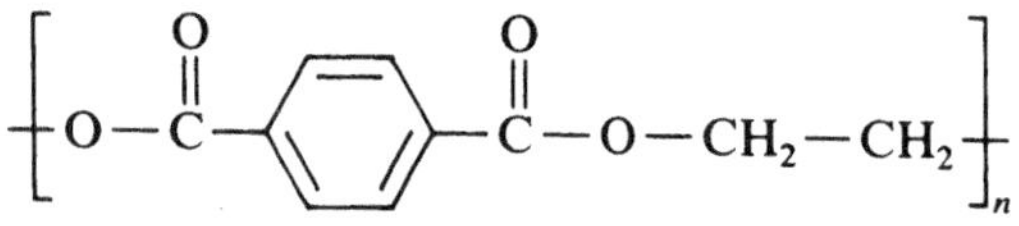

Polyethylene glycol terephthalate (Dacron)

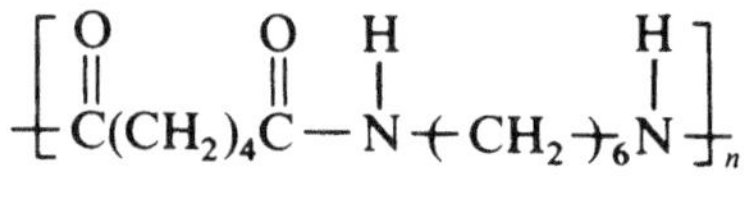

Nylon

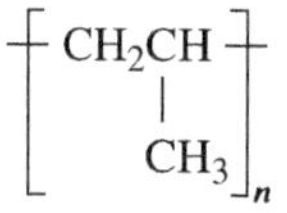

Polypropylene

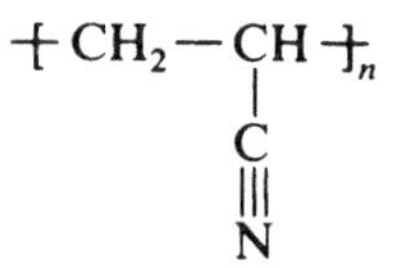

Polyacrylonitrile (Orlon)

PART 1: Dyes

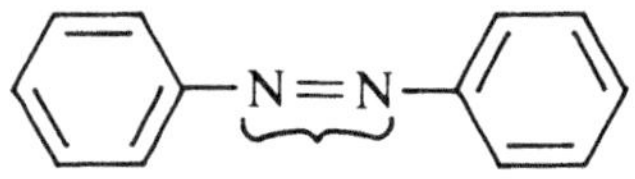

Azo group

The most common dyes are the azo dyes, formed by coupling diazotized amines to phenols. The dye can be made in bulk; as we shall see, the dye molecule can also be developed on and in the fiber by combining the reactants in the presence of the fiber.

One dye, Orange II, is made by coupling diazotized sulfanilic acid with 2-naphthol in alkaline solution; another, Methyl Orange, is prepared by coupling the same diazonium salt with N,N-dimethylaniline in a weakly acidic solution. Methyl Orange is used as an indicator because it changes color at pH 3.2–4.4. The change in color is due to the transition from one chromophore (azo group) to another (quinonoid system).

You will prepare one of these two dyes and then exchange samples with a classmate and do the tests with both dyes. Both substances dye wool, silk, and skin, so you must work carefully to avoid getting these on your hands or clothes. The dye will eventually wear off your hands; alternatively, you can clean your hands by soaking them for a few seconds in a warm, slightly acidic (H_2SO_4) permanganate solution until heavily stained with manganese dioxide and then removing the stain in a bath of warm, dilute bisulfite solution. (Follow the same procedure if dye is accidentally spilled on your clothing.)

EXPERIMENTS

1. Diazotization of Sulfanilic Acid
Microscale Procedure

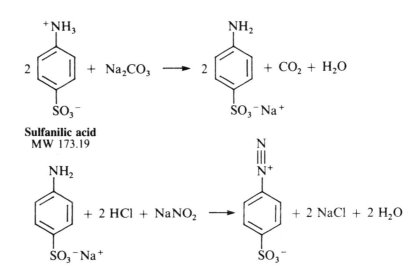

Sulfanilic acid
MW 173.19

In a 10×100-mm reaction tube (Fig. 46.2) dissolve, by boiling, 120 mg of sulfanilic acid monohydrate in 1.25 mL of 2.5% sodium carbonate solution (or use 35 mg of anhydrous sodium carbonate and 1.25 mL of water). Cool the solution to room temperature, add 50 mg of sodium nitrite, and stir until it is dissolved. Cool the tube in ice and add to it with thorough stirring a mixture of 0.75 g of ice

■ FIG. 46.2
A reaction tube with a boiling stick to promote even boiling.

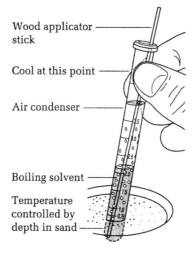

Wood applicator stick

Cool at this point

Air condenser

Boiling solvent

Temperature controlled by depth in sand

and 0.125 mL of concentrated hydrochloric acid. A powdery white precipitate of the diazonium salt should separate in 1–2 min; the material is then ready for use. The product is not collected but is used in the preparation of Orange II and/or Methyl Orange while in suspension. It is more stable than most diazonium salts and will keep for a few hours.

Macroscale Procedure

In a 50-mL Erlenmeyer flask dissolve, if necessary by boiling, 2.4 g of sulfanilic acid monohydrate in 25 mL of 2.5% sodium carbonate solution (or use 0.66 g of anhydrous sodium carbonate and 25 mL of water). Cool the solution under the tap, add 0.95 g of sodium nitrite, and stir until it is dissolved. Pour the solution into a flask containing about 15 g of ice and 2.5 mL of concentrated hydrochloric acid. A powdery white precipitate of the diazonium salt should separate in 1–2 min; the material is then ready for use. The product is not collected but is used in the preparation of Orange II and/or Methyl Orange while in suspension. It is more stable than most diazonium salts and will keep for a few hours.

2. Orange II (1-*p*-Sulfobenzeneazo-2-Naphthol, Sodium Salt)

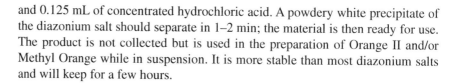

Orange II

In a 10-mL Erlenmeyer flask dissolve 90 mg of 2-naphthol in 0.5 mL of 3 *M* sodium hydroxide solution and transfer to this solution—with *thorough* stirring—the suspension of diazotized sulfanilic acid prepared in the microscale procedure of Experiment 1. Rinse all of the diazonium salt into the naphthol solution with a few drops of cold water. Coupling occurs very rapidly, and the dye, being a sodium salt, separates easily from the solution because a considerable excess of sodium ion from the carbonate, the nitrite, and the alkali is present. Stir the crystalline paste thoroughly to effect good mixing and, after 5–10 min, heat the mixture in a beaker of boiling water until the solid dissolves. Add 0.25 g of sodium chloride to further decrease the solubility of the product and bring this all into solution by heating and stirring. Allow the flask to cool to near room temperature undisturbed; then cool it in ice. Collect the product on a Hirsch funnel bearing a filter paper (Fig. 46.3). Use saturated sodium chloride solution rather than water to rinse out the flask and to wash the filter cake free of the dark-colored mother liquor. The filtration is somewhat slow, and the transfer to a Hirsch funnel is difficult because it is a paste.

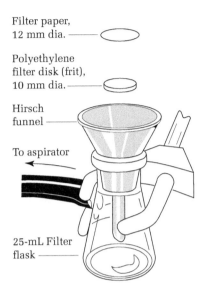

Filter paper, 12 mm dia.

Polyethylene filter disk (frit), 10 mm dia.

Hirsch funnel

To aspirator

25-mL Filter flask

Online Study Center

Photo: Vacuum Filtration into Reaction Tube through Hirsch Funnel

The product dries slowly and contains about 20% sodium chloride. Thus the crude yield is not significant, and the material need not be dried before being purified. This particular azo dye is too soluble to be crystallized from water; it can be obtained in a fairly satisfactory form by adding saturated sodium chloride solution to a hot, filtered solution in water and cooling, but the best crystals are obtained from aqueous ethanol.

Transfer the filter cake to a reaction tube, wash the material from the filter paper and funnel with 1 mL of water, and bring all of the solid into solution at the boiling point. It may be necessary to add another 0.25 mL (no more) of water in the course of filtering this hot solution through a Hirsch funnel equipped with a polyethylene frit (Fig. 46.3). Collect the filtrate in a reaction tube, add 2.5–3 mL of ethanol, and allow crystallization to proceed as the tube cools slowly to room temperature. Cool well in ice before collecting the product. Rinse the tube with some of the filtrate and, finally, wash the product with a small quantity of ethanol. The yield of pure crystalline material should be about 150 mg. Orange II separates from aqueous alcohol as the dihydrate, containing two molecules of water of crystallization; allowance for this should be made in calculating the yield. If the water of hydration is eliminated by drying at 120°C, the material will become fiery red.

Cleaning Up. The filtrate from the reaction, although highly colored, contains little dye but is highly soluble in water. It can be diluted with a large quantity of water and flushed down the drain. Alternatively, with the volume kept as small as possible, the filtrate can be placed in the aromatic amines hazardous waste container, or it can be reduced with tin(II) chloride (*see* Experiment 5). The crystallization filtrate should go into the organic solvents container. Wash the Hirsch funnel *thoroughly*.

3. Orange II (1-*p*-Sulfobenzeneazo- 2-Naphthol, Sodium Salt)

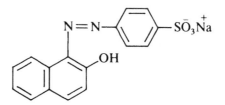

Orange II

CAUTION: Handle 2-naphthol with care, in the hood. Do not breathe the dust or allow skin contact. It is a carcinogen.

In a 250-mL beaker, dissolve 1.8 g of 2-naphthol in 10 mL of cold 3 *M* sodium hydroxide solution; pour into this solution, with stirring, the suspension of diazotized sulfanilic acid prepared in the macroscale procedure of Experiment 1. Rinse the Erlenmeyer flask with a small amount of water and add it to the beaker. Coupling occurs very rapidly, and the dye, being a sodium salt, separates easily from the solution because a considerable excess of sodium ion from the carbonate, the nitrite, and the alkali is present. Stir the crystalline paste thoroughly to effect good mixing and, after 5–10 min, heat the mixture until the solid dissolves. Add 5 g of sodium chloride to further decrease the solubility of the product, bring this all into solution by heating and stirring, set the beaker in a pan of ice and water, and let the solution cool undisturbed. When near room temperature, cool further by stirring and collect the product on a Büchner funnel. Use saturated sodium chloride solution rather than water for rinsing the material from the beaker and for washing the filter cake free of the dark-colored mother liquor. The filtration is somewhat slow.[3]

The product dries slowly; it contains about 20% of sodium chloride. Thus, the crude yield is not significant, and the material need not be dried before being purified. This particular azo dye is too soluble to be crystallized from water; it can be obtained in a fairly satisfactory form by adding saturated sodium chloride solution to a hot, filtered solution in water and cooling, but the best crystals are obtained from aqueous ethanol.

Transfer the filter cake to a beaker, wash the material from the filter paper and funnel with water, and bring the cake into solution at the boiling point. Avoid a large excess of water but use enough to prevent separation of the solid during filtration (use about 25 mL). Filter by suction through a Büchner funnel that has been preheated on a steam bath. Pour the filtrate into an Erlenmeyer flask, rinse the filter flask with a small quantity of water, add it to the flask, estimate the volume, and if greater than 30 mL evaporate by boiling. Cool to 80°C, add 50–60 mL of ethanol, and allow crystallization to proceed. Cool the solution well before collecting the product. Rinse the beaker with the mother liquor and wash finally with a little ethanol. The yield of pure crystalline material is about 3.4 g. Orange II separates from aqueous alcohol with two molecules of water of crystallization; allowance for

---

3. If the filtration must be interrupted, fill the funnel and close the rubber suction tubing (while the aspirator is still running) with a screw pinchclamp placed near the filter flask. Then disconnect the tubing from the trap and set the unit aside. Thus, suction will be maintained, and filtration will continue.

this should be made in calculating the yield. If the water of hydration is eliminated by drying at 120°C, the material will become fiery red.

Cleaning Up. The filtrate from the reaction, although highly colored, contains little dye but is highly soluble in water. It can be diluted with a large quantity of water and flushed down the drain. Alternatively, with the volume kept as small as possible, the filtrate can be placed in the aromatic amines hazardous waste container, or it can be reduced with tin(II) chloride (see Experiment 6). The crystallization filtrate should go into the organic solvents container.

4. Methyl Orange (p-Sulfobenzeneazo-4-Dimethylaniline, Sodium Salt)

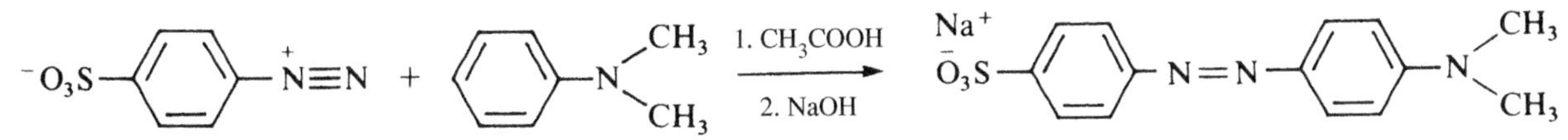

Diazotized sulfanilic acid
4-Diazobenzenesulfonic acid

N,N-**Dimethylaniline**
MW 121.18
bp 194°C

Methyl Orange
4-[4-(Dimethylamino)phenylazo]
benzenesulfonic acid, sodium salt
MW 327.34
λ_{max} 507 nm

Methyl Orange is an acid-base indicator.

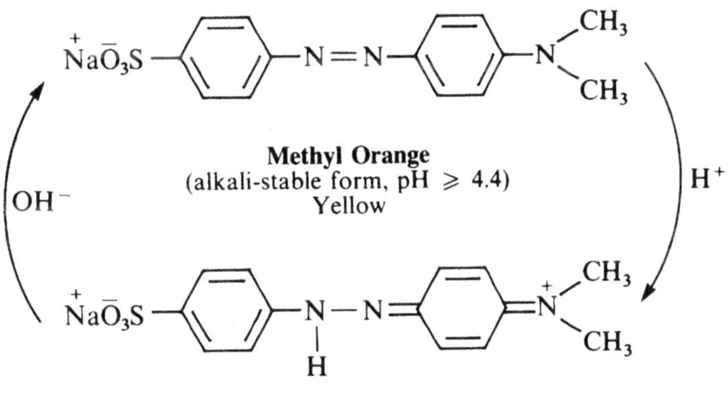

Methyl Orange
(alkali-stable form, pH ≥ 4.4)
Yellow

Methyl Orange
(acid-stable form, pH ≤ 3.2)
Red

Microscale Procedure

In a 10-mL Erlenmeyer flask, mix 75 mg of dimethylaniline and 65 mg of acetic acid. Add the suspension of diazotized sulfanilic acid prepared in the microscale procedure in Experiment 1, with stirring, to the solution of dimethylaniline acetate, rinsing out the last portions with a few drops of water. Stir the mixture thoroughly and, within a few minutes, the red acid-stable form of the dye should separate. A stiff paste should result in 5–10 min, at which time 1 mL of 3 *M* sodium hydroxide

is added to produce the orange sodium salt. Heat the mixture to the boiling point with constant stirring (to avoid bumping), at which time a large part of the dye should dissolve. A better way to do this would be to place the flask in a small beaker of boiling water. Allow the flask to cool slowly to room temperature; then cool it thoroughly in ice before collecting the product by vacuum filtration on a Hirsch funnel bearing a piece of filter paper. Use saturated sodium chloride solution rather than water to rinse the flask and to wash the dark mother liquor from the filter cake.

The crude product does not need to be dried; it can be crystallized from water after performing preliminary solubility tests to determine the proper conditions. The yield should be between 125 mg and 150 mg. Recrystallized material need not necessarily be used for the dyeing process.

Cleaning Up. The highly colored filtrates from the reaction and crystallization are very water soluble. After dilution with a large quantity of water, they can be flushed down the drain because the amount of solid is small. Alternatively, the combined filtrates should be placed in the hazardous waste container, or the mixture can be reduced with tin(II) chloride (*see* Experiment 5). Wash the Hirsch funnel *thoroughly*.

 ## Macroscale Procedure

In a test tube, thoroughly mix 1.6 mL of dimethylaniline and 1.25 mL of glacial acetic acid. To the suspension of diazotized sulfanilic acid prepared in the macroscale procedure in Experiment 1 contained in a 250-mL beaker add, with stirring, the solution of dimethylaniline acetate. Rinse the test tube with a small quantity of water and add it to the beaker. Stir and mix thoroughly; within a few minutes the red, acid-stable form of the dye should separate. A stiff paste should result in 5–10 min; 18 mL of 3 *M* sodium hydroxide solution is then added to produce the orange sodium salt. Stir well and heat the mixture to the boiling point, when a large part of the dye should dissolve. Place the beaker in a pan of ice and water and allow the solution to cool undisturbed. When cooled thoroughly, collect the product on a Büchner funnel, using saturated sodium chloride solution rather than water to rinse the flask and to wash the dark mother liquor from the filter cake.

The crude product does not need to be dried; it can be crystallized from water after making preliminary solubility tests to determine the proper conditions. The yield is about 2.5–3 g.

Cleaning Up. The highly colored filtrates from the reaction and crystallization are highly water soluble. After dilution with a large quantity of water, they can be flushed down the drain because the amount of solid is small. Alternatively, the combined filtrates should be placed in the hazardous waste container, or the mixture can be reduced with tin(II) chloride (*see* Experiment 5).

 ## 5. For Further Investigation

Following the procedure for the synthesis of Orange II, couple diazotized sulfanilic acid with one or more of the following naphthols: 4-amino-5-hydroxy-2,7-naphthalenedisulfonic acid, monosodium salt; 6-amino-4-hydroxy-2-naphthalenesulfonic acid; 7-amino-4-hydroxy-2-naphthalenesulfonic acid; 4-hydroxy-1-naphthalenesulfonic acid, sodium salt; or 6-hydroxy-2-naphthalenesulfonic acid, sodium salt. Despite the apparent complexity of these molecules, they are not expensive; they are used for the commercial synthesis of dyes. The many sulfonic acid groups have been added to make the dyes water soluble; the addition of solid sodium chloride "salts out" the dye, rendering it insoluble.

6. Tests

Solubility and Color

Compare the solubility in water of Orange II to Methyl Orange and account for the difference in terms of structure. Treat the first solution with alkali and note the change in shade due to salt formation; to the other solution alternately add acid and alkali.

Reduction

A characteristic of an azo compound is the ease with which the molecule is cleaved at the double bond by reducing agents to give two amines. Because amines are colorless, the reaction is easily followed by the color change. The reaction is useful in preparing hydroxyamino and similar compounds in analyzing azo dyes by titration with a reducing agent and in identifying azo compounds from an examination of the cleavage products.

This reaction can, if necessary, be run on five times the indicated scale. Dissolve about 0.1 g of tin(II) chloride in 0.2 mL of concentrated hydrochloric acid, add a small quantity of the azo compound (20 mg), and heat. A colorless solution should result, and no precipitate should form on adding water. The aminophenol or the diamine products are present as the soluble hydrochlorides; the other product of cleavage, sulfanilic acid, is sufficiently soluble to remain in solution.

$$\overset{+}{Na}\overset{-}{O_3}S-\!\!\!\raisebox{0pt}{⬡}\!\!\!-N\!=\!N-\!\!\!\raisebox{0pt}{⬡}\!\!\!-N\!\!\begin{smallmatrix}CH_3\\CH_3\end{smallmatrix}$$

$$\xrightarrow[\text{SnCl}_2]{\text{HCl}} HO_3S-\!\!\!\raisebox{0pt}{⬡}\!\!\!-NH_2 + Cl^-\,H_3N^+\!\!-\!\!\!\raisebox{0pt}{⬡}\!\!\!-N\!\!\begin{smallmatrix}CH_3\\CH_3\end{smallmatrix}$$

Cleaning Up. Dilute the reaction mixture with water, neutralize with sodium carbonate, and remove the solids by vacuum filtration. The solids go in the aromatic amines hazardous waste container, and the filtrate can be flushed down the drain.

PART 2: Dyeing

EXPERIMENTS

Using good laboratory techniques, your hands will not be dyed; use care.

The following experiments can, if necessary, be run on up to 10 times the indicated amounts. Wearing disposable gloves (polyethylene or vinyl) might be advisable.

 ### 1. Direct Dyes

The sulfonate groups on the Methyl Orange and Orange II molecules are polar and thus enable these dyes to combine with polar sites in fibers. Wool and silk have many polar sites on their polypeptide chains and hence bind strongly to a dye of this type. Martius Yellow, picric acid, and eosin are also highly polar dyes and thus dye directly to wool and silk.

Orange II or Methyl Orange

The dye bath is prepared from 50 mg of Orange II or Methyl Orange, 0.5 mL of 3 M sodium sulfate solution, 15 mL of water, and 5 drops of 3 M sulfuric acid in a 30-mL beaker. Place a piece of test fabric, a strip 3/4-in. wide, in the bath for 5 min at a temperature near the boiling point. Remove the fabric from the dye bath, allow it to cool, and then wash it thoroughly with soap under running water before drying it.

Dye untreated test fabric and one or more of the pieces of test fabric that have been treated with a mordant following this same procedure. *See* Experiment 3 in this part for the application of mordants.

Picric acid

Martius Yellow

Picric Acid or Martius Yellow

Dissolve 50 mg of one of these acidic dyes in 15 mL of hot water to which a few drops of dilute sulfuric acid have been added. Heat a piece of test fabric in this bath for 1 min; then remove it with a stirring rod, rinse well, scrub with soap and water, and dry. Describe the results.

Eosin

Dissolve 10 mg of sodium eosin in 20 mL of water and dye a piece of test fabric by heating it with the solution for about 10 min. Eosin is the dye used in red ink. Also dye pieces of mordanted cloth in eosin (*see* Experiment 3 in this part). Wash and rinse the dyed cloth in the usual way.

Eosin A
(λ_{max} 516,483 nm)

Cleaning Up. In each case dilute the dye bath with a large quantity of water and flush it down the drain.

2. Substantive Dyes: Congo Red

Cotton and rayons do not have the anionic and cationic carboxyl and amine groups of wool and silk and hence do not dye well with direct dyes; however, cotton and rayons can be dyed with substances of rather high molecular weight showing colloidal properties. Such dyes probably become fixed to the fiber by hydrogen bonding: Congo Red is a substantive dye.

Dissolve 10 mg of Congo Red in 40 mL of water, add about 0.1 mL each of 3 M solutions of sodium carbonate and sodium sulfate, heat to a temperature just below the boiling point, and introduce a piece of test fabric. At the end of 10 min, remove the fabric and wash in warm water with soap as long as the dye continues to be removed. Place pieces of the dyed material in very dilute hydrochloric acid solution and observe the result. Rinse and wash the material with soap.

Congo Red
(λ_{max} 497 nm)

Cleaning Up. The dye bath should be diluted with water and flushed down the drain.

3. Mordant Dyes

One of the oldest known methods of producing wash-fast dyes involves using metallic hydroxides, which form a link, or mordant (Latin *mordere*, to bite), between the fabric and the dye. Other substances, such as tannic acid, also function as mordants. The color of the final product depends on both the dye used and the mordant. For instance, the dye Turkey Red (alizarin) is red with an aluminum mordant, violet with an iron mordant, and brownish-red with a chromium mordant. Some commercially important mordant dyes possess a structure based on triphenylmethane, as well as Crystal Violet and Malachite Green.

Chromium functioning
as a mordant

Alizarin
1,2-Dihydroxyanthraquinone

Applying Mordants–Tannic Acid, Iron, Tin, Chromium, Copper, and Aluminum

Mordant pieces of test fabric or wool yarn by allowing them to stand in a hot (nearly boiling) solution of 0.1 g of tannic acid in 50 mL of water for 30 min. The tannic acid mordant must now be fixed to the cloth; otherwise it would wash out. For this purpose, transfer the cloth or yarn to a hot bath made from 20 mg of potassium antimonyl tartrate [$K(SbO)C_4H_4O_6$; tartar emetic] in 20 mL of water. After 5 min, wring or blot the cloth and dry it as much as possible over a warm hot plate.

Mordant 1/2-in. strips of test cloth or yarn in the following mordants, which are 0.1 M solutions of the indicated salts: ferrous sulfate, stannous chloride, potassium dichromate, copper sulfate, and potassium aluminum sulfate (alum). (The alum and dichromate solutions should also contain 0.05 M oxalic acid.) Immerse pieces of cloth in the solutions, which are kept near the boiling point, for about 15–20 min or longer. These mordanted pieces of cloth or yarn can then be dyed with alizarin (1,2-dihydroxyanthraquinone, Turkey Red) and either Methyl Orange or Orange II in the usual way.

Cleaning Up. Mix the mordant baths. The Fe^{2+} and Sn^{2+} will reduce the Cr^{6+} to Cr^{3+}. The mixture can then be diluted with water and flushed down the drain because the quantity of metal ions is extremely small. Alternatively, precipitate the ions as the hydroxides, collect by vacuum filtration, and place the solid in the hazardous waste container.

Dyeing with a Triphenylmethane Dye: Crystal Violet or Malachite Green

A dye bath is prepared by dissolving 10 mg of either Crystal Violet or Malachite Green in 20 mL of boiling water. Dye the mordanted cloth or yarn in this bath for 5–10 min at a temperature just below the boiling point. Dye another piece of cloth that has not been mordanted and compare the two. In each case allow as much of the dye to drain back into the beaker as possible then, using glass rods, wash the dyed cloth or yarn under running water with soap, blot, and dry.

Cleaning Up. The stains on glass produced by triphenylmethane dyes can be removed with a few drops of concentrated hydrochloric acid and washing with water, as HCl forms a di- or trihydrochloride more soluble in water than the original monosalt.

The dye bath and acid washings are diluted with water and flushed down the drain because the quantity of dye is extremely small.

Malachite Green
(λ_{max} 617 nm)

Crystal Violet
(λ_{max} 591, 540 nm)

4. Developed Dyes

A superior method of applying azo dyes to cotton, patented in England in 1880, is that in which cotton is soaked in an alkaline solution of a phenol and then in an ice-cold solution of a diazonium salt; the azo dye is developed directly on the fiber. The reverse process (ingrain dyeing) of impregnating cotton with an amine, which is then diazotized and developed by immersion in a solution of the phenol, was introduced in 1887. The first ingrain dye was Primuline Red, obtained by coupling the sulfur dye Primuline, after application to the cloth and diazotization, with 2-naphthol. Primuline (substantive to cotton) is a complex thiazole prepared by heating *p*-toluidine with sulfur and then introducing a solubilizing sulfonic acid group.

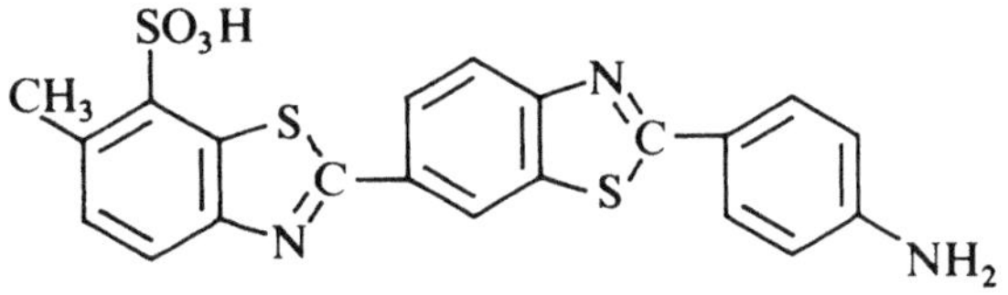

Primuline

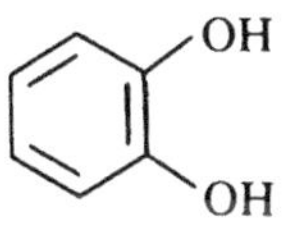

Primuline Red

Resorcinol

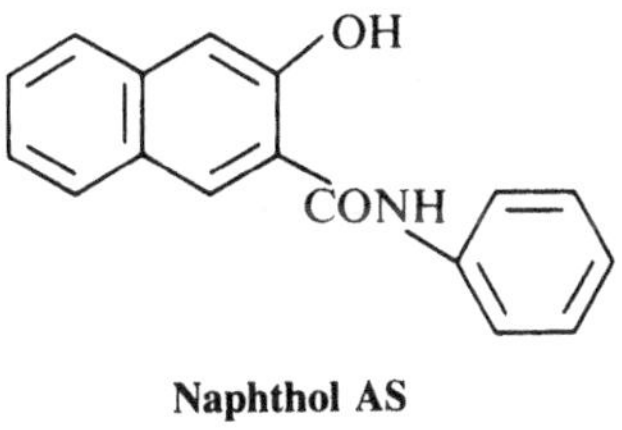

Naphthol AS

Dye three pieces of cotton cloth in a solution of 20 mg of Primuline and 0.5 mL of sodium carbonate solution in 50 mL of water at a temperature just below the boiling point for 15 min. Wash the cloth twice in about 50 mL of water. Prepare a diazotizing bath by dissolving 20 mg of sodium nitrite in 50 mL of water containing a little ice and, just before using the bath, add 0.5 mL of concentrated hydrochloric acid. Allow the cloth dyed with Primuline to stay in this diazotizing bath for about 5 min. Next, prepare three baths for the coupling reaction. Dissolve 10 mg of 2-naphthol in 0.2 mL of 1.5 *M* sodium hydroxide solution and dilute with 10 mL of water; prepare similar baths from phenol, resorcinol, Naphthol AS, or other phenolic substances.

Transfer the cloth from the diazotizing bath to a beaker containing about 50 mL of water and stir. Put one piece of cloth in each of the developing baths and allow them to remain for 5 min. Primuline coupled to 2-naphthol gives the dye called Primuline Red. Draw the structure of the dye.

Para Red, an Ingrain Color

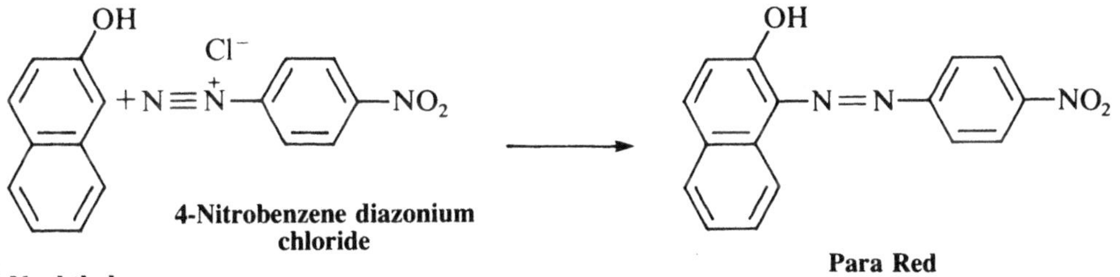

2-Naphthol

4-Nitrobenzene diazonium chloride

Para Red

A solution is prepared by suspending 300 mg of 2-naphthol in 10 mL of water, stirring well, and adding 3 *M* sodium hydroxide solution, a drop at a time, until

Para Red is the red dye used for
the American flag.

the naphthol just dissolves. Do not add excess alkali. The material to be dyed is first soaked in or painted with this solution and then dried, preferably in an oven.

Prepare a solution of 4-nitrobenzenediazonium chloride as follows: Dissolve 140 mg of 4-nitroaniline in a mixture of 3 mL of water and 0.6 mL of 3 M hydrochloric acid by heating. Cool the solution in ice (the hydrochloride of the amine may crystallize), add all at once a solution of 80 mg of sodium nitrite in about 0.5 mL of water, and stir. In about 10 min a clear solution of the diazonium salt will be obtained. Just before developing the dye on the cloth, add a solution of 80 mg of sodium acetate in 0.5 mL of cold water. Stir in the acetate well, add 30 mL of water, and immediately add the cloth. The diazonium chloride solution may also be painted onto the cloth.

Good results can be obtained by substituting Naphthol AS for 2-naphthol; in this case it is necessary to warm the Naphthol AS with alkali and to break the lumps with a flattened stirring rod to bring the naphthol into solution.

Cleaning Up. The dye baths should be diluted with water and flushed down the drain because the quantity of dye is extremely small.

5. Dyeing with Indigo: The Denim Dye

Indigo structure with reaction: Na₂S₂O₄ + NaOH / O₂ (air) converting to Indigo white

Indigo
Insoluble in water
Deep blue

Indigo white
The leuco form, soluble in water
Colorless or light yellow

Denim, de Nimes, from Nimes,
in southern France

Look around you. No matter where in the world you are carrying out these experiments, you will find people wearing cotton clothes dyed with indigo. Indigo, "the king of dyes," has been used for more than 3500 years, first in India and then in ancient Egypt. It is a unique dye in that it can be abraded from the surface of a fabric, where other dyes penetrate the fiber. This explains why the knees and other parts of blue jeans (dyed exclusively with indigo) will gradually turn white. An advantage of this fact has been taken by manufacturers who sell "stonewashed" denim blue jeans that have been tumbled with pumice to wear off some of the dye (which can then be reused).

This dye originally came from the leaves of the indigo plant, but it is now produced by chemical synthesis. It is one of the *vat dyes*, a term applied to dyes that are reduced to a colorless (leuco) form that is then oxidized on the surface of the fiber. Formerly, the reduction was carried out by fermenting plant leaves; now, chemical reducing agents, most commonly sodium hydrosulfite, are used.

Experimental Considerations

Indigo is a dark-blue powder that is completely insoluble in water and is not easily wet by water. To disperse the dye in water, a bit of soap is added to the dye bath. The liquid will appear blue, but the dye is not dissolved; it is merely suspended. Sodium hydrosulfite ($Na_2S_2O_4$) is then added to the hot dye bath. Because this reducing agent decomposes on storage, it is difficult to state exactly how much should be used. Add the required amount, stir, and then look through the side of the beaker. If the solution is not transparent and light yellow in color, add more of the reducing agent. From the top the liquid will still appear to be dark blue because of a film of the oxidized dye on the surface of the solution.

Procedure

Use 100 mg of indigo and a drop of detergent or a pinch of soap powder. Boil the dye with 50 mL of water, 2.5 mL of 3 *M* sodium hydroxide solution, and about 0.5 g of sodium hydrosulfite until the dye is reduced. At this point a clear solution will be seen through the side of the beaker. Add more sodium hydrosulfite if necessary to produce a clear solution. Introduce a piece of cloth and boil the solution gently for 10 min. Rinse the cloth well in water and then allow it to dry. To increase the intensity of the dye, repeat the process several times with no drying. Describe what happens during the drying process.

Other dyes that can be used in this procedure are Indanthrene Brilliant Violet and Indanthrene Yellow.

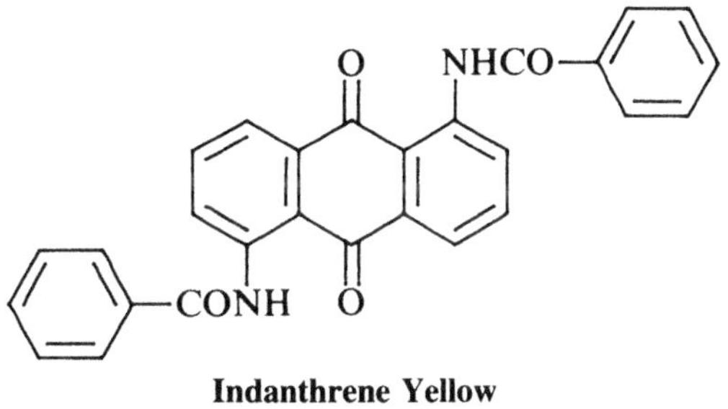

Indanthrene Yellow

Cleaning Up. Add household bleach (5.25% sodium hypochlorite solution) to the dye bath to oxidize it to the starting material. The mixture can be diluted with water and flushed down the drain; alternatively, the small amount of solid can be removed by filtration, with the solid placed in the aromatic amines hazardous waste container.

6. Disperse Dyes: Disperse Red

Fibers such as Dacron, acetate rayon, nylon, and polypropylene are difficult to dye with conventional dyes because they contain so few polar groups. These fibers are dyed with substances that are insoluble in water but that, at elevated temperatures

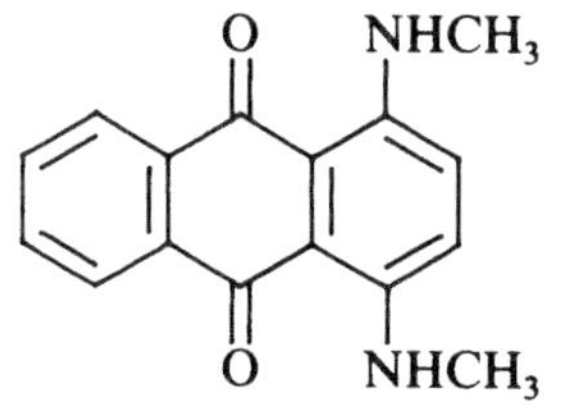

Celliton Fast Blue B

Celliton Fast Pink B

(pressure vessels), are soluble in the fiber as true solutions. They are applied to the fiber in the form of a dispersion of finely divided dye (hence the name). The Cellitons are typical disperse dyes.

Disperse Red, a brilliant red dye used commercially, is synthesized in this experiment.

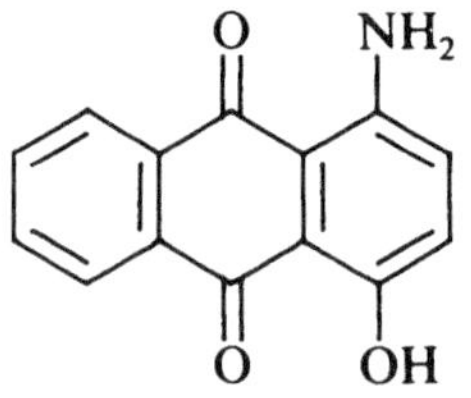

Diazotization of 2-Amino-6-methoxybenzothiazole

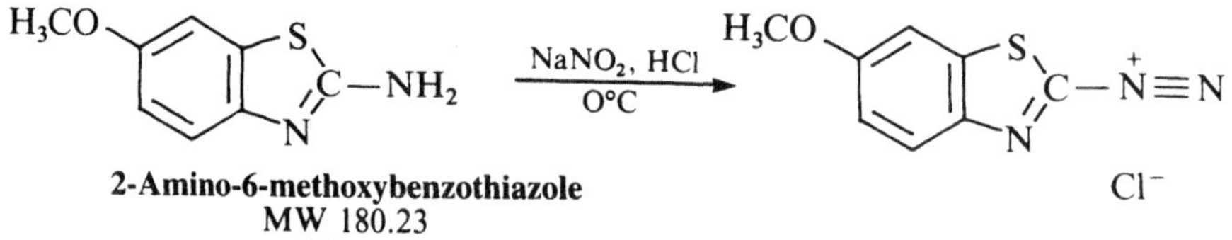

2-Amino-6-methoxybenzothiazole
MW 180.23
mp 165–167°C

To 135 mg (0.75 mmol) of 2-amino-6-methoxybenzothiazole in 1.5 mL of water in a test tube, add 0.175 mL of concentrated hydrochloric acid; then cool the solution to 0°C–5°C. To this mixture add, dropwise, an *ice-cold* solution of 55 mg of sodium nitrite that has been dissolved in 0.75 mL of water. The reaction mixture changes color, and some of the diazonium salt crystallizes out, but it should not foam. Foaming, caused by the evolution of nitrogen, is an indication the mixture is too warm. Keep the mixture ice-cold until used in the coupling reaction.

Disperse Red

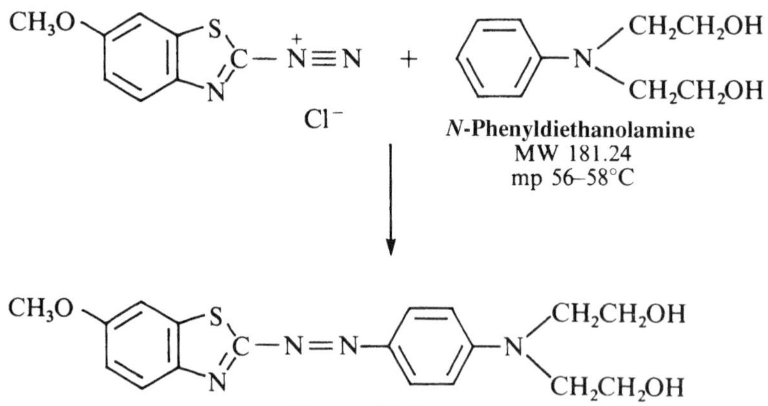

N-Phenyldiethanolamine
MW 181.24
mp 56–58°C

Disperse Red

To 135 mg (0.75 mmol) of N-phenyldiethanolamine in 0.75 mL of hot water, add just enough 3 M hydrochloric acid to bring the amine into solution. This amount is less than 0.5 mL. Cool the resulting solution to 0°C in ice, add to it, dropwise and with *very thorough mixing*, the diazonium chloride solution. Mix the solution well by drawing it into a Pasteur pipette and expelling it into a cold reaction tube. Allow the mixture to come to room temperature over a period of 10 min; then add 225 mg of sodium chloride and heat the mixture to boiling. The sodium chloride decreases the solubility of the product in water. Allowing the hot solution to cool slowly to room temperature should afford easily filterable crystals. Collect the dye on a Hirsch funnel, wash it with a few drops of saturated sodium chloride solution, and press it dry on the funnel.

Online Study Center
Video: Microscale Filtration
on the Hirsch Funnel

This reaction often produces a noncrystalline product that looks like purple tar, which is the dye. It can be used to dye the Multifiber Fabric, so do not discard the reaction mixture.

Save the filtrate. Add the material on the filter paper to 50 mL of boiling water and dye a piece of test cloth for 5 min. Do the same with the filtrate. Wash the cloth with soap and water, rinse and dry it, and compare the results.

Cleaning Up. The dye baths should be diluted with water and flushed down the drain because the quantity of dye is extremely small.

7. Test Identification Stain

In the dye industry, a proprietary mixture of three dyes called *Test Identification Stain*[4] is used to dye cloth. Prepare a dye bath by dissolving 25 mg of the dye in 25 mL of water, adding 1 drop of acetic acid, and bringing the mixture to a boil. Dye a piece of test cloth in the mixture for 5 min, remove it, immediately wash it under running water, and then scrub it *thoroughly* with soap and water before rinsing and drying. Describe the rather extraordinary result and suggest how this mixture could be used industrially. Several pieces of test cloth can be dyed in this one dye bath. Analyze the dye mixture by thin-layer chromatography, using 40:60 mixture of ethanol and hexane as the eluent.

Cleaning Up. Dilute the dye bath with copious amounts of water and flush it down the drain. Dyeing cloth removes much of the dye from the dye bath.

8. Fluorescein and Eosin

Fluorescein, as its name implies, is fluorescent; in fact, it is so fluorescent that the sodium salt in water can be detected under ultraviolet (UV) light at concentrations of 0.02 ppm. Because of this property, fluorescein is used to trace the paths of underground rivers such as those found in Mammoth Cave in Kentucky, to trace sources of contamination of drinking water, to find leaks in the huge condensers found in power plants, and so on. It is also used to visualize scratches on the cornea of the eye. The tetrabromo derivative is eosin, a dye used in lipstick, nail polish, red ink, and a biological stain.

4. The *Test Identification Stain* is available from Kontes.

The synthesis of fluorescein is very straightforward; however, in this experiment we have deliberately omitted the equations for the synthesis, the mechanism of the reaction, the structures of the molecules in acid and base, and the structures of the tetrabromo derivatives. These are left for you to solve or look up in the library.[5]

Synthesis of Fluorescein

In a reaction tube heat 100 mg of zinc chloride in a hot sand bath until no more moisture comes from the tube; then add 200 mg of resorcinol and 140 mg of phthalic anhydride. Place a thermometer into the mixture and heat the mixture to 180°C. Stir with a stirring rod and continue heating for 10 min, at which time the mixture should cease bubbling and become stiff. Cool the mixture to below 100°C, add 2 mL of water and 0.2 mL of concentrated hydrochloric acid, then stir and grind the solid (this process is called *trituration*) while heating it on a steam bath or boiling water bath (not on a sand bath because it will bump). Remove the liquid with a Pasteur pipette. After again triturating with another similar portion of water and acid, collect the solid on a Hirsch funnel, wash it well with water, and allow it to dry. A typical yield is 240 mg.

Fluorescence

Add 50 mg of fluorescein to 0.5 mL of 3 *M* sodium hydroxide and 0.5 mL of water; then dilute the mixture to 100 mL. Examine the solution by transmitted light and then by reflected sunlight or UV light. Make the solution acidic and again examine it under sunlight or UV light.

Tetrabromofluorescein, Eosin: The Red Ink Dye

In a reaction tube place 100 mg of fluorescein and 0.4 mL of ethanol. To this mixture add 215 mg of bromine in 0.5 mL of carbon tetrachloride, dropwise, mixing thoroughly after each drop. After half the bromine is added, a solution of soluble dibromofluorescein results. When all the bromine has been added, the tetrabromo compound will precipitate. After about 15 min, cool the reaction mixture in ice and remove the solvent with a Pasteur pipette. Wash the product in the reaction tube with a few drops of ice-cold ethanol; then dry it under vacuum. The yield should be about 160 mg.

Prepare the ammonium salt of eosin by exposing the salt to ammonia vapors. Moisten about 80 mg of fluorescein with a few drops of ethanol on a small filter paper. Fold up the paper and push it into a reaction tube above 0.5 mL of ammonium hydroxide. Cap the tube with a septum. After about an hour, a test portion of the derivative should be completely soluble in water. Dye a piece of test cloth with this dye. Eosin is used commercially to dye silk.

Online Study Center

Videos: Filtration of Crystals Using the Pasteur Pipette, Microscale Filtration on the Hirsch Funnel

---

5. If you would like to read about the original synthesis, see *J. Prakt. Chem.* **1922**, *104*, 123, which will require a reading knowledge of German, and the structure proof in *Comp. Rendu.* **1937**, *205*, 864, which will require a reading knowledge of French. Further searching in the library may disclose other papers in English.

Answer the following questions. The answers for some will come from your experimental observations:

1. Write a balanced equation for the reaction of resorcinol with phthalic anhydride.
2. Calculate the number of moles of starting material, determine the limiting reagent, and calculate the theoretical yield.
3. In view of the balanced equation, why is it necessary to dehydrate zinc chloride, a very hygroscopic substance?
4. To which general category of substances does zinc chloride belong that is relevant to its use in this reaction? What two functions does zinc chloride serve?
5. Write a possible mechanism for the synthesis of fluorescein.
6. Write an equation showing the reaction of fluorescein with sodium hydroxide. *Hint*: A quinoid structure is involved.
7. How do you account for the changes in fluorescence in going from an acidic to a basic medium?
8. Compare the color and fluorescence changes of fluorescein to those of the related compound, phenolphthalein.
9. Write a balanced equation for the bromination of fluorescein to give first the dibromo and then the tetrabromo compound. Explain your choices for the positions of bromination.
10. Write a balanced equation for the reaction of tetrabromofluorescein with ammonia.

9. Optical Brighteners: Fluorescent White Dyes

Most modern detergents contain a blue-white fluorescent dye that is adsorbed on the cloth during the washing process. These dyes fluoresce; that is, absorb UV light and reemit light in the visible blue region of the spectrum. This blue color counteracts the pale-yellow color of white goods, which develops because of a buildup of insoluble lipids. The modern-day usage of optical brighteners has replaced a past custom of using bluing (ferriferrocyanide).

Blankophor B
An optical brightener

Dyeing with Detergents

Immerse a piece of test fabric in a hot solution (0.5 g of detergent, 200 mL of water) of a commercial laundry detergent that you suspect may contain an optical brightener (e.g., Tide and Cheer) for 15 min. Rinse the fabric thoroughly, dry, and compare it to an untreated fabric sample under a UV lamp.

Cleaning Up. Dilute the solution with water and flush down the drain.

QUESTIONS

1. Write reactions showing how nylon can be synthesized so that it will react with (a) basic dyes and (b) acidic dyes.
2. Draw the resonance form of dimethylaniline that is most prone to react with diazotized sulfanilic acid.
3. Draw a resonance form of indigo that would be present in base.
4. Draw a resonance form of indigo that has been reduced and is therefore colorless.

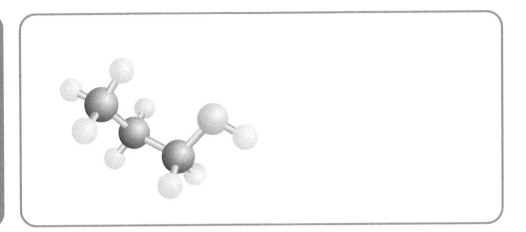

CHAPTER

47

Martius Yellow

PRELAB EXERCISE: Prepare a time line for this experiment, clearly indicating which experiments can be carried out simultaneously.

The experiments in this chapter were introduced by Louis Fieser of Harvard University over 70 years ago. Although the initial procedure starts with macroscale quantities of materials, the later steps often become microscale experiments and benefit greatly from using microscale equipment. It has long been the basis for an interesting laboratory competition. Ken Williamson, one of the present authors, was a winner over 50 years ago.

Starting with 5 g of 1-naphthol, a skilled chemist familiar with the procedures can prepare pure samples of the seven compounds in this chapter in 3–4 h. In a first trial of the experiments here, a particularly competent student, one who plans his or her work in advance, can complete the program in two laboratory periods (6 h).

The first compound of the series, Martius Yellow, a mothproofing dye for wool (1 g of Martius Yellow dyes 200 g of wool) discovered in 1868 by Karl Alexander von Martius, is the ammonium salt of 2,4-dinitro-1-naphthol (**1**). This compound is obtained by sulfonating 1-naphthol with sulfuric acid and treating the resulting disulfonic acid with nitric acid in an aqueous medium. The exchange of groups occurs with remarkable ease, and it is not necessary to isolate the disulfonic acid. The advantage of introducing the nitro groups in this indirect way is that 1-naphthol is highly sensitive to oxidation and would be partially destroyed by direct nitration. Martius Yellow is prepared by reacting the acidic phenolic group of compound **1** with ammonia to form the ammonium salt. A small portion of this salt (Martius Yellow) is converted by acidification and crystallization into pure 2,4-dinitro-1-naphthol, a sample of which is saved. The remainder is

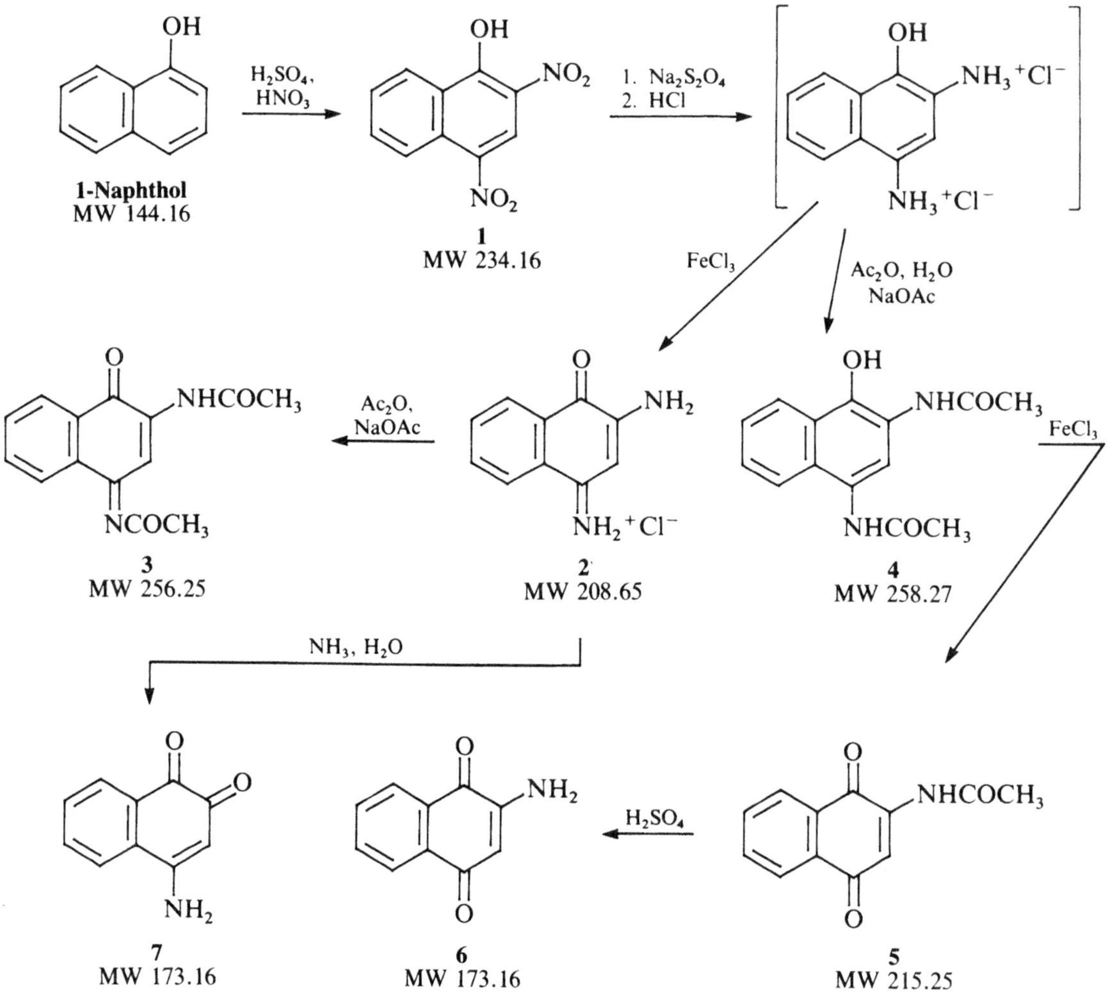
OH
1-Naphthol
MW 144.16
H_2SO_4, HNO_3
OH
NO_2
NO_2
1
MW 234.16
1. $Na_2S_2O_4$
2. HCl
OH
$NH_3^+Cl^-$
$NH_3^+Cl^-$
$FeCl_3$
Ac_2O, H_2O
NaOAc
O
$NHCOCH_3$
$NCOCH_3$
3
MW 256.25
Ac_2O, NaOAc
O
NH_2
$NH_2^+Cl^-$
2
MW 208.65
OH
$NHCOCH_3$
$NHCOCH_3$
4
MW 258.27
$FeCl_3$
NH_3, H_2O
O
O
NH_2
7
MW 173.16
O
NH_2
O
6
MW 173.16
H_2SO_4
O
$NHCOCH_3$
O
5
MW 215.25

suspended in water and reduced to diaminonaphthol with sodium hydrosulfite according to the following equation:

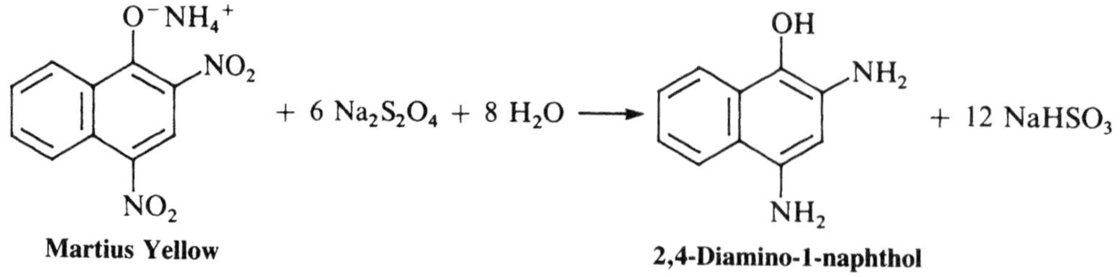
$O^-NH_4^+$
NO_2
NO_2
Martius Yellow
+ 6 $Na_2S_2O_4$ + 8 H_2O
OH
NH_2
NH_2
2,4-Diamino-1-naphthol
+ 12 $NaHSO_3$

The diaminonaphthol separates in the free condition, rather than as an ammonium salt. The diamine, unlike the dinitro compound, is a very weak acidic substance that does not readily form salts.

Because 2,4-diamino-1-naphthol is exceedingly sensitive to air oxidation as a free base, it is at once dissolved in dilute hydrochloric acid. The solution of diaminonaphthol dihydrochloride is clarified with decolorizing charcoal and divided into two equal parts. One part on oxidation with iron(III) chloride affords the fiery-red 2-amino-1,4-naphthoquinonimine hydrochloride (**2**). This substance, like many other salts, has no melting point; it is therefore converted for identification to the yellow diacetate (**3**). Compound **2** is remarkable in that it is stable enough to be isolated. On hydrolysis it affords the orange 4-amino-1,2-naphthoquinone (**7**).

The other part of the diaminonaphthol dihydrochloride solution is treated with acetic anhydride and then sodium acetate; the reaction in aqueous solution effects selective acetylation of the amino groups and affords 2,4-diacetylamino-1-naphthol (**4**). Oxidation of compound **4** by Fe^{3+} and oxygen from the air is attended with cleavage of the acetylamino group at the 4-position, and the product is 2-acetylamino-1,4-naphthoquinone (**5**). This yellow substance is hydrolyzed by sulfuric acid to the red 2-amino-1,4-naphthoquinone (**6**), the last member of the series. The reaction periods are brief, and the yields high; however, remember to scale down the quantities of the reagents and the solvents if the quantity of the starting material is less than that called for.[1]

EXPERIMENTS

If necessary, the quantities in the following seven experiments can be doubled.

1. Preparation of 2,4-Dinitro-1-Naphthol (1)

Use care when working with hot concentrated sulfuric and nitric acids.

Place 2.5 g of pure 1-naphthol[2] in a 50-mL Erlenmeyer flask, add 5 mL of concentrated sulfuric acid, and heat the mixture with swirling on a steam bath or a hot plate for 5 min, at which time the solid should have dissolved and an initial red color should be discharged. Cool in an ice bath, add 13 mL of water, and cool the solution rapidly to 15°C. Measure 3 mL of concentrated nitric acid into a test

---

1. This series of reactions lends itself to a laboratory competition, the rules for which might be as follows: (a) No practice or advance preparation is allowed except the collection of reagents not available at the contestant's bench [ammonium chloride, sodium hydrosulfite, iron(III) chloride solution, and acetic anhydride]. (b) The time scored is the actual working time, including that required for cleaning the apparatus and bench; labels for ziplock bags can be prepared outside of the working period. (c) Time is not charged during an interim period (overnight) when solutions are let stand to crystallize or solids are let to dry, on the condition that during this period no adjustments are made and no cleaning or other work is done. (d) Melting-point and color-test characterizations are omitted. (e) Successful completion of the contest requires preparing authentic and macroscopically crystalline samples of all seven compounds. (f) Judgment of the winners among the successful contestants is based on quality and quantity of samples, technique and neatness, and working time. (Superior performance is 3–4 h.)

2. If the 1-naphthol is dark, it can be purified by distillation at atmospheric pressure in the hood. The colorless distillate is most easily pulverized (also in the hood) before it has completely cooled and hardened.

OH
NO₂
NO₂

1

Avoid contact of the yellow product and its orange ammonium salt with the skin.

O⁻ NH₄⁺
NO₂
NO₂

Martius Yellow

tube and transfer it with a Pasteur pipette in small portions (0.25 mL each) to the chilled aqueous solution while maintaining the temperature between 15°C and 20°C by swirling the flask vigorously in the ice bath. When the addition is complete and the exothermic reaction has subsided (1–2 min), warm the mixture gently to 50°C (1 min), whereupon the nitration product should separate as a stiff yellow paste. Apply heat (the full heat of the steam bath for an additional 1 min if using a steam bath), fill the flask with water, break up the lumps and stir to an even paste, collect the product (**1**) on a small Büchner funnel, wash it well with water, and then wash it into a 250-mL beaker with water (50 mL). Add 75 mL of hot water and 2.5 mL of concentrated ammonium hydroxide solution (den. 0.90), heat to the boiling point, and stir to dissolve the solid. Filter the hot solution by suction if it is dirty, add 5 g of ammonium chloride to the filtrate to salt out the ammonium salt (Martius Yellow), cool in an ice bath, collect the orange salt, and wash it with water containing 1%–2% of ammonium chloride. The salt does not have to be dried (the dry weight yield is 3.8 g—a percent yield of 88%).

Set aside an estimated 150 mg of the moist ammonium salt. This sample is to be dissolved in hot water, the solution is acidified with HCl, and the free 2,4-dinitro-1-naphthol (**1**) is crystallized from methanol or ethanol (use decolorizing charcoal if necessary); it forms yellow needles (mp 138°C).

The proton nuclear magnetic resonance (¹H NMR) and infrared (IR) spectra of 1-naphthol are shown in Figures 47.1 and 47.2.

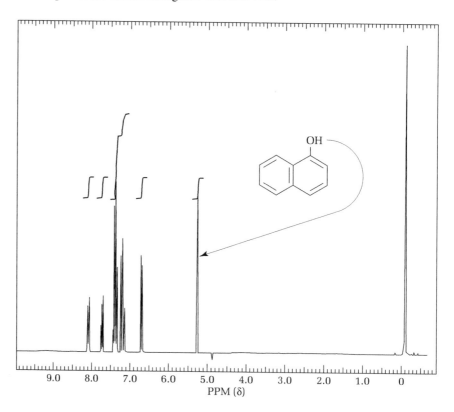

■ FIG. 47.1
The ¹H NMR spectrum of 1-naphthol (250 MHz).

■ **FIG. 47.2**
The IR spectrum of 1-naphthol
(KBr disk).

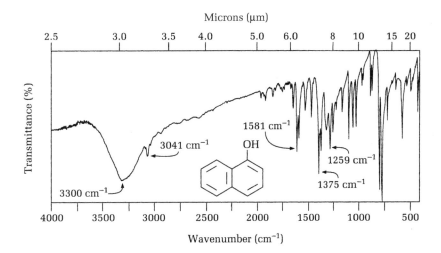

Cleaning Up. Combine aqueous filtrates, dilute with water, neutralize with sodium carbonate, and flush the solution down the drain. Recrystallization solvents go in the organic solvents container.

Preparation of Unstable 2,4-Diamino-1-Naphthol

$Na_2S_2O_4$
Sodium hydrosulfite

2,4-Diamino-1-naphthol dihydrochloride

Wash the remaining ammonium salt into a beaker with about 100 mL of water, add 20 g of sodium hydrosulfite, stir until the original orange color has disappeared and a crystalline tan precipitate has formed (5–10 min), and then cool in ice. Prepare two separate solutions: (1) 1 g of sodium hydrosulfite in 50 mL of water for use in washing and (2) a 250-mL beaker containing 3 mL of concentrated hydrochloric acid and 12 mL of water. When collecting the precipitate by suction filtration on a small Büchner funnel, use the hydrosulfite solution for rinsing and washing, be sure to avoid even briefly sucking air through the cake after the reducing agent has been drained away, and immediately wash the solid into the beaker containing the dilute hydrochloric acid and stir to convert all the diamine to the dihydrochloride.

   The acid solution, often containing suspended sulfur and filter paper, is clarified using suction filtration through a moist charcoal bed made by shaking 1 g of powdered decolorizing charcoal (carbon) with 13 mL of water in a stoppered flask to produce a slurry and pouring this on the filter paper in a 50-mm Büchner funnel. Pour out the water from the filter flask and then filter the solution of dihydrochloride. Divide the pink or colorless filtrate into approximately two equal parts and immediately add the reagents for converting one part to compound **2** (Experiment 2) and the other part to compound **4** (Experiment 4).

Cleaning Up. Neutralize the filtrate with sodium carbonate and dilute with water. In the hood, cautiously add household bleach (aqueous sodium hypochlorite solution) to the mixture until a test with 5% silver nitrate proves that no more hydrosulfite is present (the absence of a black precipitate). Neutralize the solution and filter it through Celite to remove any suspended solids. Dilute the filtrate with water and flush down the drain. The solid residue goes in the nonhazardous solid waste container.

2. Preparation of 2-Amino-1,4-naphthoquinonimine Hydrochloride (2)

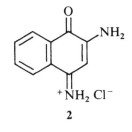

2

To one-half of the diamine dihydrochloride solution (Experiment 1), add 12.5 mL of 1.3 M iron(III) chloride solution,[3] cool in ice, and, if necessary, initiate crystallization by scratching. Rub the liquid film with a glass stirring rod at a single spot slightly above the surface of the liquid. If efforts to induce crystallization are unsuccessful, add more hydrochloric acid until crystallization occurs. Collect the red product and wash with 3 M HCl (aq). The dry weight yield is 1.2–1.35 g.

Divide the moist product into three equal parts; then spread out one part to dry for conversion to compound **3** (Experiment 3). The other two parts can be used while still moist for conversion to compound **7** (Experiment 7) and for recrystallization. Dissolve one part by warming in a little water containing 2–3 drops of hydrochloric acid, shake for a minute or two with decolorizing charcoal, filter by suction, and add concentrated hydrochloric acid to decrease the solubility. Collect the product by suction filtration.

Cleaning Up. Neutralize the filtrate with sodium carbonate and collect the iron hydroxide by vacuum filtration through Celite on a Büchner funnel. The solid goes into the nonhazardous solid waste container, whereas the filtrate is diluted with water and flushed down the drain.

3. Preparation of 2-Amino-1,4-naphthoquinonimine Diacetate (3)

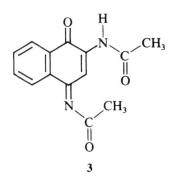

3

A mixture of 0.25 g of the dry quinonimine hydrochloride (**2**; Experiment 2), 0.25 g of sodium acetate (anhydrous), and 1.5 mL of acetic anhydride is stirred in a reaction tube and warmed gently on a sand or steam bath. With thorough stirring, the red salt should soon change into yellow crystals of the diacetate. The solution may appear red, but as soon as particles of red solid have disappeared, the mixture can be poured into about 5 mL of water. Stir until the excess acetic anhydride has either dissolved or become hydrolyzed, collect and wash the product (the dry weight yield is 250 mg) with water, and (drying is unnecessary) crystallize it from ethanol or methanol; yellow needles result (mp 189°C).

---

3. *Preparation of iron(III) chloride solution:* Dissolve 45 g of FeCl$_3$ · 6 H$_2$O (MW 270.32) in 50 mL of water and 50 mL of concentrated hydrochloric acid by warming, cooling, and filtration (produces 124 mL of solution).

Cleaning Up. The filtrate should be diluted with water and flushed down the drain. The crystallization solvent goes in the organic solvents container.

4. Preparation of 2,4-Diacetylamino-1-Naphthol (4)

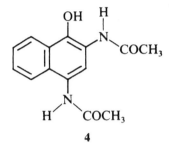

4

To one-half of the diaminonaphthol dihydrochloride solution saved from Experiment 1, add 1.5 mL of acetic anhydride, stir vigorously, and then add a solution of 1.5 g of sodium acetate (anhydrous) and about 50 mg of sodium hydrosulfite in 10–15 mL of water. The diacetate may precipitate as a white powder, or it may separate as an oil that solidifies when chilled in ice and rubbed with a stirring rod. Collect the product and, to hydrolyze any triacetate present, dissolve it in 2.5 mL of 3 M sodium hydroxide and 25 mL of water by stirring at room temperature. If the solution is colored, a few milligrams of sodium hydrosulfite may bleach it. Filter by suction and acidify by gradual addition of well-diluted hydrochloric acid (1 mL of concentrated acid in 19 mL of water). The diacetate tends to remain in a supersaturated solution; hence, either to initiate crystallization or to ensure maximum separation, it is advisable to stir well, rub the walls with a stirring rod, and cool in ice. Collect the product, wash it with water, and divide it into thirds (the dry weight yield is 1–1.3 g).

Two-thirds of the material can be converted without drying into compound **5**, and the other third can be used to prepare a crystalline sample. Dissolve the one-third reserved for crystallization (moist or dry) in enough hot acetic acid to bring about solution, add a solution of a small crystal of tin(II) chloride in a few drops of 10% HCl (aq) solution to inhibit oxidation, and dilute gradually with 5 to 6 volumes of water at the boiling point. Crystallization may be slow, and cooling and scratching may be necessary. The pure diacetate forms colorless prisms (mp 224°C, dec.).

Cleaning Up. Combine all filtrates—including the acetic acid used to crystallize the product—dilute with water, neutralize with sodium carbonate, and flush the solution down the drain.

5. Preparation of 2-Acetylamino-1,4-Napthoquinone (5)

5

Dissolve 0.75 g of the moist diacetylaminonaphthol (compound **4**; Experiment 4) in 5 mL of acetic acid (hot), dilute with 10 mL of hot water, and add 5 mL of 0.13 M iron(III) chloride solution. The product separates promptly into flat, yellow needles, which are collected (after cooling) and washed with a little alcohol; the yield is usually 0.6 g. Dry one-half of the product for conversion to compound **6** and crystallize the remaining half from 95% ethanol (mp 204°C).

Cleaning Up. Dilute the filtrate with water, neutralize it with sodium carbonate, and flush it down the drain. A negligible quantity of iron is disposed of in this way.

6. Preparation of 2-Amino-1,4-Napthoquinone (6)

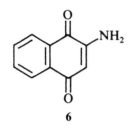

6

To 0.25 g of dry 2-acetylamino-1,4-naphthoquinone (**5**) contained in a reaction tube, add 1 mL of concentrated sulfuric acid and heat the mixture on a steam or sand bath with swirling to promote rapid solution (1–2 min). After an additional 5 min, cool the deep-red solution, dilute it with a large amount of water, and collect the precipitated product. Wash the crude material with water and recrystallize it from alcohol or an alcohol:water mixture[4] while it is still moist. The yield of red needles of the amino quinone (mp 206°C) will be about 200 mg.

Cleaning Up. The filtrate is neutralized with sodium carbonate, diluted with water, and flushed down the drain.

7. Preparation of 4-Amino-1,2-Naphthoquinone (7)

Compound **2** can be moist.

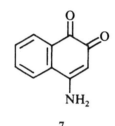

7

Dissolve 0.5 g of the aminonaphthoquinonimine hydrochloride (**2**) reserved from Experiment 2 in 12 mL of water, add 1 mL of concentrated ammonium hydroxide solution (den 0.90), and boil the mixture for 5 min. The free quinonimine initially precipitated is hydrolyzed to a mixture of the aminoquinone (**7**) and the isomeric compound **6**. Cool, collect the precipitate, suspend it in about 25 mL of water, and add 12.5 mL of 3 *M* sodium hydroxide solution. Stir well, remove the small amount of residual 2-amino-1,4-naphthoquinone (**6**) by filtration, and acidify the filtrate with acetic acid. The orange precipitate of compound **7** is collected, washed, and crystallized while still wet from 250–300 mL of hot water (the separation is slow). The yield of orange needles (dec. about 270°C) is about 200 mg.

Cleaning Up. The filtrate is neutralized with sodium carbonate, diluted with water, and flushed down the drain.

QUESTIONS

1. Write a balanced equation for the preparation of 2-amino-1,4-naphtho-quinonimine hydrochloride (**2**).
2. What part of the molecule accounts for the peak at 3300 cm^{-1} in the IR spectrum of 1-naphthol (Fig. 47.2)?

---

4. The exact ratio of alcohol:water is determined experimentally. Try different ratios, such as 50:50, 70:30, or 80:20. *See* Chapter 4 on how to choose a recrystallization solvent.

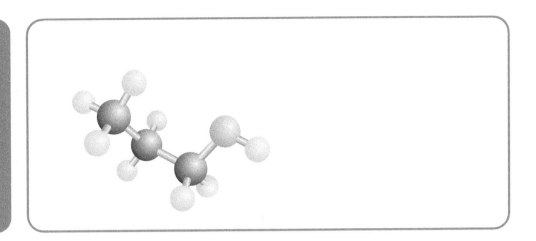

Ferrocene [Bis(cyclopentadienyl)iron]

Online Study Center

This icon will direct you to
techniques, equipment setups,
and online resources at
http://college.hmco.com/
PIC/williamsonMMOE5e.

PRELAB EXERCISE: Propose a detailed outline of the procedure
for synthesizing ferrocene, paying particular attention to the time
required for each step.

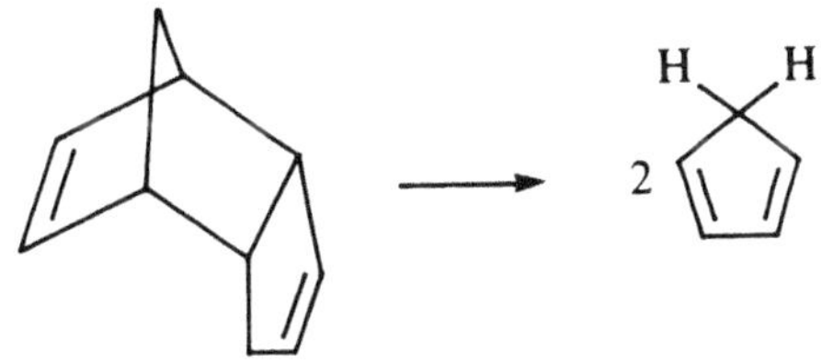

Dicyclopentadiene
den. 0.98, MW 132.20

Cyclopentadiene
bp 41°C, den. 0.80
MW 66.10

H   H

2

H   H

$+ \ KOH \longrightarrow$ $\ominus \ K^+ \ + \ H_2O$

**Potassium
cyclopentadienide**

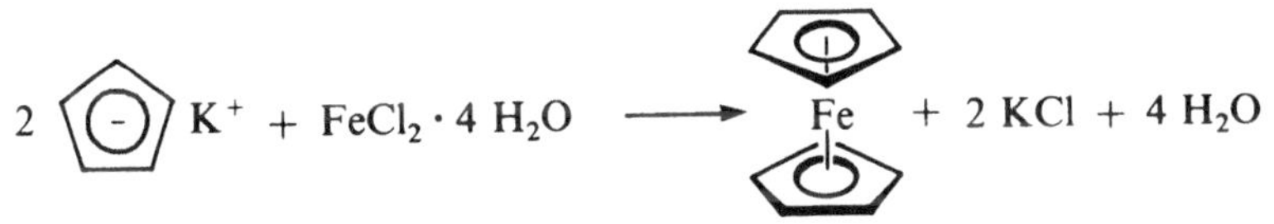

$2 \ \ominus \ K^+ \ + \ FeCl_2 \cdot 4 \ H_2O \longrightarrow$ Fe $+ \ 2 \ KCl \ + \ 4 \ H_2O$

Iron(II) chloride tetrahydrate
MW 198.81

**Ferrocene
Bis(cyclopentadienyl)iron**
MW 186.04, mp 172–174°C

he Grignard reagent is a classic organometallic compound. The magnesium ion, in Group IIA of the periodic table, needs to lose two—and only two—electrons to achieve the inert gas configuration. This metal has a strong tendency to form ionic bonds by electron transfer:

$$RBr + Mg \longrightarrow \overset{\delta^-}{R}\text{---}\overset{\delta^+}{Mg}Br$$

In the transition metals, the situation is not so simple. Consider the bonding between iron and carbon monoxide in $Fe(CO)_5$:

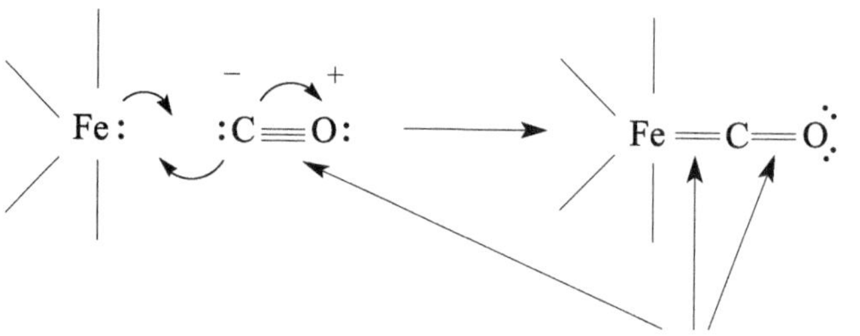

The pair of electrons on the carbon atom is shared with iron to form a σ (sigma) bond between the carbon and iron. The π (pi) bond between iron and carbon is formed from a pair of electrons in the *d* orbital of iron. The π bond is thus formed by the overlap of a *d* orbital of iron with the *p*-π bond of the carbonyl group. This mutual sharing of electrons results in a relatively nonpolar bond.

Iron has six electrons in the 3*d* orbital, two in the 4*s*, and none in the 4*p* orbital. The inert gas configuration requires 18 electrons—ten 3*d*, two 4*s*, and six 4*p* electrons. Iron pentacarbonyl enters this configuration by accepting two electrons from each of the five carbonyl groups, for a total of 18 electrons. Back-bonding of the *d*-π type distributes the excess electrons among the five carbon monoxide molecules.

Early attempts to form σ-bonded derivatives linking alkyl carbon atoms to iron were unsuccessful, but P. L. Pauson in 1951 succeeded in preparing a very stable substance, ferrocene, $C_{10}H_{10}Fe$, by reacting 2 mol of cyclopentadienyl-magnesium bromide with anhydrous ferrous chloride. Another group of chemists—Geoffrey Wilkinson, Myron Rosenblum, Mark Whiting, and Robert B. Woodward—recognized that the properties of ferrocene (its remarkable stability to water, acids, and air and its ease of sublimation) could be explained only if it had the structure depicted previously and that the bonding of the ferrous iron with its six electrons must involve all 12 of the π electrons on the two cyclopentadiene rings, with a stable 18-electron inert gas structure as the result.

In this chapter ferrocene is prepared by reacting the anion of cyclopentadiene with iron(II) chloride. Abstraction of one of the acidic allylic protons of cyclopentadiene with base gives the aromatic cyclopentadienyl anion. It is considered aromatic because it conforms to the Hückel rule in having $4n + 2\pi$

electrons (where n is 1). Two molecules of this anion will react with iron(II) chloride to give ferrocene, the most common member of the class of metal-organic compounds referred to as metallocenes. In this centrosymmetric sandwich-type π complex, all carbon atoms are equidistant from the iron atom, and the two cyclopentadienyl rings rotate more or less freely with respect to each other. The extraordinary stability of ferrocene (stable to 500°C) can be attributed to the sharing of the 12 π electrons of the two cyclopentadienyl rings with the 6 outer shell electrons of iron(II) to give the iron a stable 18-electron inert gas configuration. Ferrocene is soluble in organic solvents, can be dissolved in concentrated sulfuric acid and recovered unchanged, and is resistant to other acids and bases as well (in the absence of oxygen). This behavior is consistent with that of an aromatic compound; ferrocene can undergo electrophilic aromatic substitution reactions with ease.

Cyclopentadiene readily dimerizes at room temperature by a Diels–Alder reaction to give dicyclopentadiene. This dimer can be "cracked" by heating (an example of the reversibility of the Diels–Alder reaction) to give low-boiling cyclopentadiene. In most syntheses of ferrocene, the anion of cyclopentadiene is prepared by reacting the diene with metallic sodium. Subsequently, this anion is allowed to react with anhydrous iron(II) chloride. Here the anion is generated using powdered potassium hydroxide, which functions as both a base and a dehydrating agent.

The anion of cyclopentadiene decomposes rapidly in air and iron(II) chloride; although reasonably stable in the solid state, it is readily oxidized to the iron(III) (ferric) state in solution. Consequently, this reaction must be carried out in the absence of oxygen, which is accomplished by bubbling nitrogen gas through the solutions to displace dissolved oxygen and to flush air from the apparatus. Rather elaborate apparatus is used in research laboratories to carry out the experiment in the absence of oxygen. A very simple apparatus is used in this chapter because no gases are evolved, no heating is necessary, and the reaction is only mildly exothermic.

Ferrocene: soluble in organic solvents, stable to 500°C

EXPERIMENTS

1. Microscale Synthesis of Ferrocene

IN THIS EXPERIMENT one of the reactants and the ionic intermediate will react with atmospheric oxygen; therefore, the reaction is run under a nitrogen atmosphere. Potassium hydroxide is partially dissolved in a solvent, cyclopentadiene is added, and, on shaking and stirring, is converted to the colored anion. To this solution of the cyclopentadienyl anion is added a solution of iron(II) chloride to produce ferrocene. The mixture is poured onto ice-cold acid, and the crystalline product is collected on a Hirsch funnel. Once dry, the product is purified by sublimation.

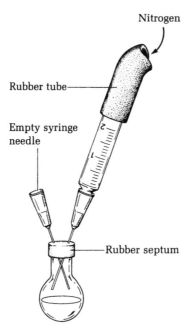

Nitrogen

Rubber tube

Empty syringe
needle

Rubber septum

■ FIG. 49.1
The apparatus for flushing air from
reaction flask. Check the nitrogen
flow rate by allowing it to bubble
through an organic solvent before
inserting the needle through the
septum.

CH₃OCH₂CH₂OCH₃

1,2-Dimethoxyethane
(Ethylene glycol dimethyl
ether, monoglyme), bp 85°C
Completely miscible with water

Online Study Center

Video: Ferrocene Synthesis

CAUTION: Potassium
hydroxide is extremely corrosive
and hygroscopic. Immediately
wash any spilled powder or
solutions from the skin and wipe
up all spills. Keep containers
tightly closed. Work in the hood.

CAUTION: Dimethyl sulfoxide
is rapidly absorbed through the
skin. Wash off spills with water.
Wear disposable gloves when
shaking the apparatus.

To a 5-mL short-necked, round-bottomed flask add a magnetic stirring bar
and then quickly add 0.75 g of finely powdered potassium hydroxide,[1] followed
by 1.25 mL of dimethoxyethane (DME). The funnel that is a part of the chromatography column makes a convenient addition funnel. Cap the flask with a
good septum and pass nitrogen into the flask or, better, through the solution for
about 1 min. This is done by connecting a tank of nitrogen via a rubber tube to a
22-gauge needle and adjusting the nitrogen flow to a few milliliters per minute by
bubbling it under a liquid such as acetone. With the nitrogen flow adjusted, insert
an empty syringe needle through the septum of the flask as an outlet and then insert the nitrogen inlet needle (Fig. 49.1). Remove the needles and shake the flask
to dislodge the solid from the bottom and to help dissolve some of it. If possible,
stir the mixture magnetically.

To a 10 × 100-mm reaction tube add 0.35 g of finely powdered green iron(II)
chloride tetrahydrate and 1.5 mL of dimethyl sulfoxide (DMSO). Cap the tube
with a good rubber septum, insert an empty syringe needle through the septum,
and pass nitrogen into the tube for about 1 min to displace the oxygen present.
Remove the needles and then shake the vial vigorously to dissolve all the iron
chloride. Some warming may be needed.

Using an accurate syringe, inject 0.300 mL of freshly prepared cyclopentadiene
(*see* Chapter 48 for the preparation of cyclopentadiene) into the flask containing the

---

1. Potassium hydroxide is easily ground to a fine powder in 25-g batches in 1 min, employing an ordinary
food blender (e.g., Waring, Osterizer). The finely powdered base is transferred in a hood to a bottle with a
tightly fitting cap. Alternatively, grind about 1 g potassium hydroxide in a mortar and transfer it rapidly to the
reaction flask.

O
‖
CH₃SCH₃

Dimethyl sulfoxide, DMSO
(Methyl sulfoxide), bp 189°C
Completely miscible with
water

Isolation of ferrocene

Purification by sublimation

potassium hydroxide. Do not grasp the body of the syringe because the heat of your hand will cause the cyclopentadiene to volatilize. Stir the mixture vigorously and note the color change as the potassium cyclopentadienide is formed. After waiting about 5 min for the anion to form, pierce the septum with an empty needle for pressure relief and inject the iron(II) chloride solution contained in the reaction tube in six 0.25-mL portions over a 10-min period. Stir the mixture well with a magnetic stirrer. If the stirring bar is immobilized, then between injections remove both needles from the septum and shake the flask vigorously. After all the iron(II) chloride solution has been added, rinse the reaction tube with an additional 0.25 mL of DMSO and add this to the flask. Continue to stir or shake the solution for about 15 min to complete the reaction.

To isolate the ferrocene, pour the dark slurry onto a mixture of 4.5 mL of 6 *M* hydrochloric acid and 5 g of ice in a 30-mL beaker. Stir the contents of the beaker thoroughly to dissolve and neutralize all the potassium hydroxide. Collect the crystalline orange ferrocene on a Hirsch funnel, wash the crystals well with water, press out excess water, squeeze the product between sheets of filter paper to complete the drying, and then purify the ferrocene by sublimation. The filtrate is blue because of dissolved ferrocinium ion. It can be reduced with a mild reducing agent, such as ascorbic acid, to regenerate ferrocene. The amount of ferrocene produced is usually negligible.

To sublime the ferrocene, add the crude dry product to a 25-mL filter flask equipped with a neoprene filter adapter (Pluro stopper) and a 15-mL centrifuge tube that is pushed to within 5 mm of the bottom of the flask (Fig. 49.2). Put a rubber bulb on the side arm of the flask; then add ice to the centrifuge tube and heat the flask on a sand bath to sublime the product. Tilting and rolling the filter flask in the hot sand will help drive ferrocene onto the centrifuge tube. Remove

■ FIG. 49.2
The apparatus for the sublimation of ferrocene.

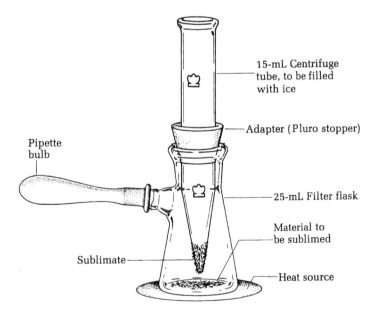

15-mL Centrifuge tube, to be filled with ice

Adapter (Pluro stopper)

25-mL Filter flask

Material to be sublimed

Heat source

Pipette bulb

Sublimate

■ **FIG. 49.3**
Evacuation of a melting point
capillary prior to sealing.

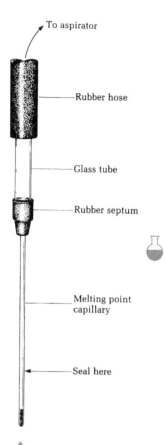

To aspirator

Rubber hose

Glass tube

Rubber septum

Melting point
capillary

Seal here

⚠

CAUTION: Potassium
hydroxide is extremely corrosive
and hygroscopic. Immediately
wash any spilled powder or
solutions from the skin and wipe
up all spills. Keep containers
tightly closed. Work in the hood.

CAUTION: Dimethoxyethane
can form peroxides. Discard
90 days after opening bottle
because of peroxide formation.

CAUTION: Dimethyl sulfoxide
is rapidly absorbed through the
skin. Wash off spills with water.

the flask from the sand bath and use a heat gun to drive the last of the ferrocene from the sides of the flask to the centrifuge tube. Ferrocene sublimes nicely at atmospheric pressure. Vacuum sublimation is not needed.

When sublimation is complete, cool the flask, remove the ice water from the centrifuge tube, and replace it with room temperature water (to prevent moisture from collecting on the tube). Transfer the product to a tared, stoppered vial, determine the weight, and calculate the percent yield. Determine the melting point in an evacuated capillary because the product sublimes at the melting point (Fig. 49.3) (*see* Chapter 3 for this technique).

Cleaning Up. The filtrate should be slightly acidic. Neutralize it with sodium carbonate, dilute it with water, and flush it down the drain. Place any unused cyclopentadiene in the recovered dicyclopentadiene or the organic solvents container. Add 0.4 mL concentrated nitric acid to the sublimation flask to clean it. After at least 24 hours, dilute the acid, neutralize it with sodium carbonate, and flush the solution down the drain.

2. Macroscale Synthesis of Ferrocene

Following the procedure described in Experiment 1 of Chapter 48, prepare 3 mL of cyclopentadiene. It need not be dry. While this distillation is taking place, rapidly weigh 12.5 g of finely powdered potassium hydroxide[2] into a 50-mL Erlenmeyer flask, add 30 mL of dimethoxyethane (DME), and immediately cool the mixture in an ice bath. Swirl the mixture in the ice bath for a minute or two; then bubble nitrogen through the solution for about 2 min. Quickly stopper the flask and shake the mixture to dislodge the cake of potassium hydroxide from the bottom of the flask and to dissolve as much of the base as possible (much will remain undissolved).

Grind 3.5 g of iron(II) chloride tetrahydrate to a fine powder, then add 3.5 g of the green salt to 12.5 mL of dimethyl sulfoxide (DMSO) in a 25-mL Erlenmeyer flask. Pass nitrogen through the DMSO mixture for about 2 min, stopper the flask, and shake it vigorously to dissolve all the iron(II) chloride. Gentle warming of the flask on a steam bath may be necessary to dissolve the last traces of iron(II) chloride. Transfer the solution rapidly to a 60-mL separatory funnel (serves as an addition funnel) equipped with a cork or stopper to fit the 50-mL Erlenmeyer flask, flush air from the funnel with a stream of nitrogen, and stopper it.

Transfer 3 mL of the freshly distilled cyclopentadiene to the slurry of potassium hydroxide in dimethoxyethane. Shake the flask vigorously and note the color change as the potassium cyclopentadienide is formed. After waiting about 5 min for the anion to form, replace the cork or stopper on the Erlenmeyer flask with the separatory funnel quickly (to avoid admission of air to the flask; Fig. 49.4). Figure 49.5 depicts a research-quality apparatus that would be used for this experiment.

■ FIG. 49.4
The apparatus for ferrocene synthesis.

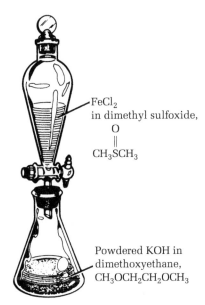

FeCl₂
in dimethyl sulfoxide,

$$CH_3 \overset{\overset{O}{\|}}{S} CH_3$$

Powdered KOH in
dimethoxyethane,
$CH_3OCH_2CH_2OCH_3$

■ FIG. 49.5
A research-quality apparatus for the preparation of ferrocene.

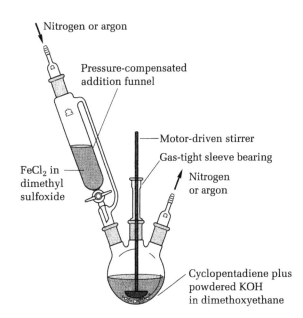

Nitrogen or argon

Pressure-compensated
addition funnel

$FeCl_2$ in
dimethyl
sulfoxide

Motor-driven stirrer

Gas-tight sleeve bearing

Nitrogen
or argon

Cyclopentadiene plus
powdered KOH
in dimethoxyethane

Add the iron(II) chloride solution to the base dropwise over a period of 20 min with vigorous swirling and shaking. Dislodge the potassium hydroxide should it cake on the bottom of the flask. The shaking will allow nitrogen to pass from the Erlenmeyer flask into the separatory funnel as the solution leaves the funnel.[3] Continue to shake and swirl the solution for 10 min after all the iron(II) chloride is added; then pour the dark slurry onto a mixture of 45 mL of 6 *M* hydrochloric acid and 50 g of ice in a 250-mL beaker. Stir the contents of the beaker thoroughly to dissolve and neutralize all the potassium hydroxide. Collect the crystalline orange ferrocene on a Büchner funnel, wash the crystals with water, press out excess water, and allow the product to dry overnight on a watch glass.

Recrystallize the ferrocene from methanol or, better, hexanes. It is also very easily sublimed. In a hood place about 0.5 g of crude ferrocene on a watch glass on a hot plate set to about 150°C. Invert a glass funnel over the watch glass. Ferrocene will sublime in about 1 h, leaving nonvolatile impurities behind. Pure ferrocene melts at 172°C–174°C. Determine the melting point in an evacuated capillary (Fig. 49.3) because the product sublimes at the melting point. Compare the melting points of your sublimed and recrystallized materials.

---

3. If the particular separatory funnel being used does not allow nitrogen to pass from the flask to the funnel, connect the two with a rubber tube leading to a glass tube and stopper at the top of the separatory funnel and to a syringe needle that pierces the rubber stopper in the flask (suggestion of D. L. Fishel).

Cleaning Up. The filtrate from the reaction mixture should be slightly acidic. Neutralize it with sodium carbonate, dilute it with water, and flush it down the drain. Place any unused cyclopentadiene in the recovered dicyclopentadiene or the organic solvents container. If the ferrocene has been crystallized from methanol or hexanes, place the mother liquor in the organic solvents container.

QUESTIONS

1. If ferrocene and all of the reagents in this experiment are stable in air before the reaction begins, why must air be so carefully excluded during the reaction?

2. What special properties do the solvents dimethoxyethane and dimethyl sulfoxide have compared to diethyl ether, for example, that make them particularly suited for this reaction?

3. What is it about ferrocene that allows it to sublime easily whereas many other compounds do not?

CHAPTER
70

Qualitative Organic Analysis

PRELAB EXERCISE: In the identification of an unknown organic compound, certain procedures are more valuable than others. For example, far more information is obtained from an IR spectrum than from a refractive index measurement. Outline, in order of priority, the steps you will employ in identifying your unknown.

Identification and characterization of the structures of unknown substances are an important part of organic chemistry. It is often, of necessity, a micro process, for example, in drug analyses. It is sometimes possible to establish the structure of a compound on the basis of spectra alone (IR, UV, and NMR), but these spectra must usually be supplemented with other information about the unknown: physical state, elementary analysis, solubility, and confirmatory tests for functional groups. Conversion of the unknown to a solid derivative of known melting point will often provide final confirmation of structure.

However, before spectra are run, other information about the sample must be obtained. Is it homogeneous (test by thin-layer, gas, or liquid chromatography)? What are its physical properties (melting point, boiling point, color, solubility in various solvents)? Is it soluble in a common NMR solvent? It might also be necessary to determine which elements are present in the sample and its percentage elemental composition (mass spectroscopy).

Nevertheless, an organic chemist can often identify a sample in a very short time by performing solubility tests and some simple tests for functional groups, coupled with spectra that have not been compared to a database. Conversion of the unknown to a solid derivative of known melting point will often provide final confirmation of structure. This chapter provides the information needed to carry out this type of qualitative analysis of an organic compound.

Procedures

All experiments in this chapter can, if necessary, be run on two to three times the indicated quantities of material.

Physical State

Check for Sample Purity

Distill or recrystallize as necessary. Constant boiling point and sharp melting point are indicators, but beware of azeotropes and eutectics. Check homogeneity by TLC, gas, HPLC, or paper chromatography.

Note the Color

Common colored compounds include nitro and nitroso compounds (yellow), α-diketones (yellow), quinones (yellow to red), azo compounds (yellow to red), and polyconjugated olefins and ketones (yellow to red). Phenols and amines are often brown to dark-purple because of traces of air oxidation products.

CAUTION: Do not taste an unknown compound. To note the odor, cautiously smell the cap of the container and do it only once.

Note the Odor

Some liquid and solid amines are recognizable by their fishy odors; esters are often pleasantly fragrant. Alcohols, ketones, aromatic hydrocarbons, and aliphatic olefins have characteristic odors. On the unpleasant side are thiols, isonitriles, and low-molecular weight carboxylic acids.

Make an Ignition Test

Heat a small sample on a spatula; first hold the sample near the side of a microburner to see if it melts normally and then burns. Heat it in the flame. If a large ashy residue is left after ignition, the unknown is probably a metal salt. Aromatic compounds often burn with a smoky flame.

Spectra

Obtain infrared and nuclear magnetic resonance spectra following the procedures of Chapters 12 and 13. If these spectra indicate the presence of conjugated double bonds, aromatic rings, or conjugated carbonyl compounds, obtain the UV spectrum following the procedures of Chapter 14. Interpret the spectra as fully as possible by reference to the sources cited at the end of the various spectroscopy chapters.

Explanation

Elementary Analysis, Sodium Fusion

This method for detection of nitrogen, sulfur, and halogen in organic compounds depends on the fact that fusion of substances containing these elements with sodium yields NaCN, Na_2S, and NaX (X = Cl, Br, I). These products can, in turn, be readily identified. The method has the advantage that the most usual elements other than C, H, and O present in organic compounds can all be detected following a single fusion, although the presence of sulfur sometimes

Rarely performed by professional chemists

interferes with the test for nitrogen. Unfortunately, even in the absence of sulfur, the test for nitrogen is sometimes unsatisfactory (nitro compounds in particular). Practicing organic chemists rarely perform this test. Either they know which elements their unknowns contain, or they have access to a mass spectrometer or atomic absorption instrument.

Place a 3-mm cube of sodium[1] (30 mg, no more)[2] in a 10 × 75-mm Pyrex test tube, and support the tube in a vertical position (Fig. 70.1). Have a microburner with small flame ready to move under the tube, place an estimated 20 mg of solid on a spatula or knife blade, put the burner in place, and heat until the sodium first melts and then vapor rises 1.5–2.0 cm in the tube. Remove the burner, and at once drop the sample onto the hot sodium. If the substance is a liquid add 2 drops of it. If there is a flash or small explosion the fusion is complete; if not, heat briefly to produce a flash or a charring. Then let the tube cool to room temperature, be sure it is cold, add a drop of methanol, and let it react. Repeat until 10 drops have been added. With a stirring rod break up the char to uncover sodium. When you are sure that all sodium has reacted, empty the tube into a 13 × 100-mm test tube, hold the small tube pointing away from you or a neighbor, and pipette into it 1 mL of water. Boil and stir the mixture, and pour the water into the larger tube; repeat with 1 mL more water. Then transfer the solution with a Pasteur pipette to a 2.5-cm funnel (fitted with a fluted filter paper) resting in a second 13 × 100-mm test tube. Portions of the alkaline filtrate are used for the tests that follow.

CAUTION: Manipulate sodium with a knife and forceps; never touch it with the fingers. Wipe it free of kerosene with a dry towel or filter paper; return scraps to the bottle or destroy scraps with methyl or ethyl alcohol, never with water. Safety glasses! Hood!

Do not use $CHCl_3$ or CCl_4 as samples in sodium fusion. They react extremely violently.

(a) Nitrogen

Run each test on a known and an unknown.

The test is done by boiling a portion of the alkaline solution from the solution fusion with iron(II) sulfate and then acidifying. Sodium cyanide reacts with iron(II) sulfate to produce ferrocyanide, which combines with iron(III) salts, inevitably formed by air oxidation in the alkaline solution, to give insoluble Prussian Blue, $NaFe[Fe(CN)_6]$. Iron(II) and iron(III) hydroxide precipitate along with the blue pigment but dissolve on acidification.

Place 50 mg of powdered iron(II) sulfate (this is a large excess) in a 10 × 75-mm test tube, add 0.5 mL of the alkaline solution from the fusion, heat the mixture gently with shaking to the boiling point, and then—without cooling—acidify with dilute sulfuric acid (hydrochloric acid is unsatisfactory). A deep-blue precipitate indicates the presence of nitrogen. If the coloration is dubious, filter through a 2.5-cm funnel and see if the paper shows blue pigment.

Cleaning Up Dilute the test solution with water and flush down the drain.

Notes for the instructor

1. Sodium spheres $\frac{1}{16}''$ to $\frac{1}{4}''$ are convenient.

2. A dummy 3-mm cube of rubber can be attached to the sodium bottle to indicate the correct amount.

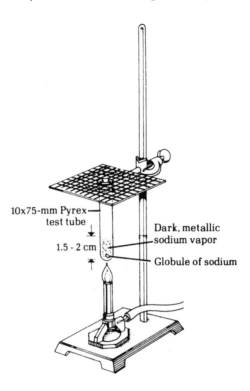

10x75-mm Pyrex test tube

1.5 - 2 cm

Dark, metallic sodium vapor

Globule of sodium

FIG. 70.1 Sodium fusion, just prior to addition of sample.

(b) Sulfur

1. Dilute 1 drop of the alkaline solution with 1 mL of water, and add a drop of sodium nitroprusside; a purple coloration indicates the presence of sulfur.
2. Prepare a fresh solution of sodium plumbite by adding 10% sodium hydroxide solution to 0.2 mL of 0.1 *M* lead acetate solution until the precipitate just dissolves, and add 0.5 mL of the alkaline test solution. A black precipitate or a colloidal brown suspension indicates the presence of sulfur.

$$Na_2(NO)Fe(CN)_6 \cdot 2H_2O$$

Sodium nitroprusside

Cleaning Up   Dilute the test solution with water and flush down the drain.

(c) Halogen

Differentiation of the halogens

Do not waste silver nitrate.

Acidify 0.5 mL of the alkaline solution from the fusion with dilute nitric acid (indicator paper) and, if nitrogen or sulfur has been found present, boil the solution (hood) to expel HCN or H_2S. On addition of a few drops of silver nitrate solution, halide ion is precipitated as silver halide. Filter with minimum exposure to light on a 2.5-cm funnel, wash with water, and then with 1 mL of concentrated ammonia solution. If the precipitate is white and readily soluble in ammonium hydroxide solution it is AgCl; if it is pale yellow and not readily soluble it is AgBr; if it is yellow and insoluble it is AgI. Fluorine is not detected in this test since silver fluoride is soluble in water.

Cleaning Up Dilute the test solution with water and flush down the drain.

> Run tests on knowns in parallel with unknowns for all qualitative organic reactions. In this way, interpretations of positive reactions are clarified and defective test reagents can be identified and replaced.

Beilstein Test for Halogens

A fast, easy, reliable test

Heat the tip of a copper wire in a burner flame until no further coloration of the flame is noticed. Allow the wire to cool slightly, then dip it into the unknown (solid or liquid), and again heat it in the flame. A green flash is indicative of chlorine, blue-green of bromine, and blue of iodine; fluorine is not detected because copper fluoride is not volatile. The Beilstein test is very sensitive; halogen-containing impurities may give misleading results. Run the test on a compound known to contain halogen for comparison to your unknown.[3]

Solubility Tests

Weigh and measure carefully.

Like dissolves like; a substance is most soluble in that solvent to which it is most closely related in structure. This statement serves as a useful classification scheme for all organic molecules. The solubility measurements are done at room temperature with 1 drop of a liquid, or 5 mg of a solid (finely crushed), and 0.2 mL of solvent. The mixture should be rubbed with a rounded stirring rod and shaken vigorously. Lower members of a homologous series are easily classified; higher members become more like the hydrocarbons from which they are derived.

If a very small amount of the sample fails to dissolve when added to some of the solvent, it can be considered insoluble; and, conversely, if several portions dissolve readily in a small amount of the solvent, the substance is obviously soluble.

If an unknown seems to be more soluble in dilute acid or base than in water, the observation can be confirmed by neutralization of the solution; the original material will precipitate if it is less soluble in a neutral medium.

If both acidic and basic groups are present, the substance may be amphoteric and therefore soluble in both acid and base. Aromatic aminocarboxylic acids are amphoteric, like aliphatic ones, but they do not exist as zwitterions. They are soluble in both dilute hydrochloric acid and sodium hydroxide, but not in bicarbonate solution. Aminosulfonic acids exist as zwitterions; they are soluble in alkali but not in acid.

The solubility tests are not infallible and many borderline cases are known. Carry out the tests according to the scheme of Fig. 70.2 and the following "Notes to Solubility Tests," and tentatively assign the unknown to one of the groups I–X.

---

3. http://odin.chemistry.uakron.edu/organic_lab/beil/
With seven very good color photos from the University of Akron, the Beilstein test is clearly demonstrated on this Web site. The dramatic differences among chlorine (green), bromine (blue-green), and iodine (blue) are quite clearly seen.

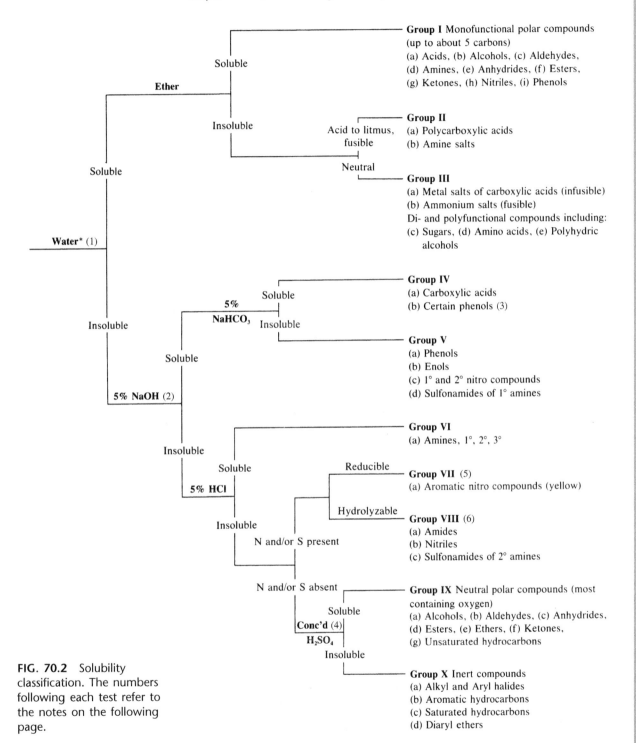

FIG. 70.2 Solubility classification. The numbers following each test refer to the notes on the following page.

Cleaning Up Because the quantities of material used in these tests are extremely small, and because no hazardous substances are handed out as unknowns, it is possible to dilute the material with a large quantity of water and flush it down the drain.

Notes to Solubility Tests

See Fig. 70.2.

1. Groups I, II, III (soluble in water). Test the solution with pH paper. If the compound is not easily soluble in cold water, treat it as water insoluble but test with indicator paper.
2. If the substance is insoluble in water but dissolves partially in 5% sodium hydroxide, add more water; the sodium salts of some phenols are less soluble in alkali than in water. If the unknown is colored, be careful to distinguish between the *dissolving* and the *reacting* of the sample. Some quinones (colored) *react* with alkali and give highly colored solutions. Some phenols (colorless) *dissolve and then* become oxidized to give colored solutions. Some compounds (e.g., benzamide) are hydrolyzed with such ease that careful observation is required to distinguish them from acidic substances.
3. Nitrophenols (yellow), aldehydophenols, and polyhalophenols are sufficiently strongly acidic to react with sodium bicarbonate.
4. Oxygen- and nitrogen-containing compounds form oxonium and ammonium ions in concentrated sulfuric acid and dissolve.
5. On reduction in the presence of hydrochloric acid, these compounds form water-soluble amine hydrochlorides. Dissolve 250 mg of tin(II) chloride in 0.5 mL of concentrated hydrochloric acid, add 50 mg of the unknown, and warm. The material should dissolve with the disappearance of the color and give a clear solution when diluted with water.
6. Most amides can be hydrolyzed by short boiling with 10% sodium hydroxide solution; the acid dissolves with evolution of ammonia. Reflux 100 mg of the sample and 10% sodium hydroxide solution for 15–20 min. Test for the evolution of ammonia, which confirms the elementary analysis for nitrogen and establishes the presence of a nitrile or amide.

Classification Tests

After the unknown is assigned to one of the solubility groups (Fig. 70.2) on the basis of solubility tests, the possible type should be further narrowed by application of classification tests; for example, for alcohols, or methyl ketones, or esters.

Complete Identification—Preparation of Derivatives

Once the unknown has been classified by functional group, the physical properties should be compared with those of representative members of the group (see tables at the end of this chapter). Usually, several possibilities present themselves, and the choice can be narrowed by preparation of derivatives. Select derivatives that distinguish most clearly among the possibilities.

Classification Tests

Group I. Monofunctional Polar Compounds (up to ca. 5 carbons)

(a) Acids

(Table 70.1, page 784; Derivatives, page 779)
No classification test is necessary. Carboxylic and sulfonic acids are detected by testing aqueous solutions with litmus. Acyl halides may hydrolyze during the solubility test.

(b) Alcohols

(Table 70.2, page 786; Derivatives, pages 779–780)

Jones' Oxidation.   Dissolve 5 mg of the unknown in 0.5 mL of pure acetone in a test tube, and add to this solution 1 small drop of Jones' reagent (chromic acid in sulfuric acid). A positive test is formation of a green color within 5 sec upon addition of the orange-yellow reagent to a primary or secondary alcohol. Aldehydes also give positive tests, but tertiary alcohols do not.

CAUTION: Cr^{+6} dust is toxic.

   Reagent: Dissolve/suspend 13.4 g of chromium trioxide in 11.5 mL of concentrated sulfuric acid, and add this carefully with stirring to enough water to bring the volume to 50 mL.

Cleaning Up   Place the test solution in the hazardous waste container.

Handle dioxane with care. It is a suspected carcinogen.

Cerium(IV) Nitrate Test [Ammonium Hexanitratocerium(IV) Test]. Dissolve 15 mg of the unknown in a few drops of water or dioxane in a reaction tube. Add to this solution 0.25 mL of the reagent, and mix thoroughly. Alcohols cause the reagent to change from yellow to red.

   Reagent: Dissolve 22.5 g of ammonium hexanitratocerium(IV), $Ce(NH_4)_2(NO_3)_6$, in 56 mL of 2 *N* nitric acid.

Cleaning Up   Dilute the solution with water and flush down the drain.

(c) Aldehydes

(Table 70.3, page 787; Derivatives, pages 442 and 444; ch. 36)

2,4-Dinitrophenylhydrazones.   All aldehydes and ketones readily form bright-yellow to dark-red 2,4-dinitrophenylhydrazones. Yellow derivatives are formed from isolated carbonyl groups and orange-red to red derivatives from aldehydes or ketones conjugated with double bonds or aromatic rings.

   Dissolve 10 mg of the unknown in 0.5 mL of ethanol, and then add 0.75 mL of 2,4-dinitrophenylhydrazine reagent. Mix thoroughly and let sit for a few minutes. A yellow to red precipitate is a positive test.

Reagent: Dissolve 1.5 g of 2,4-dinitrophenylhydrazine in 7.5 mL of concentrated sulfuric acid. Add this solution, with stirring, to a mixture of 10 mL of water and 35 mL of ethanol.

Cleaning Up   Place the test solution in the hazardous waste container.

Schiff Test.   Add 1 drop (30 mg) of the unknown to 1 mL of Schiff's reagent. A magenta color will appear within 10 min with aldehydes. Compare the color of your unknown with that of a known aldehyde.

Reagent: Prepare 50 mL of a 0.1% aqueous solution of *p*-rosaniline hydrochloride (fuchsin). Add 2 mL of a saturated aqueous solution of sodium bisulfite. After 1 h add 1 mL of concentrated hydrochloric acid.

Bisulfite Test.   Follow the procedure in Chapter 36. Nearly all aldehydes and most methyl ketones form solid, water-soluble bisulfite addition products.

Destroy used Tollens' reagent promptly with nitric acid. It can form explosive fulminates.

Tollens' Test.   Follow the procedure in Chapter 36. A positive test, deposition of a silver mirror, is given by most aldehydes, but not by ketones.

(d) Amides and Amines

(Tables 70.4, 70.5, and 70.6, pages 788–791; Derivatives of amines, pages 780–781)

Hinsberg Test.   Follow the procedure in Chapter 43, using benzenesulfonyl chloride to distinguish between primary, secondary, and tertiary amines.

(e) Anhydrides and Acid Halides

(Table 70.7, page 791; Derivatives, page 781–782) Anhydrides and acid halides will react with water to give acidic solutions, detectable with litmus paper. They easily form benzamides and acetamides.

Acidic Iron(III) Hydroxamate Test.   With iron(III) chloride alone a number of substances give a color that can interfere with this test. Dissolve 2 drops (or about 30 mg) of the unknown in 1 mL of ethanol, and add 1 mL of 1 *N* hydrochloric acid followed by 1 drop of 10% aqueous iron(III) chloride solution. If any color except yellow appears you will find it difficult to interpret the results from the following test.

Add 2 drops (or about 30 mg) of the unknown to 0.5 mL of a 1 *N* solution of hydroxylamine hydrochloride in alcohol. Add 2 drops of 6 *M* hydrochloric acid to the mixture, warm it slightly for 2 min, and boil it for a few seconds. Cool the solution, and add 1 drop of 10% ferric chloride solution. A red-blue color is a positive test.

Cleaning Up   Neutralize the reaction mixture with sodium carbonate, dilute with water, and flush down the drain.

(f) Esters

(Table 70.8, page 792. Derivatives are prepared from component acid and alcohol obtained on hydrolysis.)

Esters, unlike anhydrides and acid halides, do not react with water to give acidic solutions and do not react with acidic hydroxylamine hydrochloride. They do, however, react with alkaline hydroxylamine.

Alkaline Iron(III) Hydroxamate Test.   First test the unknown with iron(III) chloride alone. [See under Group I(e), Acidic Iron(III) Hydroxamate Test.]

To a solution of 1 drop (30 mg) of the unknown in 0.5 mL of 0.5 N hydroxylamine hydrochloride in ethanol, add 2 drops of 20% sodium hydroxide solution. Heat the solution to boiling, cool slightly, and add 1 mL of 1 N hydrochloric acid. If cloudiness develops add up to 1 mL of ethanol. Add 10% iron(III) chloride solution dropwise with thorough mixing. A red-blue color is a positive test. Compare your unknown with a known ester.

Cleaning Up   Neutralize the solutions with sodium carbonate, dilute with water, and flush down the drain.

(g) Ketones

(Table 70.14, page 796; Derivatives, pages 442 and 444)

2,4-Dinitrophenylhydrazone.   See under Group I(c), Aldehydes. All ketones react with 2,4-dinitrophenylhydrazine reagent.

Iodoform Test for Methyl Ketones.   Follow the procedure in Chapter 36.

A positive iodoform test is given by substances containing the $CH_3\overset{\displaystyle O}{\overset{\displaystyle \|}{C}}-$ group or by compounds easily oxidized to this group, e.g., CH_3COR, CH_3CHOHR, CH_3CH_2OH, CH_3CHO, $RCOCH_2COR$. The test is negative for compounds of the structure CH_3COOR, CH_3CONHR, and other compounds of similar structure that give acetic acid on hydrolysis. It is also negative for $CH_3COCH_2CO_2R$, CH_3COCH_2CN, $CH_3COCH_2NO_2$.

Bisulfite Test.   Follow the procedure in Chapter 36. Aliphatic methyl ketones and unhindered cyclic ketones form bisulfite addition products. Methyl aryl ketones, such as acetophenone, $C_6H_5COCH_3$, fail to react.

(h) Nitriles

(Table 70.15, page 797. Derivatives prepared from the carboxylic acid obtained by hydrolysis.)

At high temperature nitriles (and amides) are converted to hydroxamic acids by hydroxylamine:

$$RCN + 2\ H_2NOH \longrightarrow RCONHOH + NH_3$$

The hydroxamic acid forms a red-blue complex with iron(III) ion. The unknown must first give a negative test with hydroxylamine at lower temperature [Group I(f), Alkaline Iron(III) Hydroxamate Test] before trying this test.

Hydroxamic Acid Test for Nitriles (and Amides).

To 1 mL of a 1 M hydroxylamine hydrochloride solution in propylene glycol add 15 mg of the unknown dissolved in the minimum amount of propylene glycol. Then add 0.5 mL of 1 N potassium hydroxide in propylene glycol, and boil the mixture for 2 min. Cool the mixture, and add 0.1 to 0.25 mL of 10% iron(III) chloride solution. A red-blue color is a positive test for almost all nitrile and amide groups, although benzanilide fails to give a positive test.

Cleaning Up   Because the quantity of material is extremely small, the test solution can be diluted with water and flushed down the drain.

(i) Phenols

(Table 70.17, page 798; Derivatives, page 783)

Iron(III) Chloride Test.

Dissolve 15 mg of the unknown compound in 0.5 mL of water or water-alcohol mixture, and add 1–2 drops of 1% iron(III) chloride solution. A red, blue, green, or purple color is a positive test.

Cleaning Up   Because the quantity of material is extremely small, the test solution can be diluted with water and flushed down the drain.

Caution: $CHCl_3$ is a carcinogen.

A more sensitive test for phenols consists of dissolving or suspending 15 mg of the unknown in 0.5 mL of chloroform and adding 1 drop of a solution made by dissolving 0.1 g of iron(III) chloride in 10 mL of chloroform. Addition of a drop of pyridine, with stirring, will produce a color if phenols or enols are present.

Group II. Water-Soluble Acidic Salts, Insoluble in Ether

Amine Salts

[Table 70.5 (1° and 2° amines), pages 789–790; Table 70.6 (3° amines), page 791]

The free amine can be liberated by addition of base and extraction into ether. Following evaporation of the ether, the Hinsberg test, Group I(d), can be applied to determine if the compound is a primary, secondary, or tertiary amine.

The acid iron(III) hydroxamate test, Group I(d), can be applied directly to the amine salt (see the Hinsburg test, page 770).

Group III. Water-Soluble Neutral Compounds, Insoluble in Ether

(a) Metal Salts of Carboxylic Acids

(Table 70.1, carboxylic acids, page 784; Derivatives, page 779)

The free acid can be liberated by addition of acid and extraction into an appropriate solvent, after which the carboxylic acid can be characterized by mp or bp before proceeding to prepare a derivative.

(b) Ammonium Salts

(Table 70.1, carboxylic acids, page 784; Derivatives, page 779)

Ammonium salts on treatment with alkali liberate ammonia, which can be detected by its odor and the fact that it will turn red litmus to blue. A more sensitive test utilizes the copper(II) ion, which is blue in the presence of ammonia [see Group VIII a(i)]. Ammonium salts will not give a positive hydroxamic acid test (Ih) as given by amides.

(c) Sugars

See Chapter 36 for Tollens' test and Chapter 63 for phenylosazone formation.

(d) Amino Acids

Add 2 mg of the suspected amino acid to 1 mL of ninhydrin reagent, boil for 20 sec, and note the color. A blue color is a positive test.

Reagent: Dissolve 0.2 g of ninhydrin in 50 mL of water.

Cleaning Up   Because the quantity of material is extremely small, the test solution can be diluted with water and flushed down the drain.

(e) Polyhydric Alcohols

(Table 70.2, page 786; Derivatives, pages 779–780)

Periodic Acid Test for vic-Glycols.[4]   Vicinal glycols (hydroxyl groups on adjacent carbon atoms) can be detected by reaction with periodic acid. In addition to 1,2-glycols, a positive test is given by α-hydroxy aldehydes, α-hydroxy ketones, α-hydroxy acids, and α-amino alcohols, as well as 1,2-diketones.

To 2 mL of periodic acid reagent add 1 drop (no more) of concentrated nitric acid and shake. Then add 1 drop or a small crystal of the unknown. Shake

---

4. R. L. Shriner, R. C. Fuson, D. Y. Curtin, and T. C. Morill. *The Systematic Identification of Organic Compounds,* 6th ed., JohnWiley & Sons, Inc., New York, 1980.

for 15 sec, and add 1–2 drops of 5% aqueous silver nitrate solution. Instantaneous formation of a white precipitate is a positive test.

Reagent: Dissolve 0.25 g of paraperiodic acid (H_5IO_6) in 50 mL of water.

Cleaning Up   Because the quantity of material is extremely small, dilute the test solution with water and flush down the drain.

Group IV. Certain Carboxylic Acids, Certain Phenols, and Sulfonamides of 1° Amines

(a) Carboxylic Acids

Solubility in both 5% sodium hydroxide and sodium bicarbonate is usually sufficient to characterize this class of compounds. Addition of mineral acid should regenerate the carboxylic acid. The neutralization equivalent can be obtained by titrating a known quantity of the acid (ca. 50 mg) dissolved in water-ethanol with 0.1 N sodium hydroxide to a phenolphthalein end point.

(b) Phenols

Negatively substituted phenols such as nitrophenols, aldehydrophenols, and polyhalophenols are sufficiently acidic to dissolve in 5% sodium bicarbonate. See Group I(i) for the iron(III) chloride test for phenols; however, this test is not completely reliable for these acidic phenols.

Group V. Acidic Compounds, Insoluble in Bicarbonate

(a) Phenols

See Group I(i).

(b) Enols

See Group I(i).

(c) 1° and 2° Nitro Compounds

(Table 70.16, page 797; Derivatives, page 780)

Iron(II) Hydroxide Test.   To a small vial (capacity (1–2 mL) add 5 mg of the unknown to 0.5 mL of freshly prepared ferrous sulfate solution. Add 0.4 mL of a 2 N solution of potassium hydroxide in methanol, cap the vial, and shake it. The appearance of a red-brown precipitate of iron(III) hydroxide within 1 min is a positive test. Almost all nitro compounds give a positive test within 30 sec.

Reagents: Dissolve 2.5 g of ferrous ammonium sulfate in 50 mL of deoxygenated (by boiling) water. Add 0.2 mL of concentrated sulfuric acid and a piece of iron to prevent oxidation of the ferrous ion. Keep the bottle tightly stoppered. The potassium hydroxide solution is prepared by dissolving 5.6 g of potassium hydroxide in 50 mL of methanol.

Cleaning Up   Because the quantity of material is extremely small, the test solution can be diluted with water and flushed down the drain after neutralization with dilute hydrochloric acid.

(d) Sulfonamides of 1° Amines

An extremely sensitive test for sulfonamides (Feigl, *Spot Tests in Organic Analysis*) consists of placing a drop of a suspension or solution of the unknown on sulfonamide test paper followed by a drop of 0.5% hydrochloric acid. A red color is a positive test for sulfonamides.

   The test paper is prepared by dipping filter paper into a mixture of equal volumes of a 1% aqueous solution of sodium nitrite and a 1% methanolic solution of *N,N*-dimethyl-1-naphthylamine. Allow the filter paper to dry in the dark.

CAUTION: Handle N,N-dimethyl-1-naphthylamine with care. As a class, aromatic amines are quite toxic and many are carcinogenic. Handle them all with care—in a hood if possible.

Cleaning Up   Place the test paper in the solid hazardous waste container.

Group VI. Basic Compounds, Insoluble in Water, Soluble in Acid

Amines

See Group I(d).

Group VII. Reducible, Neutral *N*- and *S*-Containing Compounds

Aromatic Nitro Compounds

See Group V(c).

Group VIII. Hydrolyzable, Neutral *N*- and *S*-Containing Compounds (identified through the acid and amine obtained on hydrolysis)

(a) Amides

Unsubstituted amides are detected by the hydroxamic acid test, Group I(h).

(1) Unsubstituted Amides.   Upon hydrolysis, unsubstituted amides liberate ammonia, which can be detected by reaction with cupric ion [Group III(b)].

   To 1 mL of 20% sodium hydroxide solution, add 25 mg of the unknown. Cover the mouth of the reaction tube with a piece of filter paper moistened with a few drops of 10% copper(II) sulfate solution. Boil for 1 min. A blue color on the filter paper is a positive test for ammonia.

Cleaning Up   Neutralize the test solution with 10% hydrochloric acid, dilute with water, and flush down the drain.

(2) Substituted Amides.

The identification of substituted amides is not easy. There are no completely general tests for the substituted amide groups and hydrolysis is often difficult.

Hot sodium hydroxide solution is corrosive; use care.

Hydrolyze the amide by refluxing 250 mg with 2.5 mL of 20% sodium hydroxide for 20 min. Isolate the primary or secondary amine produced, by extraction into ether, and identify as described under Group I(d). Liberate the acid by acidification of the residue, isolate by filtration or extraction, and characterize by bp or mp and the mp of an appropriate derivative.

Use care in shaking concentrated sulfuric acid.

Cleaning Up   Dilute the test solution with water and flush down the drain.

(3) Anilides.

Dichromate dust is carcinogenic, when inhaled. Cr^{+6} is not a carcinogen when applied to the skin or ingested.

Add 50 mg of the unknown to 1.5 mL of concentrated sulfuric acid. Carefully stopper the reaction tube with a rubber stopper, and shake vigorously. (*Caution!*) Add 25 mg of finely powdered potassium dichromate. A blue-pink color is a positive test for an anilide that does not have substituents on the ring (e.g., acetanilide).

Cleaning Up   Carefully add the solution to water, neutralize with sodium carbonate, and flush down the drain.

(b) Nitriles

See Group I(h).

(c) Sulfonamides

See Group V(d).

Group IX. Neutral Polar Compounds, Insoluble in Dilute Hydrochloric Acid, Soluble in Concentrated Sulfuric Acid (most compounds containing oxygen)

(a) Alcohols

See Group I(b).

(b) Aldehydes

See Group I(c).

(c) Anhydrides

See Group I(e).

(d) Esters

See Group I(f).

(e) Ethers

(Table 70.9, page 793)

    Ethers are very unreactive. Care must be used to distinguish ethers from those hydrocarbons that are soluble in concentrated sulfuric acid.

$Fe[Fe(SCN)_6]$

Iron(III) hexathiocyanato-ferrate(III)

Ferrox Test.   In a dry test tube grind together, with a stirring rod, a crystal of iron(III) ammonium sulfate (or iron(III) chloride) and a crystal of potassium thiocyanate. Iron(III) hexathiocyanatoferrate(III) will adhere to the stirring rod. In a clean tube place 3 drops of a liquid unknown or a saturated toluene solution of a solid unknown, and stir with the rod. The salt will dissolve if the unknown contains oxygen to give a red-purple color, but it will not dissolve in hydrocarbons or halocarbons. Diphenyl ether does not give a positive test.

    Alkyl ethers are generally soluble in concentrated sulfuric acid; alkyl aryl and diaryl ethers are not soluble.

Cleaning Up   Place the mixture in the hazardous waste container.

(f) Ketones

(Table 70.14, page 796; Derivatives, pages 442 and 444).

(g) Unsaturated Hydrocarbons

(Table 70.12, page 794)

Use care in working with the bromine solution

Bromine in Carbon Tetrachloride.   Dissolve 1 drop (20 mg) of the unknown in 0.5 mL of carbon tetrachloride. Add a 2% solution of bromine in carbon tetrachloride dropwise with shaking. If more than 2 drops of bromine solution are required to give a permanent red color, unsaturation is indicated. The bromine solution must be fresh.

Cleaning Up   Place the mixture in the halogenated solvents container.

Potassium Permanganate Solution.   Dissolve 1 drop (20 mg) of the unknown in reagent grade acetone and add a 1% aqueous solution of potassium permanganate dropwise with shaking. If more than one drop of reagent is required to give a purple color to the solution, unsaturation or an easily oxidized functional group is present. Run parallel tests on pure acetone and, as usual, a compound known to be an alkene.

Cleaning Up   Dilute the solution with water and flush down the drain.

Group X. Inert Compounds. Insoluble in Concentrated Sulfuric Acid

(a) Alkyl and Aryl Halides

Alkyl Halides (Table 70.10, page 793)
Aryl Halides (Table 70.11, page 794)

Do not waste silver nitrate.

Alcoholic Silver Nitrate.   Add 1 drop of the unknown (or saturated solution of 10 mg of unknown in ethanol) to 0.2 mL of a saturated solution of silver nitrate. A precipitate that forms within 2 min is a positive test for an alkyl bromide, or iodide, or a tertiary alkyl chloride, as well as alkyl halides.

If no precipitate forms within 2 min, heat the solution to boiling. A precipitate of silver chloride will form from primary and secondary alkyl chlorides. Aryl halides and vinyl halides will not react.

Cleaning Up   Because the quantity of material is extremely small, it can be diluted with water and flushed down the drain.

(b) Aromatic Hydrocarbons

(Table 70.13, page 795; Derivatives, pages 780 and 782)

Aromatic hydrocarbons are best identified and characterized by UV and NMR spectroscopy, but the Friedel–Crafts reaction produces a characteristic color with certain aromatic hydrocarbons.

Keep moisture away from aluminum chloride.
CAUTION: Chloroform is carcinogenic. Carry out this test in a hood.

Friedel–Crafts Test.   Heat a test tube containing about 50 mg of anhydrous aluminum chloride in a hot flame to sublime the salt up onto the sides of the tube. Add a solution of about 10 mg of the unknown dissolved in a drop of chloroform to the cool tube in such a way that it comes into contact with the sublimed aluminum chloride. Note the color that appears.

Nonaromatic compounds fail to give a color with aluminum chloride, benzene and its derivatives give orange or red colors, naphthalenes a blue or purple color, biphenyls a purple color, phenanthrene a purple color, and anthracene a green color.

Cleaning Up   Place the test mixture in the halogenated organic solvents container.

(c) Saturated Hydrocarbons

Saturated hydrocarbons are best characterized by NMR and IR spectroscopy, but they can be distinguished from aromatic hydrocarbons by the Friedel–Crafts test [Group X(b)].

(d) Diaryl Ethers

Because they are so inert, diaryl ethers are difficult to detect and may be mistaken for aromatic hydrocarbons. They do not give a positive Ferrox test (see p. 777) for ethers and do not dissolve in concentrated sulfuric acid. Their infrared spectra, however, are characterized by an intense C—O single-bond, stretching vibration in the region 1270–1230 cm^{-1}.

Derivatives

1. Acids

(Table 70.1)

CAUTION: p-Toluidine is a highly toxic irritant.

p-Toluidides and Anilides.   Reflux a mixture of the acid (100 mg) and thionyl chloride (0.5 mL) in a reaction tube for 0.5 h. Cool the reaction mixture, and add 0.25 g of aniline or *p*-toluidine in 3 mL of toluene. Warm the mixture on the steam bath for 2 min, and then wash with 1-mL portions of water, 5% hydrochloric acid, 5% sodium hydroxide, and water. The toluene is dried briefly over anhydrous calcium chloride pellets and evaporated in the hood; the derivative is recrystallized from water or ethanol–water.

Cleaning Up   Dilute the aqueous layers with water and flush down the drain. Place the drying agent in the hazardous waste container.

Thionyl chloride is an irritant. Use it in a hood.

Amides.   Reflux a mixture of the acid (100 mg) and thionyl chloride (0.5 mL) for 0.5 h. Transfer the cool reaction mixture into 1.4 mL of ice-cold concentrated ammonia. Stir until reaction is complete, collect the product by filtration, and recrystallize it from water or water–ethanol.

Cleaning Up   Neutralize the aqueous filtrate with 10% hydrochloric acid, dilute with water, and flush down the drain.

2. Alcohols

(Table 70.2)

Note to instructor: Check to ascertain that the 3,5-dinitrobenzoyl chloride has not hydrolyzed. The mp should be >65°C. Reported mp is 68–69°C.

3,5-Dinitrobenzoates.   Gently boil 100 mg of 3,5-dinitrobenzoyl chloride and 25 mg of the alcohol for 5 min. Cool the mixture, pulverize any solid that forms, and add 2 mL of 2% sodium carbonate solution. Continue to grind and stir the solid with the sodium carbonate solution (to remove 3,5-dinitrobenzoic acid) for about a minute, filter, and wash the crystals with water. Dissolve the product in about 2.5–3 mL of hot ethanol, add water to the cloud point, and allow crystallization to proceed. Wash the 3,5-dinitrobenzoate with water–alcohol and dry.

Phenyl isocyanate

CAUTION: *Lachrymator*

Cleaning Up   Dilute the aqueous filtrate with water and flush down the drain.

Phenylurethanes.   Mix 100 mg of anhydrous alcohol (or phenol) and 100 mg of phenyl isocyanate (or α-naphthylurethane), and heat on the steam bath for 5 min. (If the unknown is a phenol add a drop of pyridine to the reaction mixture.) Cool, add about 1 mL of ligroin, heat to dissolve the product, filter hot to remove a small amount of diphenylurea which usually forms, and cool the filtrate in ice, with scratching, to induce crystallization.

Cleaning Up   Place the ligroin filtrate in the organic solvents container.

3. Aldehydes

(Table 70.3)

Semicarbazones.   See Chapter 36. Use 0.5 mL of the stock solution and an estimated 1 mmol of the unknown aldehyde (or ketone).

2,4-Dinitrophenylhydrazones.   See Chapter 36. Use 1 mL of the stock solution of 0.1 M 2,4-dinitrophenylhydrazine and an estimated 0.1 mmol of the unknown aldehyde (or ketone).

4. Primary and Secondary Amines

(Table 70.5)

Benzamides.   Add about 0.25 g of benzoyl chloride in small portions with vigorous shaking and cooling to a suspension of 0.5 mmol of the unknown amine in 0.5 mL of 10% aqueous sodium hydroxide solution. After about 10 min of shaking the mixture is made pH 8 (pH paper) with dilute hydrochloric acid. The lumpy product is removed by filtration, washed thoroughly with water, and recrystallized from ethanol–water.

Cleaning Up   Dilute the filtrate with water and flush down the drain.

**Picric acid
(2,4,6-Trinitrophenol)**

Handle pure acid with care (explosive). It is sold as a moist solid. Do not allow to dry out.

Picrates.   Add a solution of 30 mg of the unknown in 1 mL of ethanol (or 1 mL of a saturated solution of the unknown) to 1 mL of a saturated solution of picric acid (2,4,6-trinitrophenol, a strong acid) in ethanol, and heat the solution to boiling. Cool slowly, remove the picrate by filtration, and wash with a small amount of ethanol. Recrystallization is not usually necessary; in the case of hydrocarbon picrates the product is often too unstable to be recrystallized.

Cleaning Up   See page 523 for the treatment of solutions containing picric acid.

Acetamides.   Reflux about 0.5 mmol of the unknown with 0.2 mL of acetic anhydride for 5 min, cool, and dilute the reaction mixture with 2.5 mL of water.

Acetic anhydride is corrosive. Work with this in a hood.

Initiate crystallization by scratching, if necessary. Remove the crystals by filtration, and wash thoroughly with dilute hydrochloric acid to remove unreacted amine. Recrystallize the derivative from alcohol–water. Amines of low basicity, e.g., *p*-nitroaniline, should be refluxed for 30–60 min with 1 mL of pyridine as a solvent. The pyridine is removed by shaking the reaction mixture with 5 mL of 2% sulfuric acid solution; the product is isolated by filtration and recrystallized.

Cleaning Up   Neutralize the filtrate from the usual reaction with sodium carbonate. Then dilute it with water and flush down the drain. If pyridine is used as the solvent, neutralize the filtrate with sodium carbonate and extract it with ligroin. Place the ligroin/pyridine in the organic solvents container; dilute the aqueous layer with water and flush down the drain.

5. Tertiary Amines

(Table 70.6)

Picrates.   See under Primary and Secondary Amines.

$$R_3N + CH_3I$$
$$\downarrow$$
$$R_3\overset{+}{N}CH_3I^-$$

Methyl iodide is a suspected carcinogen.

Methiodides.   Reflux 100 mg of the amine and 100 mg of methyl iodide for 5 min on the steam bath. Cool, scratch to induce crystallization, and recrystallize the product from ethyl alcohol or ethyl acetate.

Cleaning Up   Because the filtrate may contain some methyl iodide, place it in the halogenated solvents container.

6. Anhydrides and Acid Chlorides

(Table 70.7)

Acids.   Reflux 40 mg of the acid chloride or anhydride with 1 mL of 5% sodium carbonate solution for 20 min or less. Extract unreacted starting material with 1 mL of ether, if necessary, and acidify the reaction mixture with dilute sulfuric acid to liberate the carboxylic acid.

Cleaning Up   Place the ether in the organic solvents container; dilute the aqueous layer with water and flush it down the drain.

Amides.   Because the acid chloride (or anhydride) is already present, simply mix the unknown (50 mg) and 0.7 mL of ice-cold concentrated ammonia until reaction is complete, collect the product by filtration, and recrystallize it from water or ethanol–water.

Cleaning Up   Neutralize the filtrate with dilute hydrochloric acid and flush it down the drain.

Anilides. Reflux 40 mg of the acid halide or anhydride with 100 mg of aniline in 2 mL of toluene for 10 min. Wash the toluene solution with 5-mL portions each of water, 5% hydrochloric acid, 5% sodium hydroxide, and again with water. The toluene solution is dried over anhydrous calcium chloride and evaporated; the anilide is recrystallized from water or ethanol–water.

Cleaning Up   Dilute the combined aqueous layers with water and flush down the drain. Place the soldium sulfate in the aromatic amines hazardous waste container.

7. Aryl Halides

(Table 70.11)

Nitration. Add 0.4 mL of concentrated sulfuric acid to 100 mg of the aryl halide (or aromatic compound) and stir. Add 0.4 mL of concentrated nitric acid dropwise with stirring and shaking while cooling the reaction mixture in water. Then heat and shake the reaction mixture in a water bath at about 50°C for 15 min, pour into 2 mL of cold water, and collect the product by filtration. Recrystallize from methanol to constant melting point.

    To nitrate unreactive compounds, use fuming nitric acid in place of concentrated nitric acid.

Use great care when working with fuming nitric acid.

Cleaning Up   Dilute the filtrate with water, neutralize with sodium carbonate, and flush the solution down the drain.

Sidechain Oxidation Products. Dissolve 0.2 g of sodium dichromate in 0.6 mL of water, and add 0.4 mL of concentrated sulfuric acid. Add 50 mg of the unknown and boil for 30 min. Cool, add 0.4–0.6 mL of water, and then remove the carboxylic acid by filtration. Wash the crystals with water and recrystallize from methanol–water.

Cleaning Up   Place the filtrate from the reaction, after neutralization with sodium carbonate, in the hazardous waste container.

8. Hydrocarbons: Aromatic

(Table 70.13)

Nitration. See preceding, under Aryl Halides.

Picrates. See preceding, under Primary and Secondary Amines.

9. Ketones

(Table 70.14)

Semicarbazones and 2,4-dinitrophenylhydrazones. See preceding directions under Aldehydes.

10. Nitro Compounds

(Table 70.16)

Reduction to Amines. Place 100 mg of the unknown in a reaction tube, add 0.2 g of tin, and then—in portions—2 mL of 10% hydrochloric acid. Reflux for 30 min, add 1 mL of water, then add slowly, with good cooling, sufficient 40% sodium hydroxide solution to dissolve the tin hydroxide. Extract the reaction mixture with three 1-mL portions of *t*-butyl methyl ether, dry the ether extract over anhydrous calcium chloride pellets, wash the drying agent with ether, and evaporate the ether to leave the amine. Determine the boiling point or melting point of the amine and then convert it into a benzamide or acetamide as described under the section on Primary and Secondary Amines.

Cleaning Up Neutralize the aqueous layer with 10% hydrochloric acid, remove the tin hydroxide by filtration, and discard it in the nonhazardous solid waste container. Dilute the filtrate with water and flush down the drain. After the ether evaporates from the calcium chloride, place it in the nonhazardous waste container.

11. Phenols

(Table 70.17)

α-Naphthylurethane. Follow the procedure for preparation of a phenylurethane under the Alcohols section.

Use great care when working with bromine. Should any touch the skin wash it off with copious quantities of water. Work in a hood and wear disposable gloves.

Bromo Derivative. In a reaction tube dissolve 160 mg of potassium bromide in 1 mL of water. *Carefully* add 100 mg of bromine. In a separate flask dissolve 20 mg of the phenol in 0.2 mL of methanol, and add 0.2 mL of water. Add about 0.3 mL of the bromine solution with swirling (hood); continue the addition of bromine until the yellow color of unreacted bromine persists. Add 0.6–0.8 mL of water to the reaction mixture, and shake vigorously. Remove the product by filtration, and wash well with water. Recrystallize from methanol-water.

Cleaning Up Destroy any unreacted bromine by adding sodium bisulfite solution dropwise until the color dissipates. Then dilute the solution with water and flush it down the drain.

TABLE 70.1 Acids

			Derivatives		
			p-*Toluidide*[a]	*Anilide*[b]	*Amide*[c]
bp	mp	Compound	mp	mp	mp
101		Formic acid	53	47	43
118		Acetic acid	126	106	79
139		Acrylic acid	141	104	85
141		Propionic acid	124	103	81
162		*n*-Butyric acid	72	95	115
163		Methacrylic acid		87	102
165		Pyruvic acid	109	104	124
185		Valeric acid	70	63	106
186		2-Methylvaleric acid	80	95	79
194		Dichloroacetic acid	153	118	98
202–203		Hexanoic acid	75	95	101
237		Octanoic acid	70	57	107
254		Nonanoic acid	84	57	99
	31–32	Decanoic acid	78	70	108
	43–45	Lauric acid	87	78	100
	47–49	Bromoacetic acid		131	91
	47–49	Hydrocinnamic acid	135	92	105
	54–55	Myristic acid	93	84	103
	54–58	Trichloroacetic acid	113	97	141
	61–62	Chloroacetic acid	162	137	121
	61–62.5	Palmitic acid	98	90	106
	67–69	Stearic acid	102	95	109
	68–69	3,3-Dimethylacrylic acid		126	107
	71–73	Crotonic acid	132	118	158
	77–78.5	Phenylacetic acid	136	118	156
	101–102	Oxalic acid dihydrate		257	400 (dec)
	98–102	Azelaic acid (nonanedioic)	164 (di)	107 (mono)	93 (mono)
				186 (di)	175 (di)
	103–105	*o*-Toluic acid	144	125	142
	108–110	*m*-Toluic acid	118	126	94
	119–121	DL-Mandelic acid	172	151	133
	122–123	Benzoic acid	158	163	130
	127–128	2-Benzoylbenzoic acid		195	165
	129–130	2-Furoic acid	107	123	143
	131–133	DL-Malic acid	178 (mono)	155 (mono)	
			207 (di)	198 (di)	163 (di)
	131–134	Sebacic acid	201	122 (mono)	170 (mono)
				200 (di)	210 (di)

(continued)

a. For preparation, see page 779.
b. For preparation, see page 779.
c. For preparation, see page 779.

TABLE 70.1 *continued*

bp	mp	Compound	Derivatives		
			p-*Toluidide*[a]	*Anilide*[b]	*Amide*[c]
			mp	mp	mp
	134–135	*E*-Cinnamic acid	168	153	147
	134–136	Maleic acid	142 (di)	198 (mono)	260 (di)
				187 (di)	
	135–137	Malonic acid	86 (mono)	132 (mono)	
			253 (di)	230 (di)	
	138–140	2-Chlorobenzoic acid	131	118	139
	140–142	3-Nitrobenzoic acid	162	155	143
	144–148	Anthranilic acid	151	131	109
	147–149	Diphenylacetic acid	172	180	167
	152–153	Adipic acid	239	151 (mono)	125 (mono)
				241 (di)	220 (di)
	153–154	Citric acid	189 (tri)	199 (tri)	210 (tri)
	157–159	4-Chlorophenoxyacetic acid		125	133
	158–160	Salicylic acid	156	136	142
	163–164	Trimethylacetic acid		127	178
	164–166	5-Bromosalicylic acid		222	232
	166–167	Itaconic acid		190	191 (di)
	171–174	D-Tartaric acid		180 (mono)	171 (mono)
				264 (di)	196 (di)
	179–182	3,4-Dimethoxybenzoic acid		154	164
	180–182	4-Toluic acid	160	145	160
	182–185	4-Anisic acid	186	169	167
	187–190	Succinic acid	180 (mono)	143 (mono)	157 (mono)
			255 (di)	230 (di)	260 (di)
	201–203	3-Hydroxybenzoic acid	163	157	170
	203–206	3,5-Dinitrobenzoic acid		234	183
	210–211	Phthalic acid	150 (mono)	169 (mono)	149 (mono)
			201 (di)	253 (di)	220 (di)
	214–215	4-Hydroxybenzoic acid	204	197	162
	225–227	2,4-Dihydroxybenzoic acid		126	228
	236–239	Nicotinic acid	150	132	128
	239–241	4-Nitrobenzoic acid	204	211	201
	299–300	Fumaric acid		233 (mono)	270 (mono)
				314 (di)	266 (di)
	>300	Terephthalic acid		334	

a. For preparation, see page 779.
b. For preparation, see page 779.
c. For preparation, see page 779.

TABLE 70.2 Alcohols

			Derivatives	
			3,5-Dinitrobenzoate[a]	Phenylurethane[b]
bp	mp	Compound	mp	mp
65		Methanol	108	47
78		Ethanol	93	52
82		2-Propanol	123	88
83		t-Butyl alcohol	142	136
96–98		Allyl alcohol	49	70
97		1-Propanol	74	57
98		2-Butanol	76	65
102		2-Methyl-2-butanol	116	42
104		2-Methyl-3-butyn-2-ol	112	
108		2-Methyl-1-propanol	87	86
114–115		Propargyl alcohol		63
114–115		3-Pentanol	101	48
118		1-Butanol	64	61
118–119		2-Pentanol	62	oil
123		3-Methyl-3-pentanol	96(62)	43
129		2-Chloroethanol	95	51
130		2-Methyl-1-butanol	70	31
132		4-Methyl-2-pentanol	65	143
136–138		1-Pentanol	46	46
139–140		Cyclopentanol	115	132
140		2,4-Dimethyl-3-pentanol	75	95
146		2-Ethyl-1-butanol	51	
151		2,2,2-Trichloroethanol	142	87
157		1-Hexanol	58	42
160–161		Cyclohexanol	113	82
170		Furfuryl alcohol	80	45
176		1-Heptanol	47	60(68)
178		2-Octanol	32	oil
178		Tetrahydrofurfuryl alcohol	83	61
183–184		2,3-Butanediol		201 (di)
183–186		2-Ethyl-1-hexanol		34
187		1,2-Propanediol		153 (di)
194–197		Linaloöl		66
195		1-Octanol	61	74
196–198		Ethylene glycol	169	157 (di)
204		1,3-Butanediol		122
203–205		Benzyl alcohol	113	77
204		1-Phenylethanol	95	92
219–221		2-Phenylethanol	108	78
230		1,4-Butanediol		183 (mono)
231		1-Decanol	57	59

(continued)

a. For preparation, see page 779.
b. For preparation, see page 780.

TABLE 70.2 *continued*

bp	mp	Compound	3,5-Dinitrobenzoate[a] mp	Phenylurethane[b] mp
			Derivatives	
259		4-Methoxybenzyl alcohol		92
	33–35	Cinnamyl alcohol	121	90
	38–40	1-Tetradecanol	67	74
	48–50	1-Hexadecanol	66	73
	58–60	1-Octadecanol	77(66)	79
	66–67	Benzhydrol	141	139
	147	Cholesterol	195	168

a. For preparation, see page 779.
b. For preparation, see page 780.

TABLE 70.3 Aldehydes

bp	mp	Compound	Semicarbazone[a] mp	2,4-Dinitrophenylhydrazone[b] mp
			Derivatives	
21		Acetaldehyde	162	168
46–50		Propionaldehyde	89(154)	148
63		Isobutyraldehyde	125(119)	187(183)
75		Butyraldehyde	95(106)	123
90–92		3-Methylbutanal	107	123
98		Chloral	90	131
104		Crotonaldehyde	199	190
117		2-Ethylbutanal	99	95(130)
153		Heptaldehyde	109	108
162		2-Furaldehyde	202	212(230)
163		2-Ethylhexanal	254	114(120)
179		Benzaldehyde	222	237
195		Phenylacetaldehyde	153	121(110)
197		Salicylaldehyde	231	248
204–205		4-Tolualdehyde	234(215)	232
209–215		2-Chlorobenzaldehyde	146(229)	213
247		2-Ethoxybenzaldehyde	219	
248		4-Anisaldehyde	210	253

(continued)

a. For preparation, see page 444.
b. For preparation, see page 442.

TABLE 70.3   *continued*

| | | | Derivatives | |
| | | | *Semicarbazone*[a] | *2,4-Dinitrophenylhydrazone*[b] |
bp	mp	Compound	mp	mp
250–252		*E*-Cinnamaldehyde	215	255
	33–34	1-Naphthaldehyde	221	254
	37–39	2-Anisaldehyde	215	254
	42–45	3,4-Dimethoxybenzaldehyde	177	261
	44–47	4-Chlorobenzaldehyde	230	254
	57–59	3-Nitrobenzaldehyde	246	293
	81–83	Vanillin	230	271

a. For preparation, see page 444.
b. For preparation, see page 442.

TABLE 70.4   Amides

bp	mp	Name of Compound	mp	Name of Compound
153		*N,N*-Dimethylformamide	127–129	Isobutyramide
164–166		*N,N*-Dimethylacetamide	128–129	Benzamide
210		Formamide	130–133	Nicotinamide
243–244		*N*-Methylformanilide	177–179	4-Chloroacetanilide
	26–28	*N*-Methylacetamide		
	79–81	Acetamide		
	109–111	Methacrylamide		
	113–115	Acetanilide		
	116–118	2-Chloroacetamide		

TABLE 70.5   Primary and Secondary Amines

			Derivatives		
			Benzamide[a]	*Picrate*[b]	*Acetamide*[c]
bp	mp	Compound	mp	mp	mp
33–34		Isopropylamine	71	165	
46		*t*-Butylamine	134	198	
48		*n*-Propylamine	84	135	
53		Allylamine		140	
55		Diethylamine	42	155	
63		*s*-Butylamine	76	139	
64–71		Isobutylamine	57	150	
78		*n*-Butylamine	42	151	
84		Diisopropylamine		140	
87–88		Pyrrolidine	oil	112	
106		Piperidine	48	152	
111		Di-*n*-propylamine	oil	75	
118		Ethylenediamine	244 (di)	233	172 (di)
129		Morpholine	75	146	
137–139		Diisobutylamine		121	86
145–146		Furfurylamine		150	
149		*N*-Methylcyclohexylamine	85	170	
159		Di-*n*-butylamine	oil	59	
182–185		Benzylamine	105	199	60
184		Aniline	163	198	114
196		*N*-Methylaniline	63	145	102
199–200		2-Toluidine	144	213	110
203–204		3-Toluidine	125	200	65
205		*N*-Ethylaniline	60	138(132)	54
208–210		2-Chloroaniline	99	134	87
210		2-Ethylaniline	147	194	111
216		2,6-Dimethylaniline	168	180	177
218		2,4-Dimethylaniline	192	209	133
218		2,5-Dimethylaniline	140	171	139
221		*N*-Ethyl-*m*-toluidine	72		
225		2-Anisidine	60(84)	200	85
230		3-Chloroaniline	120	177	72(78)
231–233		2-Phenetidine	104		79
241		4-Chloro-2-methylaniline	142		140
242		3-Chloro-4-methylaniline	122		105
250		4-Phenetidine	173	69	137
256		Dicyclohexylamine	153(57)	173	103

(continued)

a. For preparation, see page 780.
b. For preparation, see page 780.
c. For preparation, see page 780.

TABLE 70.5   *continued*

			Derivatives		
			Benzamide[a]	*Picrate*[b]	*Acetamide*[c]
bp	mp	Compound	mp	mp	mp
	35–38	*N*-Phenylbenzylamine	107	48	58
	41–44	4-Toluidine	158	182	147
	49–51	2,5-Dichloroaniline	120	86	132
	52–54	Diphenylamine	180	182	101
	57–60	4-Anisidine	154	170	130
	57–60	2-Aminopyridine	165 (di)	216(223)	
	60–62	*N*-Phenyl-1-naphthylamine	152		115
	62–65	2,4,5-Trimethylaniline	167		162
	64–66	1,3-Phenylenediamine	125 (mono)	184	87 (mono)
			240 (di)		191 (di)
	66	4-Bromoaniline	204	180	168
	68–71	4-Chloroaniline	192	178	179(172)
	71–73	2-Nitroaniline	110(98)	73	92
	97–99	2,4-Diaminotoluene	224 (di)		224 (di)
	100–102	1,2-Phenylenediamine	301	208	185
	104–107	2-Methyl-5-nitroaniline	186		151
	107–109	2-Chloro-4-nitroaniline	161		139
	112–114	3-Nitroaniline	157(150)	143	155(76)
	115–116	4-Methyl-2-nitroaniline	148		99
	117–119	4-Chloro-2-nitroaniline			104
	120–122	2,4,6-Tribromoaniline	198(204)		232
	131–133	2-Methyl-4-nitroaniline			202
	138–140	2-Methoxy-4-nitroaniline	149		
	138–142	1,4-Phenylenediamine	128 (mono)		162 (mono)
			300 (di)		304 (di)
	148–149	4-Nitroaniline	199	100	215
	162–164	4-Aminoacetanilide			304
	176–178	2,4-Dinitroaniline	202(220)		120

a. For preparation, see page 780.
b. For preparation, see page 780.
c. For preparation, see page 780.

TABLE 70.6   Tertiary Amines

		Derivatives	
		Picrate[a]	*Methiodide*[b]
bp	Compound	mp	mp
85–91	Triethylamine	173	280
115	Pyridine	167	117
128–129	2-Picoline	169	230
143–145	2,6-Lutidine	168(161)	233
144	3-Picoline	150	92(36)
145	4-Picoline	167	149
155–158	Tri-*n*-propylamine	116	207
159	2,4-Lutidine	180	113
183–184	N,N-Dimethylbenzylamine	93	179
216	Tri-*n*-butylamine	105	186
217	N,N-Diethylaniline	142	102
237	Quinoline	203	133(72)

a. For preparation, see page 780.
b. For preparation, see page 781.

TABLE 70.7   Anhydrides and Acid Halides

			Derivatives			
			Acid[a]		*Amide*[b]	*Anilide*[c]
bp	mp	Compound	bp	mp	mp	mp
52		Acetyl chloride	118		82	114
77–79		Propionyl chloride	141		81	106
102		Butyryl chloride	162		115	96
138–140		Acetic anhydride	118		82	114
167		Propionic anhydride	141		81	106
198–199		Butyric anhydride	162		115	96
198		Benzoyl chloride		122	130	163
225		3-Chlorobenzoyl chloride		158	134	122
238		2-Chlorobenzoyl chloride		142	142	118
	32–34	*cis*-1,2-Cyclohexanedicarboxylic anhydride		192		
	35–37	Cinnamoyl chloride		133	147	151
	39–40	Benzoic anhydride		122	130	163
	54–56	Maleic anhydride		130	181 (mono)	173 (mono)
					266 (di)	187

(continued)

a. For preparation, see page 781.
b. For preparation, see page 781.
c. For preparation, see page 782.

TABLE 70.7 *continued*

			Derivatives			
			Acid[a]		*Amide*[b]	*Anilide*[c]
bp	mp	Compound	bp	mp	mp	mp
	72–74	4-Nitrobenzoyl chloride		241	201	211
	119–120	Succinic anhydride		186	157 (mono)	148 (mono)
					260 (di)	230 (di)
	131–133	Phthalic anhydride		206	149 (mono)	170 (mono)
					220 (di)	253 (di)
	254–258	Tetrachlorophthalic anhydride		250		
	267–269	1,8-Naphthalic anhydride		274		250–282 (di)

a. For preparation, see page 781.
b. For preparation, see page 781.
c. For preparation, see page 782.

TABLE 70.8 Esters

bp	mp	Compound	bp	mp	Compound
34		Methyl formate	169–170		Methyl acetoacetate
52–54		Ethyl formate	180–181		Dimethyl malonate
72–73		Vinyl acetate	181		Ethyl acetoacetate
77		Ethyl acetate	185		Diethyl oxalate
79		Methyl propionate	198–199		Methyl benzoate
80		Methyl acrylate	206–208		Ethyl caprylate
85		Isopropyl acetate	208–210		Ethyl cyanoacetate
93		Ethyl chloroformate	212		Ethyl benzoate
94		Isopropenyl acetate	217		Diethyl succinate
98		Isobutyl formate	218		Methyl phenylacetate
98		*t*-Butyl acetate	218–219		Diethyl fumarate
99		Ethyl propionate	222		Methyl salicylate
99		Ethyl acrylate	225		Dimethyl maleate
100		Methyl methacrylate	229		Ethyl phenylacetate
101		Methyl trimethylacetate	234		Ethyl salicylate
102		*n*-Propyl acetate	268		Diethyl suberate
106–113		*s*-Butyl acetate	271		Ethyl cinnamate
120		Ethyl butyrate	282		Dimethyl phthalate
127		*n*-Butyl acetate	298–299		Diethyl phthalate
128		Methyl valerate	298–299		Phenyl benzoate
130		Methyl chloroacetate	340		Dibutyl phthalate
131–133		Ethyl isovalerate		56–58	Ethyl *p*-nitrobenzoate
142		*n*-Amyl acetate		88–90	Ethyl *p*-aminobenzoate
142		Isoamyl acetate		94–96	Methyl *p*-nitrobenzoate
143		Ethyl chloroacetate		95–98	*n*-Propyl *p*-hydroxybenzoate
154		Ethyl lactate		116–118	Ethyl *p*-hydroxybenzoate
168		Ethyl caproate (ethyl hexanoate)		126–128	Methyl *p*-hydroxybenzoate

TABLE 70.9   Ethers

bp	mp	Compound	bp	mp	Compound
32		Furan	215		4-Bromoanisole
33		Ethyl vinyl ether	234–237		Anethole
65–67		Tetrahydrofuran	259		Diphenyl ether
94		*n*-Butyl vinyl ether	273		2-Nitroanisole
154		Anisole	298		Dibenzyl ether
174		4-Methylanisole		50–52	4-Nitroanisole
175–176		3-Methylanisole		56–60	1,4-Dimethoxybenzene
198–203		4-Chloroanisole		73–75	2-Methoxynaphthalene
206–207		1,2-Dimethoxybenzene			

TABLE 70.10   Halides

bp	Compound	bp	Compound
34–36	2-Chloropropane	100–105	1-Bromobutane
40–41	Dichloromethane	105	Bromotrichloromethane
44–46	Allyl chloride	110–115	1,1,2-Trichloroethane
57	1,1-Dichloroethane	120–121	1-Bromo-3-methylbutane
59	2-Bromopropane	121	Tetrachloroethylene
68	Bromochloromethane	123	3,4-Dichloro-1-butene
68–70	2-Chlorobutane	125	1,3-Dichloro-2-butene
69–73	Iodoethane	131–132	1,2-Dibromoethane
70–71	Allyl bromide	140–142	1,2-Dibromopropane
71	1-Bromopropane	142–145	1-Bromo-3-chloropropane
72–74	2-Bromo-2-methylpropane	146–150	Bromoform
74–76	1,1,1-Trichloroethane	147	1,1,2,2-Tetrachloroethane
81–85	1,2-Dichloroethane	156	1,2,3-Trichloropropane
87	Trichloroethylene	161–163	1,4-Dichlorobutane
88–90	2-Iodopropane	167	1,3-Dibromopropane
90–92	1-Bromo-2-methylpropane	177–181	Benzyl chloride
91	2-Bromobutane	197	(2-Chloroethyl)benzene
94	2,3-Dichloro-1-propene	219–223	Benzotrichloride
95–96	1,2-Dichloropropane	238	1-Bromodecane
96–98	Dibromomethane		

TABLE 70.11   Aryl Halides

			Derivatives			
			Nitration Product[a]		Oxidation Product[b]	
bp	mp	Compound	Position	mp	Name	mp
132		Chlorobenzene	2, 4	52		
156		Bromobenzene	2, 4	70		
157–159		2-Chlorotoluene	3, 5	63	2-Chlorobenzoic acid	141
162		4-Chlorotoluene	2	38	4-Chlorobenzoic acid	240
172–173		1,3-Dichlorobenzene	4, 6	103		
178		1,2-Dichlorobenzene	4, 5	110		
196–203		2,4-Dichlorotoluene	3, 5	104	2,4-Dichlorobenzoic acid	164
201		3,4-Dichlorotoluene	6	63	3,4-Dichlorobenzoic acid	206
214		1,2,4-Trichlorobenzene	5	56		
279–281		1-Bromonaphthalene	4	85		
	51–53	1,2,3-Trichlorobenzene	4	56		
	54–56	1,4-Dichlorobenzene	2	54		
	66–68	1,4-Bromochlorobenzene	2	72		
	87–89	1,4-Dibromobenzene	2, 5	84		
	138–140	1,2,4,5-Tetrachlorobenzene	3	99		
			3, 6	227		

a. For preparation, see page 782.
b. For preparation, see page 782.

TABLE 70.12   Hydrocarbons: Alkenes

bp	Compound	bp	Compound
34	Isoprene	149–150	1,5-Cyclooctadiene
83	Cyclohexene	152	DL-α-Pinene
116	5-Methyl-2-norbornene	160	Bicyclo[4.3.0]nona-3,7-diene
122–123	1-Octene	165–167	(−)-β-Pinene
126–127	4-Vinyl-1-cyclohexene	165–169	α-Methylstyrene
132–134	2,5-Dimethyl-2,4-hexadiene	181	1-Decene
141	5-Vinyl-2-norbornene	181	Indene
143	1,3-Cyclooctadiene	251	1-Tetradecene
145	4-Butylstyrene	274	1-Hexadecene
145–146	Cyclooctene	349	1-Octadecene
145–146	Styrene		

TABLE 70.13 Hydrocarbons: Aromatic

| | | | Melting Point of Derivatives | | |
| | | | *Nitro*[a] | | *Picrate*[b] |
bp	mp	Compound	Position	mp	mp
80		Benzene	1, 3	89	84
111		Toluene	2, 4	70	88
136		Ethylbenzene	2, 4, 6	37	96
138		*p*-Xylene	2, 3, 5	139	90
138–139		*m*-Xylene	2, 4, 6	183	91
143–145		*o*-Xylene	4, 5	118	88
145		4-*t*-Butylstyrene	2, 4	62	
145–146		Styrene			
152–154		Cumene	2, 4, 6	109	
163–166		Mesitylene	2, 4	86	97
			2, 4, 6	235	
165–169		α-Methylstyrene			
168		1,2,4-Trimethylbenzene	3, 5, 6	185	97
176–178		*p*-Cymene	2, 6	54	
189–192		4-*t*-Butyltoluene			
197–199		1,2,3,5-Tetramethylbenzene	4, 6	181(157)	
203		*p*-Diisopropylbenzene			
204–205		1,2,3,4-Tetramethylbenzene	5, 6	176	92
207		1,2,3,4-Tetrahydronaphthalene	5, 7	95	
240–243		1-Methylnaphthalene	4	71	142
	34–36	2-Methylnaphthalene	1	81	116
	50–51	Pentamethylbenzene	6	154	131
	69–72	Biphenyl	4, 4′	237(229)	
	80–82	1,2,4,5-Tetramethylbenzene	3, 6	205	
	80–82	Naphthalene	1	61(57)	149
	90–95	Acenaphthene	5	101	161
	99–101	Phenanthrene			144(133)
	112–115	Fluorene	2	156	87(77)
			2, 7	199	
	214–217	Anthracene			138

a. For preparation, see page 782.
b. For preparation, see page 780.

TABLE 70.14   Ketones

| | | | Derivatives | |
| | | | Semicarbazone[a] | 2,4-Dinitrophenylhydrazone[b] |
bp	mp	Compound	mp	mp
56		Acetone	187	126
80		2-Butanone	136, 186	117
88		2,3-Butanedione	278	315
100–101		2-Pentanone	112	143
102		3-Pentanone	138	156
106		Pinacolone	157	125
114–116		4-Methyl-2-pentanone	132	95
124		2,4-Dimethyl-3-pentanone	160	88, 95
128–129		5-Hexen-2-one	102	108
129		4-Methyl-3-penten-2-one	164	205
130–131		Cyclopentanone	210	146
133–135		2,3-Pentanedione	122 (mono) 209 (di)	209
145		4-Heptanone	132	75
145		5-Methyl-2-hexanone	147	95
145–147		2-Heptanone	123	89
146–149		3-Heptanone	101	81
156		Cyclohexanone	166	162
162–163		2-Methylcyclohexanone	195	137
169		2,6-Dimethyl-4-heptanone	122	66, 92
169–170		3-Methylcyclohexanone	180	155
173		2-Octanone	122	58
191		Acetonylacetone	185 (mono) 224 (di)	257 (di)
202		Acetophenone	198	238
216		Phenylacetone	198	156
217		Isobutyrophenone	181	163
218		Propiophenone	182	191
226		4-Methylacetophenone	205	258
231–232		2-Undecanone	122	63
232		*n*-Butyrophenone	188	191
232		4-Chloroacetophenone	204	236
235		Benzylacetone	142	127
	35–37	4-Chloropropiophenone	176	223
	35–39	4-Phenyl-3-buten-2-one	187	227
	36–38	4-Methoxyacetophenone	198	228
	48–49	Benzophenone	167	238
	53–55	2-Acetonaphthone	235	262
	60	Desoxybenzoin	148	204
	76–78	3-Nitroacetophenone	257	228
	78–80	4-Nitroacetophenone		257
	82–85	9-Fluorenone	234	283
	134–136	Benzoin	206	245
	147–148	4-Hydroxypropiophenone		240

a. For preparation, see page 444.
b. For preparation, see page 442.

TABLE 70.15   Nitriles

bp	mp	Compound	bp	mp	Compound
77		Acrylonitrile	212		3-Tolunitrile
83–84		Trichloroacetonitrile	217		4-Tolunitrile
97		Propionitrile	233–234		Benzyl cyanide
107–108		Isobutyronitrile	295		Adiponitrile
115–117		*n*-Butyronitrile		30.5	4-Chlorobenzyl cyanide
174–176		3-Chloropropionitrile		32–34	Malononitrile
191		Benzonitrile		38–40	Stearonitrile
205		2-Tolunitrile		46–48	Succinonitrile
				71–73	Diphenylacetonitrile

TABLE 70.16   Nitro Compounds

bp	mp	Compound	Amine Obtained by Reduction of Nitro Groups			
			bp	mp	*Acetamide*[a] mp	*Benzamide*[b] mp
210–211		Nitrobenzene	184		114	160
225		2-Nitrotoluene	200		110	146
225		2-Nitro-*m*-xylene	215		177	168
230–231		3-Nitrotoluene	203		65	125
245		3-Nitro-*o*-xylene	221		135	189
245–246		4-Ethylnitrobenzene	216		94	151
	34–36	2-Chloro-6-nitrotoluene	245		157(136)	173
	36–38	4-Chloro-2-nitrotoluene		21	139(131)	
	40–42	3,4-Dichloronitrobenzene		72	121	
	43–50	1-Chloro-2,4-dinitrobenzene		91	242 (di)	178 (di)
	52–54	4-Nitrotoluene		45	147	158
	55–56	1-Nitronaphthalene		50	159	160
	83–84	1-Chloro-4-nitrobenzene		72	179	192
	88–90	*m*-Dinitrobenzene		63	87 (mono)	125 (mono)
					191 (di)	240 (di)

a. For preparation, see page 780.
b. For preparation, see page 780.

TABLE 70.17   Phenols

| | | | Derivatives | |
| | | | α-*Naphthylurethane*[a] | *Bromo*[b] |
bp	mp	Compound	mp	mp
175–176		2-Chlorophenol	120	48 (mono)
				76 (di)
181	42	Phenol	133	95 (tri)
202	32–34	*p*-Cresol	146	49 (di)
				108 (tetra)
203		*m*-Cresol	128	84 (tri)
228–229		3,4-Dimethylphenol	141	171 (tri)
	32–33	*o*-Cresol	142	56 (di)
	42–43	2,4-Dichlorophenol		68
	42–45	4-Ethylphenol	128	
	43–45	4-Chlorophenol	166	90 (di)
	44–46	2,6-Dimethylphenol	176	79
	44–46	2-Nitrophenol	113	117 (di)
	49–51	Thymol	160	55
	62–64	3,5-Dimethylphenol		166 (tri)
	64–68	4-Bromophenol	169	95 (tri)
	74	2,5-Dimethylphenol	173	178 (tri)
	92–95	2,3,5-Trimethylphenol	174	
	95–96	1-Naphthol	152	105 (di)
	98–101	4-*t*-Butylphenol	110	50 (mono)
				67 (di)
	104–105	Catechol	175	192 (tetra)
	109–110	Resorcinol	275	112 (tri)
	112–114	4-Nitrophenol	150	142 (di)
	121–124	2-Naphthol	157	84
	133–134	Pyrogallol	173	158 (di)

a. For preparation, see page 783.
b. For preparation, see page 783.